T0212391

SURGERY AND SOCIETY IN PEACE AND WAR

SCIENCE, TECHNOLOGY AND MEDICINE IN MODERN HISTORY

General Editor: John V. Pickstone, Centre for the History of Science, Technology and Medicine, University of Manchester, England

One purpose of historical writing is to illuminate the present. In the late twentieth century, science, technology and medicine are enormously important, yet their development is little studied. Histories of politics and literature abound, and historical biography is established as an effective way of setting individuals in context. But the historical literature on science, technology and medicine is relatively small, and the better studies are rarely accessible to the general reader. Too often one finds mere chronicles of progress, or scientific biographies which do little to illuminate either the science or the society in which it was produced, let alone their interactions.

The reasons for this failure are as obvious as they are regrettable. Education in many countries, not least in Britain, draws deep divisions between the sciences and the humanities. Men and women who have been trained in science have too often been trained away from history, or from any sustained reflection on how societies work. Those educated in historical or social studies have usually learned so little of science that they remain thereafter suspicious, overawed, or both.

Such a diagnosis is by no means novel, nor is it particularly original to suggest that good historical studies of science may be peculiarly important for understanding our present. Indeed this series could be seen as extending research undertaken over the last half-century, especially by American historians. But much of that work has treated science, technology and medicine separately; this series aims to draw them together, partly because the three activities have become ever more intertwined. This breadth of focus and the stress on the relationships of knowledge and practice are particularly appropriate in a series which will concentrate on modern history and on industrial societies. Furthermore, while much of the existing historical scholarship is on American topics, this series aims to be international, encouraging studies on European material. The intention is to present science, technology and medicine as aspects of modern culture, analysing their economic, social and political aspects, but not neglecting the expert content which tends to distance them from other aspects of history. The books will investigate the uses and consequences of technical knowledge, and how it was shaped within particular economic, social and political structures.

Such analyses should contribute to discussions of present dilemmas and to assessments of policy. 'Science' no longer appears to us as a triumphant agent of Enlightenment, breaking the shackles of tradition, enabling command over nature. But neither is it to be seen as merely oppressive and dangerous. Judgement requires information and careful analysis, just as intelligent policy-making requires a community of discourse between men and women trained in technical specialties and those who are not.

This series is intended to supply analysis and to stimulate debate. Opinions will vary between authors; we claim only that the books are based on searching historical study of topics which are important, not least because they cut across conventional academic boundaries. They should appeal not just to historians, nor just to scientists, engineers and doctors, but to all who share the view that science, technology and medicine are far too important to be left out of history.

Published

Roger Cooter
SURGERY AND SOCIETY IN PEACE AND WAR: Orthopaedics and the Organization of Modern Medicine, 1880–1948

David Edgerton
ENGLAND AND THE AEROPLANE: An Essay on a Militant and Technological Nation

John V. Pickstone (*editor*)
MEDICAL INNOVATIONS IN HISTORICAL PERSPECTIVE

Surgery and Society in Peace and War

Orthopaedics and the Organization of Modern Medicine, 1880–1948

Roger Cooter
Wellcome Unit for the History of Medicine
University of Manchester

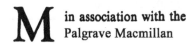 in association with the
Palgrave Macmillan

First published 1993 by
THE MACMILLAN PRESS LTD
Houndmills, Basingstoke, Hampshire RG21 2XS
and London
Companies and representatives
throughout the world

A catalogue record for this book is available
from the British Library.

Copy-edited and typeset by Povey–Edmondson
Okehampton and Rochdale, England

ISBN 978-1-349-64283-0   ISBN 978-1-137-10235-5 (eBook)
DOI 10.1007/978-1-137-10235-5

Series Standing Order (Science, Technology and Medicine in Modern History)

If you would like to receive future titles in this series as they are published, you can make
use of our standing order facility. To place a standing order please contact your
bookseller or, in case of difficulty, write to us at the address below with your name
and address and the name of the series. Please state with which title you wish to
begin your standing order. (If you live outside the United Kingdom we may not have
the rights for your area, in which case we will forward your order to the publisher
concerned.)

Customer Services Department, Macmillan Distribution Ltd
Houndmills, Basingstoke, Hampshire RG21 2XS, England

For Harrison & Helen
In memory of Sir Harry Platt

To attempt to isolate the history of medicine, and to comprehend its curious ebbs and flows of doctrine from medical writing only, is like cutting a narrow strip from the center of a piece of tapestry and speculating upon the origin and purpose of the cut threads of patterns that may be found in it.

John Shaw Billings (1838–1913)

CONTENTS

List of Tables		viii
List of Maps, Figures and Illustrations		ix
Acknowledgements		x
List of Abbreviations		xii
1	Introduction	1
2	The Medical Context of Bones	11
3	Politics and Professionalization	29
4	The Cause of the Crippled Child	53
5	Happenings by Accident	79
6	The Great War	105
7	Industry and Labour, Part I Britain and America, 1920s	137
8	Colonization Among Cripples	152
9	The Fracture Movement	180
10	Industry and Labour, Part II Rehabilitation and the Assault on Trauma, 1930s	199
11	The Phoney War	218
12	An End to 'Adolescence'	234
Notes		250
Bibliography		357
Index		374

LIST OF TABLES

2.1 Age of patients in Royal Orthopaedic Hospital, 1849, 1900   15

2.2 Inpatient cases, Royal Orthopaedic Hospital, 1849   16

2.3 Inpatient cases, Royal Orthopaedic Hospital, 1900   16

4.1 Causes of crippling in London, 1909, 1914   67

4.2 Major causes of crippling in Birmingham, 1911   68

8.1 Tuberculosis mortality in children under 15, 1898–1927   173

8.2 Poliomyelitis cases in London and in England and Wales, 1926–37   174

8.3 Percentage of 'defects' among Manchester schoolchildren at routine inspection periods, 1919, 1931   175

8.4 Malnutrition and rickets among Manchester schoolchildren, 1929–33   175

9.1 Fatal and non-fatal street accidents caused by vehicles, 1913–32   190

12.1 British Orthopaedic Association members, 1918–48   241

LIST OF MAPS, FIGURES
AND PLATES

Map

8.1 Orthopaedic schemes in England and Wales, 1934   163

Figures

2.1 Thomas splints   26
2.2 Thomas knee splint   27
2.3 Thomas knee bed splint   27
11.1 Structure for handling orthopaedic cases during the
      Second World War   226

Plates

3.1 Hugh Owen Thomas and Robert Jones, *c.* 1885   32
4.1 An open-air 'ward' at Baschurch *c.* 1910   75
4.2 Child patients at Baschurch *c.* 1910   75
5.1 Constructing the Manchester Ship Canal, 1888–93   101
6.1 Sir Robert Jones and Lady Jones with American
      orthopaedic surgeons, 1917   126
6.2 British Orthopaedic Association Inaugural Meeting,
      Roehampton, 1918   128
8.1 Biddulph Grange Orthopaedic Hospital, Staffordshire,
      in the 1930s   162

ACKNOWLEDGEMENTS

All history is social, but some contexts for its writing are more social than others. I was fortunate to enter one such place when I was invited to join the core of what has since become the University of Manchester's Wellcome Unit for the History of Medicine, in the Centre for the History of Science, Technology and Medicine. A more vibrant, committed and companionable environment would be difficult to find. To John Pickstone, the Director of the Centre and the Unit, I owe my first and foremost debts. He commissioned the project, invited me to undertake it, and arranged for its funding from the Wellcome Trust. Generous with his counsel throughout the book's preparation, he has also been patient. To the Wellcome Trust we are both enormously grateful, for without its support this study would not have been possible. Similar is our feeling of debt to the late Sir Harry Platt (1886–1986). His local presence was instrumental in the conception of the project; moreover, in interview after interview his great age proved no obstacle to great memory, wisdom and enthusiasm for serious history.

Many others have been liberal with their help and encouragement. Besides those acknowledged individually in the notes, I owe special debts to the orthopaedic surgeons Norman Roberts, Geoffrey Osborne, John Cholmeley and William Waugh, who were unfailingly cordial with all my inquiries. Lady Reginald Watson-Jones, with uncalled-for hospitality, shared with me the memory of her husband's work. The staff of the British Orthopaedic Association kindly gave me access to their archives and xeroxing facilities, as did Mr Peter Smith, Librarian of the Royal National Orthopaedic Hospital, and Mr E. Cornelius, the now retired Librarian of the Royal College of Surgeons. Dr David Kerr and Mr West of Manor House Hospital, Golders Green, the staff of the Liverpool Medical Institution, and the Cripples' Aid Society, Manchester, are also to be thanked for placing their abundant resources at my disposal.

It is impossible to mention by name all the others who made the research for this book pleasurable. The staff of the Public Record Office, the Greater London Record Office, the King's Fund Library, the library of the Wellcome Institute for the History of Medicine, the British Library, the John Rylands University Library of Manchester, and the Central Reference libraries of Manchester and Liverpool will know what repayments are due. Three archivists in particular, however, deserve to be singled out for their sterling professionalism: Adrian Allen of the University Archives of Liverpool; Derek Dow, until recently of the Greater Glasgow Health Board and University of Glasgow Archives; and Jonathan Peppler, until recently the Archivist of the Tower Hamlets Health Authority at the London Hospital.

Early drafts of some of the chapters were much improved by invitations to speak at conferences and seminars in Leeds, Bristol, Glasgow, London, Liverpool, Oxford, and at Queen's University, Kingston, Ontario. For all the insights and hospitality received, I remain warmly thankful. Various chapters also profited from careful readings by friends and colleagues, among them, Peter Bartrip, David Edgerton, Mary Fissell, Stephen Jacyna, Joan Mottram, and Mick Worboys. David Cantor, John Pickstone and Steve Sturdy read the entire manuscript with great care and improved it enormously through their expertise. Many of these friends and colleagues were also the victims of off-the-wall inquiries during countless lunchtime and evening meals. Always, their intellectual nourishments vastly compensated for the notorious cuisine of the University of Manchester. In the more comfortable surroundings of the Trevor Arms, Bill Luckin not only tolerated all my intellectual distractions, but committed himself to (and then discerningly carried out) the task of reading and refining the penultimate draft. Such generous support defies mere words of thanks. Nor can I forget the expert help of Yvonne Aspinall who at an earlier stage in the book's preparation saw to various essential practical matters. To Stella Butler I also owe many thanks, not least for her foresight and care in the preservation of the Platt Papers.

Finally, to Helen Cooter, whose magnificent patience and cosmic sense of proportion never failed, I owe more than I can say.

ROGER COOTER

LIST OF ABBREVIATIONS

AMA   American Medical Association
AJOS   *American Journal of Orthopaedic Surgery* (1902–19)
AMS   Army Medical Service
AOA   American Orthopaedic Association*
BMA   British Medical Association
BMJ   *British Medical Journal*
BOA   British Orthopaedic Association (est. 1918)
BOS   British Orthopaedic Society (est. 1894)
CCCC   Central Council for the Care of Cripples (est. 1919)
CMO   Chief Medical Officer
COS   Charity Organization Society
DNB   *Dictionary of National Biography*
EMS   Emergency Medical Service
GLRO   Greater London Record Office
ICAA   Invalid Children's Aid Association
JAMA   *Journal of the American Medical Association*
JBJS   *Journal of Bone and Joint Surgery* (1922–)
JOS   *Journal of Orthopaedic Surgery* (1919–22)
LCC   London County Council
MAB   Metropolitan Asylums Board
MOsH   Medical Officers of Health
NHI   National Health Insurance
NHS   National Health Service
PRO   Public Record Office
RNOH   Royal National Orthopaedic Hospital
ROH   Royal Orthopaedic Hospital
RSM   Royal Society of Medicine
SMO   School Medical Officer
SMS   School Medical Service
SSDC   Society for the Study of Diseases of Children
TUC   Trades Union Congress

* *Note on spellings*: The now-conventional American spelling of *orthopedics* was inconsistently used in the USA during the period under study, even within journals so entitled. The 'ae' in *orthopaedics* tended to be torn between aspirants to new world modernism and those inclined to the elitism of classical pretentions. Throughout this book, in quotations and in the titles of books and articles, spellings are faithful to the original.

1

INTRODUCTION

It is not immediately obvious why a history of orthopaedics should interest anyone other than orthopaedists themselves. As a well-established surgical specialism, now mainly concerned with the treatment of fractures, low back pain, and the replacement of hip joints,[1] orthopaedics ranks among the least controversial and most socially uninteresting areas of medicine. Its highly remunerated practitioners may not always be characterized in the most flattering terms,[2] and in Britain certain problems still surround its place in undergraduate medical education and in accident and emergency services.[3] But on the whole, its autonomy is unthreatened, it is not racked by problems of definition, and its professional organizations are secure. To an outsider, its most intriguing feature is probably its association with special types of shoes and mattresses.

But appearances deceive. Today's 'mature' specialism not only veils a turbulent and contentious 'adolescence', it also conceals an important and neglected pathway into the history of modern medicine as a whole. As it emerged in Britain between the 1880s and 1940s, orthopaedics did not depend on new sets of techniques and tools, but rather on crucial transformations in the organization of medicine. To focus on the 'adolescence' of the specialism is both to illuminate the politics of those transformations and to reveal the ways in which they were bound to wider changes in society and economy during peace and war. Such is the purpose of this book.

This study thus differs from triumphalist 'tunnel' histories of medicine, such as those written by orthopaedists themselves.[4] Although it follows a narrative structure and utilizes metaphors of growth which are evaluatively laden (hence the inverted commas), it is not concerned with celebrating 'progress' or 'maturity'. Its purpose, rather, is to investigate and analyze the nature of changes in the organization, provision, and use of health-care services. As such, this study also differs from academic accounts of medical specialization, most of which have been the product of American researchers

1

concerned with contemporary questions about professionalization and divisions of expert labour.[5] Written within a medical culture dominated by specialists, the latter studies have tended to conceive of medical specialization as among 'the most significant of human activities'.[6] Sociologically motivated accounts have generally taken as their starting point 'the assertion that the rise of medical specialties . . . is a variant of the phenomenon social scientists call professionalization',[7] and have approached the specialties as specimens of a prescribed general concept, rather than as individual empirical cases.[8] More historically-orientated studies have often been less concerned with the social, economic and political contexts for specialization than with the identification of specific determining factors. Moreover, as in George Rosen's pioneering study of ophthalmology, privileged epistemological status has usually been granted to science and technology in developments after 1800 – though Rosen, unlike many subsequent scholars, also sought to expose social and other factors in specialty formation. Several recent studies have further endeavoured to emphasize the role of market forces in the creation of the occupational divisions in American medicine.[9] But these works, too, tend to share the same basic object: to account for the division of spoils within the existing medical system, or to show, retrospectively, how the cake got cut.

A British standpoint encourages a different perspective, not least because here specialization has been less open to market forces. There has been no tradition of dependence on private beds for hospital income, and there has been little competition between hospitals for patients. Thus the compulsion for new techniques and specialty services within hospitals in order to attract patients has been far less intense than in America.[10] Equally important in working against specialization, or at least in not compelling it, has been the structure of the profession itself. At least since the mid-nineteenth century, the dominant groups have been the 'consultant' physicians and surgeons, organized through the Royal Colleges in London, and sharply differentiated in status from general practitioners (GPs).[11] In America, most doctors in urban areas have had access to hospitals to attend their patients, and many doctors, if not the majority, have had hospital appointments at some point in their career. In Britain, however, control over hospital beds (and medical teaching) has been the province of consultants. GPs in rural areas have sometimes had access to small cottage hospitals, but the majority have only been allowed to 'refer' patients to hospital consultants. With the introduction of National Health Insurance in 1911 this division of practice tended to be reinforced, and in 1948, with the National Health Service (NHS), the closed market referral system was written into law. Before the NHS most consultants held honorary positions in charity – 'voluntary' – hospitals. Obtaining such an appointment was the recognized means of gaining the status necessary to enter into private practice among the middle and upper classes – private practice remaining the object of most medical, and almost all specialist, work before the NHS. Although it was not unknown in the nineteenth century for general practitioners to establish their own specialist charity hospitals, by the end of the century this short-cut to elite private practice had become less possible. Increasingly, the major voluntary hospitals established their own specialty departments and recruited from their own generalist-trained staff. Thus to acquire the reputation necessary to enter into successful private practice as a specialist entailed following the same career pattern as becoming a consultant physician or surgeon. This in itself was restrictive, for a fellowship qualification from one of the Royal Colleges was a prerequisite by the end of the century, and considerable time and money were needed to put in long periods as an unpaid hospital staff member building up experience and status. Whereas in America it was not uncommon for doctors to cultivate specialties whilst living off the proceeds of general practice, in Britain this was exceedingly rare. To pursue the career of a consultant was to cut oneself off from general practice.

The structure of the medical profession in Britain before the NHS, then, and the fact that consultancy work in general medicine and surgery constituted a form of specialization in itself, did little to encourage the development of independent specialty careers of the American sort. Because of the status attached to voluntary hospital appointments, and because the large voluntary hospitals were virtually unrivalled as centres of excellence, few consultants who had specialty interests felt the need either to style themselves specialists or to pursue such careers independent of the voluntary hospitals. By 1900 the term 'consultant' had begun to be applied mainly to those who were distinguishable from an older generation of hospital 'generalists' by virtue of having restricted their private practice to either surgery or medicine.[12] To be more specialized than this was to run the risk not only of losing sufficient custom, but of being identified by one's peers as following in the footsteps of the single-nostrum 'quacks' of old or, indeed, of following the path of many an American specialist of late.[13] Thus within the less open and more structured British medical market, for a whole range of interrelated economic, intellectual and professional reasons, there was neither the need nor the desire for the certifying specialty boards characteristic of early twentieth-century American medicine.[14]

In view of these differences, it is not surprising that attention to professionalization in the history of British medicine has focused less on specialization *per se* than on the division between GPs and hospital consultants, and on the role of the state in reinforcing such divisions.[15] The most systematic study of British medical specialization, Rosemary Stevens's *Medical Practice in Modern England: the impact of specialization and state medicine* (1966), is typical in being largely taken up with the history of the NHS. In fact, Stevens sees the NHS not only as having 'formalized and exaggerated' the division between specialists and GPs, but as actually predicated upon 'the vast and sudden strides in medical research that compelled a rapid growth of specialization'.[16] Her chronological emphasis is justified on the grounds that specialization was simply not a characteristic feature of British medicine prior to the Second World War.[17] Equally important in determining Stevens's orientation, however, is her interest in 'policy alternatives for future developments in medical care'.[18] In this, as in her assumptions about the impact of medical science on specialization, her work is further typical of most writing on medicine and the British state.[19]

The present study is not primarily concerned with social policy, nor with the political formation of the NHS and the roles of specialists within it. Neither is it principally directed to the history of medical specialization in terms of the organization and consolidation of elite medical activity around particular bodies of therapeutic knowledge and technical skill. It concentrates more on a particular historical episode, and addresses what tends to be overlooked both in accounts of specialization and in studies of the politics and institutions of state medicine – the socioeconomic interests, professional relations, conceptual formulations and ideologies that underlay and historically permitted the divisions of labour that came to be embedded in the structure of the NHS. In short, the concern is with the interrelations between society, economy and medical politics, including the expression of those relations in the pursuit (or absence of pursuit) of scientific research and development.

The potential of the history of orthopaedics to uncover these connections over the crucial decades for the shaping of modern medicine between the 1880s and 1940s is considerable, in large part because of the specialism's own diffuse origins and controversial political establishment. Concerned with the prevention and cure of the dysfunctions of the locomotor system, modern orthopaedics not only straddled both medicine and surgery, but it also crossed the boundaries between public health and hospitals. In doing so, it encroached upon the territory of general practitioners and hospital general surgeons, as well as on that of state-employed medical officers of health (MOsH). At the same time, the would-be specialism intersected the organ geographies of certain specialisms and the age and gender specificities of yet others. Hence much of the rhetoric of orthopaedists was to be cast less in terms of the value of specialist expertise than in terms of the virtues of 'generalism', and of the need for holistic comprehensions of the body both in itself and in its relation to its physical and social environment. Furthermore, within the context of late nineteenth-century physiological thought, on the one hand, and the scientific management of business and industry, on the other, both the conception of modern orthopaedics and its practice came to exemplify an organization for medicine as a whole: elaborated through the physiological notion of the locomotor *system*, and through the practice of restoring efficient *function* to injured or diseased limbs, was a managerial ideal in which the various functions of medicine would be integrated for the greater efficiency and economy of the whole.

Orthopaedics was not unique in all these respects, but the fact that it epitomized them, whilst lacking any persuasive scientific or technological foci for specialization (comparable, say, to the ophthalmoscope for ophthalmology) renders it particularly well-suited to illuminate aspects of the social, economic, political and cultural milieus so frequently obscured in studies of specialisms and/or the NHS. Indeed, to the extent that the integrative philosophy of organization represented by orthopaedics was undermined around the time of the NHS, studies of the latter must almost inevitably, if unwittingly, draw a veil over the process of elaboration, whether or not science and technology are wheeled out as the invincible – and in themselves ahistorical – forces of change. A history of orthopaedics written narrowly from the perspective of its consolidation and integration into the NHS would similarly conceal more than it reveals. As shall be shown in the last few chapters of this book, the inherent contradiction between the organizational philosophy of modern orthopaedists and its struggle to succeed as a specialism within an increasingly specialized and research-orientated medical order – a contradiction which was opened up during and after the Second World War – was resolved in the NHS in favour of specialist autonomy. Accepting what it was to be 'of age', orthopaedists negotiated their autonomy, not through the unmediated power of scientific research, but rather through the professionally strategic use of the image and power accorded to it.

That prior to the NHS orthopaedics was in a state of 'making' is of course central to its historical potentiality, for the marginality of its would-be practitioners is partly what rendered them unusually open to the possibilities for reforming the whole structure of medicine. Two factors in particular intensified this marginality and encouraged a reform disposition. One was the deliberate obstructions that were thrown in the specialism's path by power-wielding general surgeons in the medical establishment. This was particularly the case after the well-publicized achievements of orthopaedics during the First World War. The other factor was that the majority of patients for orthopaedists – physically handicapped children from urban slums, wounded soldiers, and the victims of industrial accidents – were heavily dependent on medical welfare. Both these sets of conditions placed constraints on orthopaedic surgeons achieving status and wealth in the manner of most other specialists. Specialty legitimation and survival had to be sought from less conventional sources: from new types of charity organizations concerned with children and, increasingly during the interwar period, from local authorities, industry, trade unions, and the state. But to obtain support from these sources necessitated strong arguments for economy and efficiency above all else. Thus orthopaedists came increasingly to contribute to an ideology of cost-conscious rationalized health-care systems. To some extent making a virtue of necessity, they became architects and exemplars of local, regional and nationally integrated medical services. Although this book is less concerned with measuring the impact of orthopaedics on these restructurings of medicine than with exposing how the making of the specialty was informed by and reflects upon the restructuring process, it should not be overlooked that as a result of their ideological and architectural efforts orthopaedists came to be regarded by influential medical planners in the late 1930s as among the 'darlings of the Gods'.[20]

In this study, then, medical specialists are to be seen participating in the politics of health-care planning neither as science- or technology-wielding professionalizers, nor as policy-informing medical politicians, nor, least of all, as the familiar self-interested opponents of 'socialized' medicine. Rather, they are to be seen as ideologues of corporatism, rationalization, and statism: at a time when the state was reaching out to embrace professionals in its effort to efficiently manage its welfare, these were professionals embracing the state.[21]

The category of patients that orthopaedic surgeons focused on constitutes, in itself, a further important reason for pursuing orthopaedics as an object of social historical investigation. Although children, soldiers and industrial workers have been recognized as patient groups fundamental to the making of modern medicine, and each has vast resources available for study, they remain – paradoxically – among the least attended to in histories of medicine and society. This applies most blatantly to wounded soldiers and to the assumption – perhaps the greatest untested item of medical folklore – that the theatre of war has acted as a 'handmaid to medical progress'. As yet, no one really knows whether the major holocausts of this century 'fundamentally challenged' inherited priorities and professional structures in medicine,[22] or compelled radical reconceptualizations and reorganizations of civilian medical practice. If the two world wars *were* major turning points in health and social policy, it is as yet unclear *how* precisely such changes were effected. Nor is it immediately obvious how one might strike the historical balance between continuities and discontinuities over time. Indeed, to what extent can one generalize on the relations between war and medicine? The history of modern orthopaedics sheds light on such questions, for the specialism largely arrived on the map of medicine as a result of the First World War and was in various ways reconstituted through the Second.

So too, many of the uncertainties surrounding the history of child health and welfare can be addressed through the engagement of orthopaedists with physically handicapped children. Despite a plethora of recent works on infant mortality and infant feeding, and on childbearing, childrearing and childhood, contextual studies of child morbidity and child welfare focused specifically on medical involvement are still thin on the ground. This is particularly true for Britain for the period covered by this study.[23] How and why did 'the child' become a focus of medical attention towards the end of the nineteenth century? How were medical and lay understandings, interests and interventions related? Even the role of professional interests in shaping new institutions, such as children's hospitals, remains largely unknown.[24] The distinct and crucial role of 'crippled children' in the formulation of health-care policy and provision has also gone unexplored, despite the fact that this was one of the main areas for initiatives by voluntary organizations and local and national government in interwar Britain. It was also one of the arenas within which the vital concept of regional planning was refined.

Even less remarked upon has been the other object for specifically orthopaedic formulations – the victims of accidents. Although by the 1880s the accident victim was the archetypal patient in large voluntary hospitals, the problems of provision that this raised, and the associated spatial and conceptual dilemmas in hospital planning have remained almost entirely neglected. Nor has there been adequate discussion of the closely connected subject of the relations between medicine and industry, and the role of workers and working-class organizations in and around health care and safety.[25] Shop-floor collections are known to have been one of the major sources of revenue for many voluntary hospitals between the 1880s and the 1940s; hospital outpatient departments were often preoccupied with the treatment of industrial casualties; and workers' friendly societies exercised considerable pressure on the medical market. Yet few studies have dealt specifically with any of these topics.[26] Indeed, one of the most important issues affecting workers' lives and livelihoods – compensation for industrial accidents – has until recently 'languished in a historical backwater, the subject of no more than a couple of specialised studies'.[27] Such neglect is all the more surprising given that in Britain the Trades Union Congress and the Labour Party were involved in shaping health policy over much of this period – particularly in relation to the organization of trauma and rehabilitation services for industrial injuries.

In their different ways, each of these sites for the formation of modern orthopaedics provokes consideration of what is perhaps the most crucial but most obscure question in the history of medicine: the relations of life to medicine, as opposed to the importance of medicine to life. Though the time may be past when the history of medicine could be accused of only dwelling on great institutions, great discoveries and great men[28] – 'legends of saints' and 'court historiographies'[29] – the charge still holds that the history of the experience of illness has been little studied.[30] As Christopher Lawrence has recently pointed out, even one of the commonest encounters with medicine in the past – surgical practice of the simplest sort – has scarcely begun to find its place in the social history of medicine.[31] The study of hospitals has been conducted mostly without reference to patients, and has tended to focus on the work conducted in wards and operating theatres to the exclusion of outpatient departments, clinics and emergency services. Prevalent forms of care by the community have also been eclipsed.[32] As accidents have been ignored – despite now being the largest single cause of death in the age group 1–35[33] – so too have quotidian ills.

This book does not purport to be a corrective history of medicine from 'the patient's point of view'. Nevertheless, it embraces at least three essential components for any such history of individual encounters with illness and medicine. It draws attention to the sub-acute and sub-chronic muscular and skeletal conditions from which few people have died, but from which countless numbers have suffered extensive and incalculable physical, emotional and financial hardship. It reveals how the politics of professionalization have helped determine the provision of general and specialist health services. And it gives consideration to how the body itself came to be reconstructed in accordance with emergent concepts of labour management and economic rationality. In these respects at least, this study is as much about the impact of specialization on patients as on doctors and their relationship with the state.

The 'adolescence' of modern orthopaedics in Britain offers a 'window', therefore, on fundamental features in the history of medicine. As such, it parallels the historical function assigned to the peculiarly amorphous specialism of Internal Medicine in America. The study of the latter, it has been said, suggests

> new ways of defining several historical themes: the nature of scientific ideology; the connections between professional perceptions, practices and organizations; the relationship between the specialty and external influences; the role of prestige structures within medicine . . . ; the relationship between knowledge and skill base of a special field and its organizational arrangements; the nature of a 'specialty' per se.[34]

But the history of modern orthopaedics could be said to do more, for it provides a means of comprehending the history of medicine as part of the development of modern society as a whole. As orthopaedists themselves came to appreciate, theirs was 'no narrow specialism'; they were not, so to speak, like proctologists '[who] wilfully imprison their narrow lives within the muddy walls of the rectum'![35] Rather, 'with Ulysses', they could proclaim themselves to be '*a part of all we have met*'.[36] The principal sites for the making of modern orthopaedics – the urban slum, the industrial workplace, and the military battlefield – were much more than simply neutral spaces to be invaded by would-be specialists. They were critical locations for the restructuring of society, and for the interactions of philanthropy, labour, capital and the state.

Themselves ideological, the sites demanded the ideological engagement of all who sought to perform their labour and extend their professional interests within them. The thought and practice of the 'modernists' in orthopaedics cannot, therefore, be separated from the temporal and spatial domains in which their expertise was developed. Whether we focus on the patients with their problems, the doctors with their tools, or the relevant aspects of social policy, key features of social reconstruction emerge which compel exploration into their causes and consequences.

2

THE MEDICAL CONTEXT
OF BONES

Concepts, institutions, pioneers and techniques comprise what are often regarded as the essential building blocks for the history of a medical specialty. This chapter considers those integral to the history of orthopaedics. In so doing, however, a part of its purpose is to indicate how insufficient they are as historical explanation for the making of the modern specialism. More significant – as subsequent chapters will show – were the politics of surgery, ideologies bearing on childhood, economic interests, and the experience of war.

But 'insufficient' does not mean 'irrelevant'. Because certain institutions and practices were labelled 'orthopaedic' long before modern orthopaedics was organized, the latter cannot be fully understood without reference to the former. No absolute divide can in fact be drawn between the 'modern' and the 'pre-modern' forms of the specialism. To pursue the early history is to cast light on a multiplicity of continuous and discontinuous traditions, practices and meanings crucial to illuminating key developments not only in orthopaedics, but in the modernization of medicine as a whole.

WHAT'S IN A NAME?

Orthopaedics may be unique in that in the beginning was the word.[1] Whereas other medical specialisms name themselves as practitioners realize their mutual interests, 'orthopaedics' existed as a named practice before any such interests had made themselves visible – indeed, before any practitioners had emerged. Coined by Nicolas Andry in his *Orthopaedia, or the art of correcting and preventing deformities in children* (1741), the word was supposedly formed from the Greek words for 'straightening' and 'child'. Andry (1659–1742), the 'Professor of Medicine in the Royal College and Senior Dean of the Faculty of Physick at Paris' (as his title page has it), was by then an old man. But his coining of 'orthopaedia' bears few signs of senility or of idle invention. 'Orthopaedia' was intended, he said, 'to teach the

11

different Methods of preventing and correcting the Deformities of children' by (as the rest of the title has it) *such means, as may easily be put in practice by parents themselves, and all such as are employed in educating children.*[2] By invoking the power of parents and education, Andry may well have been seeking to counter the recently enhanced authority of French surgeons, a subject on which he had been an outspoken critic.[3] Whether or not his emphasis on the power of education and training over natural endowment or debility also permits him to be seen as sharing company with Enlightenment philosophers such as Claude Helvétius, his book situates him with those identified by Michel Foucault as involved with 'moral orthopaedics', or the straightening of supposedly impressionable minds.[4] Andry's frontispiece illustration of a bent sapling being righted by external pressures (which Foucault reproduces in his work on penitentiaries) was both descriptive of Andry's medical concern with the bones of children, and a metaphor for his wider educational philosophy. 'Orthopaedia' had in fact been carefully chosen by him to align with *paedotrophia* or child *management*, as coined by Scevale de St Marthe in a book of 1584 which Andry professed to admire.

From the very start, therefore, 'orthopaedics' had ambiguous social and professional resonances. It might imply merely a narrow physical technique for preventing and correcting deformity, or a much broader moral and medical practice focused on the treatment of skeletal disorders. During the second half of the eighteenth century the word came to be applied to various continental institutions for the treatment of deformity in children by means of mechanical apparatus. But by the early nineteenth century it had begun to burst the bounds of its etymology – literally, in such labels for foot and spinal techniques as Isidore Bricheteau's *orthosomatique* (1824) and Jacques Delpech's *orthomorphie* (1828), or, later, J. H. Green's *orthopaedy* and H. H. Bigg's *orthopraxy*.[5] By this date, not only was the correction of deformity in children no longer the essential characteristic of orthopaedics, but the correction of physical deformity itself was not necessarily implied.[6] In the 1830s the French alienist and phrenologist Felix Voisin applied the root word 'ortho' to mental deformity, founding in Paris an 'Orthophrenic' institute as a place to remould minds.[7] Equally significant during this period was the demise of Andry's anti-surgical emphasis. In 1842 the American surgeon, Valentine Mott, wrote of his 'happy lot' at having witnessed in Paris 'the dawning as well as the meridian splendour of another new and illustrious era in the healing art, . . . that beautiful and exact science *limitedly* denominated [by Delpech's successors] ORTHOPAEDIC SURGERY'. By this Mott meant the 'surgical and mechanical means . . . successfully directed to relieve almost every description of human deformity originating in the muscular system'.[8] More specifically, as would be made clear a few years later by another distinguished American surgeon, Henry Jacob Bigelow in his prize-winning *Manual of Orthopaedic Surgery* (1845), the new orthopaedics was primarily concerned with 'the division of muscles and tendons, or other parts'. Deformities to be thus treated were no longer necessarily specific to the locomotor system; they might also include muscle-cutting operations for squints. The latter was apparently still the case in the orthopaedic department at the Liverpool Royal Southern Hospital in the late-nineteenth century.[9]

In early nineteenth-century Britain 'orthopaedics' was applied mainly to the correction of club foot (talipes). It was largely for this purpose that three small infirmaries were established in London between 1838 and 1864 – the Royal Orthopaedic (1838), the City Orthopaedic (1851), and the National Orthopaedic (1864).[10] To a degree, the application of the word 'orthopaedics' to these hospitals was arbitrary. William John Little (1810–94), the founder of the first of them, seems to have employed the word almost as an afterthought to what he had originally proposed in 1838 as an 'Infirmary for the Cure of Club Foot and other Contractions by a new and very successful operation and mode of treatment, and by every other available means'. The name of Little's institution was changed to the Orthopaedic Institution shortly after it opened in Bloomsbury in July 1840 apparently only because the ground landlord, the Duke of Bedford, objected to the fixing of a door-plate with the word 'infirmary' on it. Little therefore removed the plate and had 'Orthopaedic Institution' painted on the door.[11]

Despite the etymological confusions that were to result from the application of 'orthopaedics' to institutions dealing mainly with club foot – many believing that 'paedia' was a misspelling of 'pedis', from the Latin for foot[12] – the usage was not incorrect. Although Little may only have chosen it to lend scientific or classical lustre to an otherwise mundane-looking medical procedure and institution, its application to the cutting or dividing of the Achilles tendon in the treatment of club foot (tenotomy) was very much a part of the French orthopaedics extolled by Mott and Bigelow. Little's understanding of the procedure, which was earlier (as well as more personal, in that he himself suffered from club foot),[13] differed only in that it derived mainly from German practice – in particular, from Louis Stromeyer (1804–76) of Hanover, who had modified the procedure of Delpech for dividing the *tendo Achilles*. In 1834 Stromeyer published on the success of the operation, and it was this which led Little to travel to Hanover in 1836 in search of a cure for his own affliction. Consequent upon the success of Stromeyer's operation, Little became Stromeyer's pupil. By the time he returned to London in 1837, he had become a full-fledged 'apostle of tenotomy'. Such was the 'new and very successful operation' for which he founded the infirmary in Bloomsbury.

Except, then, for the narrowness of its application, Little's 'orthopaedics' was apiece with that developing elsewhere. It similarly deviated from Andry's idea of orthopaedics by introducing into the treatment of a specific deformity an invasive surgical operation – albeit a minor one, which was, as Mott claimed, 'in many instances, almost *free from pain*, and without a drop of blood!'. And although it did not dispense with the use of mechanical apparatus (nor massage and exercise), it relied on skills other than the traditional ones of the hospital fitters and manufacturers of such equipment.[14] Like his French and American colleagues, therefore, Little regarded the 'new orthopaedics' as emancipated from its hitherto purely mechanical constraints.[15] It was liberated, too, in part, from the domain of physicians, though, importantly, in Britain the separation of orthopaedic surgery from medicine was never absolute. Unlike Mott, who was renowned for his operative boldness, Little, like Andry, was a physician. In 1839, shortly after the publication of his *A Treatise on Club Foot and Analogous Distortions, Including Their Treatment*, Little was elected honorary assistant physician to the London Hospital, and thereafter he became increasingly suspicious of the use of surgery in orthopaedics. In the 1850s he published in the *Lancet* 'On Unnecessary Orthopaedic Operations',[16] and in 1862, commenting on 'the means employed in Orthopaedic Surgery for the rectification of deformities', was at pains to point out that 'this department of practice avails itself of much that is common with general Medicine and Surgery', namely, 'Constitutional, medicinal, and dietetic treatment, exercise, gymnastics, inunctions, &c.'.[17]

LONDON ORTHOPAEDIC INSTITUTIONS

The orthopaedic infirmaries established in London in the nineteenth century were intended, as the City Orthopaedic Hospital advertised at mid-century, for all 'Persons . . . Afflicted with Club Foot, Contractions or Distortions of the Limbs, Curvature of the Spine or other Bodily Deformities'.[18] However, since the conditions they treated were mostly those acquired or manifested during childhood, they were to all intents and purposes small-scale children's hospitals. As can be seen from the statistics on the Royal Orthopaedic Hospital (Table 2.1), the tendency as the century advanced was towards an ever-greater proportion of young patients.

Table 2.1   Age of patients in Royal Orthopaedic Hospital, London, 1849, 1900

| | 1849 | | | 1900 | | | |
| | Inpatients | | | Inpatients | | In & outpatients | |
Age	Total	%		Total	%		Total	%
0–2	2	1.5	⎫	3	1.5	⎫	99	9.0
2–4	8	6.1	⎪	20	9.9	⎪	141	12.8
5–6	8	6.1	⎬ 29%	24	11.9	⎬ 42%	100	9.0
7–8	8	6.1	⎪	20	9.9	⎪	53	4.9
9–10	12	9.2	⎭	18	8.9	⎭	62	5.7
11–15	37	28.3	⎬ 53%	23	16.3	⎬ 35.1%	160	14.6
16–20	32	24.4		38	18.8		191	17.3
21–25	6	4.6		14	6.9		91	8.2
26–30	10	7.6		12	5.9		66	6.0
31–40	3	2.2		8	4.0		55	5.0
41–50	2	1.5		9	4.5		46	4.2
51–60	1	0.7		2	1.0		25	2.0
60+	1	0.7		1	0.5		14	1.3
Totals	130	100		202	100		1103	100

Sources: Royal Orthopaedic Hospital, *10th Annual Report* (1849); Royal National Orthopaedic Hospital, *62nd Annual Report* (1900).

The inpatient records also make clear that throughout the century the treatment of club foot and other deformities of the feet remained the preoccupation of these hospitals (Tables 2.2, 2.3). In 1901 the Royal Orthopaedic Hospital could boast some 344 operations performed over the past year, but only 58 of these were for other than tenotomy.[19] Contemporary comments stressing the variety of deformities treated in the London orthopaedic hospitals refer mainly to work conducted in the outpatient departments, where cases of knock

Table 2.2   Inpatient cases, Royal Orthopaedic Hospital, London, 1849

	Total	%
Club foot	32	24.5
Contracted knee	30	22.9
Contracted feet	18	13.8
Spinal curvature	17	12.9
Other	10	7.6
Contracted knee and feet and/or ankles	10	7.6
Knock knee	7	5.3
Paralysis	4	3.1
Flat foot	2	1.5
Dislocations	1	0.8
Total	131	100

Source: Royal Orthopaedic Hospital, *10th Annual Report* (1849).

Table 2.3   Inpatient cases, Royal Orthopaedic Hospital, London, 1900

	Total	%
Club foot (talipes)	101	39.3
Other*	37	14.3
Curvature of bones of the leg (including knock knee, bow knee and curved shin)	32	12.4
Paralysis	31	12.0
Diseases of the larger joints	24	9.4
Spinal curvature	24	9.4
Hallux valgus (big toe deformity)	8	3.2
Total	257†	100

* Includes old fractures, ingrown toe-nails, sprains, ulcers, corns, contractures from burns, etc.
† The discrepancy with Table 2.1 results from some patients coming under more than one heading
Source: Royal National Orthopaedic Hospital, *62nd Annual Report* (1900).

knee and bow leg tended to predominate.[20] Adolf Lorenz, the 'father of German orthopaedics' (and of the ethologist Konrad Lorenz) visited London in the 1880s in the hope of learning something new in orthopaedic surgery, but saw only 'some ridiculous attempts at correcting congenital club-feet. That was all.'[21]

The narrowness of the focus and the techniques on offer in the orthopaedic hospitals in Britain rendered them increasingly marginal as the century progressed. When first established they had not been charged with being 'quack-like' in supplying a single-nostrum or specialty service. But by the end of the century they were being subjected to the increasingly powerful claim of general surgeons that the proper place for surgery was in general hospitals.[22] To Henry Burdett, writing in the *Encyclopaedia Britannica* in 1911 it seemed 'very doubtful whether this type of [orthopaedic] hospital is really desirable or necessary'.[23] Within the rhetoric of general surgeons, the orthopaedic institutions were, like other specialist hospitals, inimical to the 'essential unity' of surgery.[24]

Working against them, too, by the latter part of the nineteenth century, was the decline in the initial excitement surrounding tenotomy. As early as 1865 it was being said that 'the nature and treatment of clubfoot have so long been a pet theme of writers on orthopaedic surgery, that the subject is almost exhausted, little in the way of novelty being left either to be discovered or accomplished'.[25] Moreover, the *need* for tenotomy in the treatment of club foot had begun to come under review,[26] not least by Little himself in a BMA address of 1875.[27] Small wonder, then, that the orthopaedic institutions came seriously to question their objects and purpose. The managing committee of the City Orthopaedic Hospital, for example, remarked in its *Annual Report* for 1869:

> It has long been painfully evident to your Committee that the name *Orthopaedic Hospital* has by no means been happy in designating the nature of the Institution, for instead of being limited to the care of Club Foot, as the word would imply, it is for the treatment of bodily deformities generally. Unfortunately, to the great bulk of the people the word Orthopaedic has no conceivable meaning, and, therefore, the Institution has suffered by not gaining the publicity to which it is deservedly entitled. (p. 13)

Similar confusions and feelings of being trapped by the label led the governors of the National Orthopaedic Hospital in 1880 to change the name of their institution to that of the National Hospital for the Deformed (though three years later, for reasons unexplained, they reverted back to 'Orthopaedic').[28] In fact, the orthopaedic infirmaries were on a downhill slope and it was not until 1907, when they were amalgamated into the Royal National Orthopaedic Hospital (RNOH), that the decline was halted.

THE SURGICAL REVOLUTION OF THE 1860s AND 1870s

Central to the fate of the orthopaedic infirmaries in the second half of the nineteenth century was the rise of 'orthopaedic surgery' as performed by general surgeons in general hospitals. Incorporating the treatment of inherited and acquired skeletal deformities, the orthopaedic surgery conducted in this context came to be celebrated as one of the major medical advances of the century.

Conventionally, the rapid development and acclaim of this surgery has been attributed to the introduction of anaesthetics and Listerian antiseptics. On the face of it, this explanation is credible. Although one finds little discussion of the impact of anaesthetics within accounts of orthopaedic surgery at this time,[29] the kudos of Listerian antiseptics was sufficiently apparent to those in general surgery who were beginning to specialize in orthopaedics at the end of the century for it to be claimed that it was to this alone that all 'progress was due'.[30] Later generations of orthopaedists would boast that Lister's first experiments with antiseptics in the 1860s were conducted on their patch – on cases of compound fractures.[31]

However, as Lindsay Granshaw, Christopher Lawrence and others have shown, the idea that the introduction of Listerian antiseptics brought about a 'technological fix', establishing a single bridge to modern surgery, is wrong.[32] When leading medical spokesmen commented in the late 1870s on 'the daily triumphs of orthopaedic surgery', it was not to the application of antiseptic procedures that they were referring, but to the progress of subcutaneous osteotomy – the technique for cutting and dividing bones by making insertions under the skin.[33] Pioneered by general surgeons in the 1860s, the technique made possible a whole range of treatments for diseased and deformed bones.[34] The expansion in the scope of orthopaedics resulting from this is reflected in Little's contribution on 'Orthopaedic Surgery' to the 1883 edition of Timothy Holmes's *System of Surgery*. Here, 'osteotomy' competes with the space that 'tenotomy' had monopolized in the 1862 edition, and omitted is Little's earlier remark on club foot, that 'no other deformity exhibits in a greater degree the incidents of orthopaedic experience'.[35]

Subcutaneous osteotomy was devised and taken up so as to avoid the risk of exposing tissues, thus preventing dreaded forms of post-operative blood poisoning (septicaemia, erysipelas, pyaemia, and so on). Because of the usually fatal consequences of such infection, the technique impressed lay and medical men alike. Not until 1878 was the procedure superseded by open 'antiseptic osteotomy', which was introduced into Britain by Lister's pupil William Macewen.[36] Since Macewen was a past master at subcutaneous osteotomy, it is tempting to regard the pre-antiseptic subcutaneous procedure merely as a prelude to the later antiseptic procedure. But to do so is to eclipse the significance (not to mention the popular acclaim) of subcutaneous osteotomy in the 1860s and 1870s as essentially a preventive-infective measure. Turning the historical telescope the right way around, subcutaneous osteotomy can be seen to stand alongside other contemporary pre- and post-operative hygienic practices (general and local) which were introduced into hospitals to stem the tide of high mortality from post-operative infection. It was the success of these measures, before and after Lister, which provided the ammunition for the opposition to Lister's dogmatic claims for 'antisepticism', and which constituted the basis for an alternative anti-infective tradition within the medical mainstream. As revealed by the following passage, from a major article in the *Lancet* in 1882 on wound and fracture treatment, the subscription to this alternative tradition did not necessarily entail the rejection of germ theory:

> To those who have noticed the omission of reference to so-called antiseptic surgery, I beg permission to address a few remarks. Life is the great antiseptic. Preserve it, restore healthy function, control by rest, position, and pressure, nervous, vascular, and muscular action, so as to minimise the material for, and the causes of, discharge, carry it off as it is produced by drainage-tubes and absorbent dressings, and the repair of injuries proceeds like healthy nutrition, uninterruptedly and painlessly. That infection is always floating in the atmosphere, ready to settle in the shape of impalpable and implacable germs into any breach which may be made in the surface of a living body, is an idea which has never troubled me.[37]

End-of-the-century aseptic surgery, with its attention to germ-free environments, was as much the result of this alternative to Listerism as to Listerism itself.

But 'proto-asepticism' did not stand on its own; it was part of a long-existing tradition which had been undergoing revival at least since mid-century, and which quickened its pace in reaction to Lister. Conservative surgery, so-called, was important within surgery as a whole, but within orthopaedics it was to have a significance at once deeper and more enduring.

Based on the principle of sacrificing as little of the body as possible to the surgeon's scalpel, in the interest of curing rather than cutting, conservative surgery, emerged in the eighteenth century out of the teachings of John Hunter.[38] Linked to it was a belief in the healing power of nature (*vis medicatrix naturae*) which included appreciation of the curative value of fresh air, rest, the immobilization of injured limbs, and notions of the use of minimal doses of physic.[39] Hunter's principles, applied generally, were taken up by several eminent surgeons in the nineteenth century. Among them was James Syme (1799–1870) of Edinburgh, Lister's father-in-law.[40] Another was Sir William Fergusson (1808-77) of London, one of the last of the great teachers of surgical generalism and one of the first – some two decades before Lister published – to express reservations about the 'brilliancy' and 'éclat' of the operative surgery then becoming so attractive to 'the younger members of our profession'.[41]

It was Fergusson, in 1852, in the course of becoming interested in homoeopathy, who coined the term conservative surgery to refer to the practice of always 'saving the most by removing the least', hence to maximizing the integrity of the body.[42] Like others who followed in this tradition, Fergusson tended to emphasize the use of restorative manipulative techniques in the place of operative ones, and to rely heavily on the use of mechanical apparatus. Hitherto, manipulation had largely been in the hands of 'a number of ignorant charlatans', especially bonesetters and masseurs. But increasingly, practices such as medical gymnastics, massage, and mechanotherapy, electrotherapy, galvanotherapy and hydrotherapy were put 'upon a scientific basis' and entered orthodox medicine as physiotherapeutic procedures in the treatment of musculo-skeletal disorders.[43]

Another important figure in conservative surgery during this period, whose career bears in a different way on the practice and professional evolution of orthopaedics, was Frederick Howard Marsh (1839–1915) of St Bartholomew's Hospital.[44] Marsh was a colleague of George Callender (1830–79), a surgeon who, besides having a special interest in bone formation, was one of the leading anti-Listerians. Callender's concern to eliminate fatal infections on his wards led him to innovate with 'aseptic' techniques, while his attention to and curtailing of amputations accounted for drastic reductions in the prevailing hospital mortality from this source (reckoned in the 1870s to be between 25 to 50 per cent).[45] Like Callender, Marsh also was profoundly influenced by the work of John Hilton (1805–78), Hunterian Orator and a major advocate of the restorative powers of rest in the treatment of accidents and surgical diseases.[46] Marsh devoted himself to applying Hilton's principles to the treatment of children suffering from tuberculosis of the bone and joints; he had ample opportunity to do so through his appointment to the Alexandra Hospital for Children with Hip Disease. His 'conservatism' also expressed itself through a keen interest in the manipulative practices of bonesetters. In 1878, when he was appointed Demonstrator of Orthopaedics and Officer-in-Charge of orthopaedic outpatients at St Bartholomew's Hospital, Marsh published on the need to incorporate into surgery all that was useful in the bonesetter's craft.[47] In this he followed James Paget whose paper 'On the Cases that Bone-setters Cure' was first published in the *British Medical Journal* in 1867.[48] Marsh went beyond Paget and most other writers, however, in his opposition to the excision of joints and in his reliance on mechanical appliances, on both of which subjects he was apparently a persuasive teacher.[49]

In general, the principles and practices indicated by Fergusson, Callender and Marsh contributed to an appreciation of the value of good diet, clean environment, rest and gentleness on health, and to a respect for non-heroic surgery and physic. Holistic in their orientation, and inclined to prevention as much as to cure, these principles and practices owed something to popular 'alternative' medical cults and practices,[50] but only as reworked *within* orthodox medicine according to ideas of the body's own natural restorative capacities. Almost certainly these men's 'conservatism' also owed something to the mounting antivivisection movement, much of the contemporary literature on which criticized those in medicine who were, as Tennyson put it, 'Happier in using the knife than in trying to save the limb'.[51] Yet there was nothing of a 'fringe' quality to their writings; in major British and American hospitals their conservative principles were widely applied both to the treatment of chronic disorders and to acute traumatic injury.[52]

What was happening in surgery towards the end of the century, then, was far more complex than most historical accounts of surgery would lead us to believe. Rather than a simple shift from conservative to bold invasive surgery, hastened by Listerism, a range of principles and practices had been opened up, from the extreme radicalism of someone like William Halsted and his British followers, to the conservatives discussed above. Indeed, as Gert Brieger has shown, the situation is even more complex than this, for, by the end of the nineteenth century, 'radical could also mean conservative in the sense of complete or finally curative: conservation of life', and invasive operative surgery could seem the most 'conservative' or 'preventive' choice facing a surgeon and his patient. Halsted himself, one of the most influential of American surgeons, was in fact, 'the epitome of the conservator in surgery', especially in relation to a key aspect by 1900, that of preventing the loss of patients' blood.[53]

'Progress' in surgery was not therefore simply or inevitably on the side of bold operations, even if by early in the twentieth century it was already being made to appear that scientific knowledge compelled this to be so. One of Marsh's obituarists, for example, was to claim that '[Marsh was] an old-fashioned clinician with scant sympathy for the revolutionary changes which shook the world of surgery around him, least of all for the germ theory of disease'. Accordingly, Marsh's conservatism, far from appearing 'progressive', was put down to a personal shortcoming – that his talents did 'not include that skill and high capacity for bold and novel operative measures which distinguished so many of his contemporaries'.[54] But such retrospection tells us more about the social and professional needs of the surgical authors of the 'Lister myth' than about historical realities.[55] In truth, just as there was no simple or absolute divide between pre- and post-germ theory in surgical practice, so there was no corresponding germ-theory-related antipathy between conservative and radical surgery.

Radical 'heroic' surgery, which linked 'great men' to bold invasive operations, could be a resource that replenished arguments for conservative surgery. But it was not the only such resource. Further incentives to conservatism around the turn of the century emerged from a variety of sources – the treatment of tuberculosis of the bone and joints,[56] the treatment of fractures, and, not least, from the pioneering work of Adolf Lorenz in the treatment of congenital hip dislocation. The latter is worth pausing over since Lorenz's 'closed operation' for this condition emerged not from hostility to Listerism, but from an allergic reaction to Lister's carbolic spray. Because of this physical response, Lorenz turned from a promising career in 'wet' (general) operative surgery to 'dry' (manipulative) surgery, in the ambitious pursuit of which he perfected and popularized a number of conservative operations and mechanical and gymnastic corrective procedures. Although it was in the USA, where 'radical' surgery had developed most rapidly, that Lorenz's 'bloodless surgery' achieved its greatest acclaim, in Britain, too, it was well received. In the popular press it was heralded as the dawn of an age of 'bloodless hospitals'.[57]

It was not, therefore, necessarily the shared impression of late Victorians and Edwardians that the advance of surgery was due entirely to the introduction of antiseptics and the expansion of radically invasive surgical techniques. Conservative surgery was understood to be at least as impressive, whether it was linked to or separate from the growth of knowledge about the importance of cleaner, more germ-free environments. Moreover, the conceptual and professional implications of the advance of conservative surgery were clearly understood. 'It is commonly observed', noted the *Westminster Review* in 1881,

> that the tendency of modern surgery is to become more conservative – that is, to dispense with the knife and rely more upon the recuperative and compensatory capabilities of the body. In this it has distinctly approached medical practice, and the line of demarcation between the two branches of this healing art is decidedly becoming more and more faint. While surgery, on the one hand, is falling back to a greater extent upon the ordinary powers of the system, medicine, on the other, is tending towards the adoption of manipulative measures.[58]

Nowhere were these lines of demarcation more faint, and the tendency to conservative surgery more marked, than in the practice of the putative 'father' of modern orthopaedics, Hugh Owen Thomas of Liverpool.

HUGH OWEN THOMAS (1834–91)

Thomas was a general practitioner who, after qualifying at Edinburgh in 1857, built up an extensive 'club' practice – worth over £3000 a year by the 1880s – mainly among the Liverpool dockers. He also operated a small eight-bed private hospital, which he continued to use even after 1866 when he opened his larger and eventually famous Liverpool premises at 11 Nelson Street. Thomas's father had been a popular bonesetter in the same dockland area of Liverpool (the seventh bonesetter in lineal descent from Anglesey farmers), and Thomas worked with him until 1859. They then parted company, the son refusing any longer to countenance and defend his father's unlicensed practice. Thomas took away more than bitterness and some crucial experience, however; he took, too, his father's Methodist conscience (though not his Methodism), a proud sense of independence, a disdain for the establishment, and an argumentative, intolerant personality. His heroes were history's popular underdogs, John Wilkes and Tom Paine, and among his few friends was the secularist orator, Charles Bradlaugh. Thomas clearly aspired to a similar reputation, and to some extent he achieved it locally by holding free Sunday sessions at the Nelson Street Clinic.[59]

Although often depicted as a lone pioneer in orthopaedics, Thomas can scarcely be said to have been out of touch with his contemporaries. Indeed, his writings have considerable historical value as a result of embracing much of what was most recent in medicine and surgery, as well as most of what was new in the then existing practice of orthopaedics. As a student of medicine in Edinburgh, Thomas had witnessed the work of Syme, and he subsequently studied in London and Paris. Like Callender and Marsh, he was inspired by the teachings of Hilton, and like them he turned against the methods of Lister. To disturb the wounds of patients through invasive surgery he regarded as a violation of the principle of repair by rest.[60] For this reason, like Marsh, Thomas repudiated the excision of joints. Later, when he became the visiting surgeon at a children's hospital at Rhyl in North Wales, he deployed open-air therapy as a primary restorative for children suffering from the suppurating wounds of bone and joint tuberculosis. Thomas was also like Marsh in his enthusiasm for mechanical apparatus (especially splints), the design and manufacture of which he undertook in his splendidly equipped workshop at the Nelson Street Clinic. Good splinting, he believed, facilitated good conservative surgery, and the simple design of his splints encouraged this. He once remarked that 'a man who understands my principles will do better with a bandage and a broomstick than another can do with an instrument-maker's arsenal'.[61]

But, unlike Marsh and most others working in the field of orthopaedics during the last third of the nineteenth century, Thomas had little time for the manipulative techniques derived from bone-setters. The rupture with his father had hardened his attitude on the subject, even though his father had actually practised very little manipulative bonesetting (being keenly aware of the legal and medical risks involved in roughly handling undetected adhesions and tuberculous joints). It is doubly ironic, therefore, that Thomas should have been outspoken on the practice of bonesetting. Reflecting on Paget's lecture on the subject, he commented,

My contention is this, that in the practice of bone-setting nothing is to be found that can be added to our present knowledge, yet discussion of the matter will show us our own ignorance. That some of the bone-setters, who practised in past time, were in some few special matters superior to their qualified contemporaries, I know to be a fact, but this assertion does not apply to their general knowledge or practice. And concerning diseases of joints, I never met with the slightest evidence that any of them had any knowledge of the subject or a method of treatment which was not utterly wrong.[62]

For personal reasons, as Paget came to realize, Thomas hoped that the medical profession would redeem the public 'from the wiles and snares of the bone-setter'.[63] But Thomas's reaction was also consistent with his commitment to the therapeutics of rest. In effect, he was only extending to the practice of massage and extension-movement the principles that Paget, Marsh and others had begun to apply in other spheres.

Like the portrait of Thomas as the isolated pioneer, the view of him as the proud prophet 'in the wilderness ... either ignored or discountenanced' during his life, also falls short of the truth.[64] Though he was aloof and isolated, lacked significant hospital appointments, and communicated to his peers primarily through an obscure local publisher, he nevertheless became reasonably well-known. His *Diseases of the Hip, Knee, and Ankle Joints with their deformities treated by a new and efficient method (enforced, uninterrupted and prolonged rest)* (1875) reached a third edition in 1878. In a context of growing interest in bone and joint surgery, the book in fact initiated a reputation for Thomas both at home and abroad. In 1876 he was elected a member of the Liverpool Medical Institution (rising to vice-president before his death),[65] and as early as 1878 it was remarked by C. G. Wheelhouse of Leeds, in an address to the BMA on the 'Progress of Surgery', that:

even while we are still rejoicing in what excision has enabled us to do for the cure of joints disabled by disease, are we not learning from our Liverpool friends, under the leadership of Mr. Thomas, that a better method even than excision is at our disposal, and that, by an intelligent and persistent use of absolute immobility, we can procure for our patients, without the smallest risk, results far better and more perfect than follow the most skilfully performed and most successful of excisions?[66]

Shortly after this Thomas was invited to Leeds to demonstrate his splinting techniques; in 1881 his splints were extolled at the International Medical Congress; and in 1884 his writings were noticed and his splinting illustrated in the major surgical compendia by J. E. Erichsen and Thomas Bryant.[67] In November 1887, there arrived the first of the many American visitors to the Nelson Street Clinic, the 35-year-old John Ridlon, one of the founder members of the American Orthopaedic Association established earlier that year.[68] Ridlon became Thomas's leading American disciple and loudly proclaimed the virtues of his methods. Indeed, Ridlon's defence of Thomas's dogmatic advocacy of 'fixation' in the face of his New York colleagues' advocacy of 'extension and excision' was ultimately to cost him the chance of an appointment to the Chair of Orthopaedic Surgery in New York and to force him to move to Chicago.[69] Able American surgeons such as L. A. Sayre, C. F. Taylor and De Forest Willard, like their English counterparts, William Adams, Marsh, Paget and Frederick Treves, were not prepared to let pass Thomas's slashing comments on their

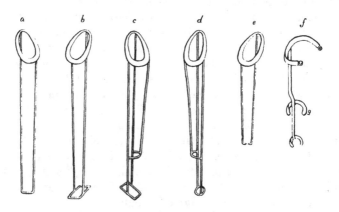

a. Ordinary Bed Knee Splint, for either side.
b. Ditto, with new end to raise it off the bed.
c. Bed Splint, with three bars for right side.
d. Walking Knee Splint, for left side. Three bars and terminal patten.
e. Calliper Walking Splint, to clip in boot heel.
f. Hip Splint, for right side.

Figure 2.1   Thomas splints

Source: Rushton Parker, 'On the Treatment of Fractured Femur', in his *Surgical Cases and Essays* (Adam Holden, Liverpool), p. 38.

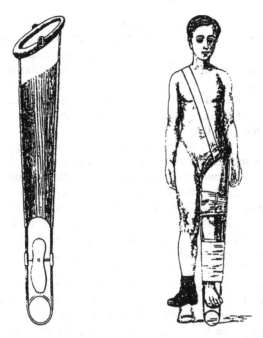

Figure 2.2   Thomas knee splint

Source: John Eric Erichsen, *The Science and Art of Surgery: a treatise on surgical injuries, diseases and operations* (Longmans, Green & Co., London, 1884), Vol. II, p. 389.

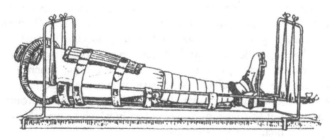

Figure 2.3   Thomas knee bed splint as adapted for hospital use during the First World War

Source: H. M. W. Gray, *The Early Treatment of War Wounds* (Henry Frowde and Hodder & Stoughton, London, 1919), p. 60.

methods.[70] Some also drew attention to the fact that Thomas made little use of the newer scientific findings in pathology, bacteriology and chemistry, and hinted at his neglect of operative methods, medication and massage and exercise.[71] But such adverse advertisement may have been advertisement nonetheless: it did not prevent Thomas from being elected to lecture before the Harveian Society of London in 1887, nor halt the spread of his reputation as an expert in the treatment of bone injuries and deformities.

Some of the Americans who recognized Thomas's expertise thought of him as a specialist in orthopaedics. But this was not his own view. He was always proud to be a general practitioner and prouder still to be the author of what he styled collectively his *Contributions to Surgery and Medicine* (1881–90), which included works on nerve inhibition, lithotomy and intestinal obstructions.[72] Although he acknowledged 'orthopaedic surgery' as it was then developing in America, he shared the contemporary prejudice in British medicine against the 'mania for specialities'.[73] British orthopaedics, as Thomas understood its origins in the work of Little and the specialist hospitals in London, bore only a slight resemblance to his own work in Liverpool. The latter embraced not only the congenital deformities of children and the treatment of tuberculosis of the bone and joint, but also the traumatic injuries of industrial workers. The latter was still outside the scope of orthopaedics.

Thomas can thus be seen as the first generalist in medicine whose practice encompassed much of the *future* clinical territory of modern orthopaedics. For this reasons he deserves the place of honour that was created for him after the First World War by his nephew and assistant, Robert Jones.[74] How Thomas's reputation might have fared without the interventions of Jones can only be surmized. But in establishing that reputation, and in planting the half-truth of Thomas's contemporary neglect, Jones was consciously seeking to sustain as rightfully 'orthopaedic' all that had been encompassed in Thomas's work. Jones was well-equipped for the task: charming, generous and unegotistical, he was quietly persuasive where Thomas had been aggressively intolerant and alienating.[75] What may have prompted Jones to carry out this task, and the context in which he was able to do so is that to which we can now turn.

3

POLITICS AND PROFESSIONALIZATION

Modern orthopaedics was not fated to exist, and Robert Jones was not preordained to play a major part in its development. Had it not been for the First World War, the specialism in its modern form might never have come into being. In Britain, at least, the treatment of fractures and other traumatic injuries involving bones and joints might have remained in the hands of general surgeons, and the aspects of child health that preoccupied orthopaedists after the First World War might have been subsumed in paediatrics. Specialization might have taken place only around the treatment of chronic skeletal deformities and muscular contractures, in which case the work of William Little, rather than that of Hugh Owen Thomas and Robert Jones, might now be regarded as more relevant to the study of its origins. In fact, shortly before the war, it was largely in terms of such a practice (applicable mainly to children) that a niche for orthopaedics came to be established in hospital medicine. In 1906, at the Charing Cross Hospital, H. A. T. Fairbank became the first orthopaedic consultant appointed to a British hospital with no responsibility for surgery outside the specialty. Thereafter several other major teaching hospitals established orthopaedic departments to which they appointed general surgeons with specialist interests in orthopaedics.[1] Similar developments took place in America and in Europe around the same time.[2]

That a different basis for orthopaedics came into being during the First World War was not, however, a consequence of the war itself. Nor was it simply a result of the specialism's wartime take-up of fracture treatment under Jones. 'Modern' orthopaedics was to be defined by more than merely an additional focus on trauma. Fundamental to its making were certain attitudes towards, and outlooks on, the organization of medicine in general. The war, as we shall see in later chapters, only accelerated or epitomized the broader politics of restructuring that had been gathering force in medicine for some time.

29

This and the following two chapters take up the changing circumstances between 1881 to 1913 that lay behind and ultimately enabled the 'modernist' construction of orthopaedics. The period is conveniently staked out by the two International Medical Congresses held in London in those years. In 1881, the application for a separate section on orthopaedics was rejected on the grounds that 'there were not, in England, enough orthopaedic surgeons of good repute to officer the section'.[3] In 1913 the same application not only succeeded, but was complemented by the formation of an orthopaedic subsection at the Royal Society of Medicine. This chapter discusses the medicopolitical and institutional factors behind that shift, while the following two consider wider aspects of the social and economic context in relation to the growth of medical interest in and organization around crippled children and the treatment of accidents and fractures. Such detail is warranted in light of the prevailing view of medical specialization as stemming simply from professional self-interest, usually seen as conditional upon advances in science and technology, and with the latter construed as more or less independent from the course of social and political history. Close attention is all the more necessary in view of the fact that there does not yet exist any readily available political history of medicine and surgery to cover these years. Medicine, and surgery in particular, was politically as well as therapeutically in a state of transition during the period, and developments in and around orthopaedics fully reflect the flux. Our discussion can begin, however, with Jones himself; as one of the central participants in these changes, his biography serves well to delineate the transformations involved.

ROBERT JONES (1857–1933)

When Jones began his formal medical education in Liverpool in 1873 at the age of 16, few could have predicted that he would be hailed in his lifetime as 'the Prime Minister of Orthopaedics and the Lord High Chancellor of Cripples'.[4] Even after he qualified in 1878, it would have been difficult for anyone to suppose that behind his apparent simplicity lurked 'the keenest brain, the most brilliant organizing ability, and the tact, the patience, the perseverance and astuteness of a diplomat'.[5] Hugh Owen Thomas, for one (with whom Jones began his apprenticeship, also in 1873), doubted that his nephew had the proper 'fighting spirit' to succeed. Perhaps it was because of this opinion that Jones tried all the harder, though it would have been almost impossible not to have learned the habits of hard work whilst apprenticing under Thomas. One of the benefits was that at an early age, when other young doctors were still waiting for patients, Jones was in full harness. After assuming control of the Nelson Street Clinic in 1888, he was also uniquely placed to enhance his reputation among his peers through the introduction of new techniques and procedures. In the 1890s he was early to adopt the 'no-touch' methods of aseptic surgery, and he was among the first in Britain to experiment with X-rays.[6] At the same time, he rationalized the procedures for dealing with patients so as to be able to handle staggeringly heavy case-loads. The latter arrangements in particular impressed the many visitors to Nelson Street before the war, the majority of whom were surgeons from America who recalled their visit as their 'introduction to modern orthopaedic surgery'.[7]

Jones's ambitions were not wholly confined to Nelson Street. In 1881, when he was 23, he secured an appointment as honorary assistant surgeon at Liverpool's Stanley Hospital (essentially an accident hospital), where he rose to the status of full surgeon five years later. In 1888, three years after he had set up his own private practice separate from Nelson Street, he was invited to act as the consulting surgeon for the construction of the 35-mile Liverpool-to-Manchester Ship Canal. In this capacity, responsible for the well-being of the 10 000–20 000 navvies annually employed on this five-year project, he devised what has come to be regarded as the world's first comprehensive accident service. Meanwhile, in addition to establishing one of the first hospitals in the country for chronically ill children, he began his life-long association with the open-air hospital at Oswestry in Shropshire. From the latter would eventually radiate a national network of over 400 orthopaedic clinics and hospitals.[8]

Jones's aspirations were further fulfilled in 1889 when he was elected Honorary Surgeon and Dean of the Clinical School at Liverpool's Royal Southern Hospital. There, in 1906, he organized one of the few hospital orthopaedic outpatient clinics in Britain – the only one at that time to deal with injured workers as well as crippled children. Run in tandem with the hospital's children's outpatient department, the orthopaedic clinic became as much of a Mecca for surgical pilgrims as the Nelson Street Clinic. Again, it was Jones's meticulous organization that left the strongest impressions. To the New York orthopaedist, Leo Mayer, for example, 'Most amazing were [Jones's]

Plate 3.1   Hugh Owen Thomas and Robert Jones, *c.* 1885.
(Probably taken by themselves with their own equipment.)

Source: By kind permission of the Wellcome Institute Library, London.

operative clinics at the Royal Southern Hospital, when at a single session he usually performed twenty or more operations. He never hurried, but he never wasted a motion, and his organization was so perfect that there was not a moment's delay between cases.'[9] Jones's colleague at the Southern, Charles Macalister, recalled in more detail how Jones:

got through an immensity of work, . . . rendered possible by the systematic preparation of the patients and by the work of the anaesthetists who had each successive patient ready by the time the operation on its predecessor had been completed. . . . He had round him a number of helpers, some of them medical men glad of the opportunity to get experience, others consisting of a nursing staff trained in the application of splints and plaster-of-paris. . . . [O]ther workers who had received some training kept an eye on the home conditions of the patients with reference to their feeding and regular attendance for massage, or other special treatment, at the Hospital.[10]

At a time when the whole idea of labour was being transformed, when the notion of the efficient 'human motor' was high on the intellectual agenda of the Western world, and when utopian visions of 'bodies without fatigue' – without waste – were burgeoning,[11] the organization of Jones's orthopaedic clinics embodied and exemplified modernity in medicine.

But it would be wrong to conclude that modern orthopaedics was simply the extension, upgrading and acceleration of traditional orthopaedics. Against such a view, first and foremost, lies the fact that Jones's work did not emerge out of 'orthopaedics', and that he himself did not aspire to be an 'orthopaedist' – to use the Americanism dating from the 1850s.[12] More so than Thomas, Jones's approach to bone and joint surgery was self-consciously generalist. From the time of his appointment at the Stanley Hospital he strove to cultivate a reputation as a general surgeon. Before 1910, he published on subjects as varied as diphtheria, trephining, colotomy, haemorrhoids, ulcers, intestinal obstructions, wounds of the abdomen and liver, and the cure of lupus, as well as on such later recognizably 'orthopaedic' subjects as fracture treatment, osteotomy, and joint injuries.[13] Only after 1910 did he devote himself almost exclusively to the treatment of locomotor complaints, but even then he remained firmly committed to generalism. In 1920, at the peak of his international reputation as an orthopaedic surgeon, he chose as the subject for his presidential address to the Liverpool Medical Institution Thomas's treatment of intestinal obstruction – thus underlining Thomas's breadth of vision, and his own.[14]

The shift to exclusively orthopaedic work was not the outcome of any new self-interested commitment to specialization. As William Mayo observed after visiting Liverpool in 1907, Jones belonged 'to that type of specialist who had been, and continues to be, a general surgeon, but has been forced by the large amount of work to become a specialist'.[15] In fact, Jones deliberately planned the steps of his career so as not to be stigmatized as a traditional (narrowly focused) orthopaedic specialist. He aspired, rather, to the newly emerging career structure of consulting general surgeon with a specialist interest, and even considered abandoning the Nelson Street Clinic once he had obtained his position as a surgeon at the Southern.[16]

SPECIALTY FORMATION

It was almost in spite of seeking not to be an 'orthopaedist' that Jones articulated the modern specialism as a branch of general surgery. But what he himself was eventually to take up as 'modern orthopaedics' was in large part defined for him by others who were simultaneously developing career structures and professional organizations in orthopaedics. Most prominent in this pursuit were the founders of the American Orthopaedic Association (AOA), whose first meeting took place in 1887 in the New York office of Lewis A. Sayre (1820–1900), the first incumbent (1861) of the first Chair of Orthopaedic Surgery in America (at Bellevue Hospital Medical College). Although Sayre's role in the AOA was soon usurped by younger men, it is worth noting that he was a radical surgeon who was abreast of the latest in general surgery as well as in the politics of American medicine. He had recently been the president of the American Medical Association and had played an important part in the founding in 1883 of the AMA's *Journal*.[17] In organizing the AOA, Sayre could hardly have been unaware of the contemporary organization of other specialist bodies both within and without the AMA. Emulating these, the AOA sought to bring together American orthopaedic practitioners in order to elevate their status both at home and abroad.[18]

Almost from the start, the founders of the AOA were concerned to emancipate orthopaedics from its association with 'braces and straps'.[19] Although the practice of most American orthopaedists before 1900 consisted mainly of treating deformities by mechanical means,[20] they aspired to a more credible surgical image. In 1891, one of the founder members of the AOA, Virgil Gibney of the New York Hospital for the Relief of the Ruptured and Crippled, redefined 'orthopaedics' as that 'department of general surgery which includes the prevention, the mechanical treatment, and the operative treatment of chronic or progressive deformities'.[21] By 1894, leading members of the AOA were referring to the 'orthopaedic surgeon' as one who 'has been thoroughly schooled in all the departments of medicine, who will have a perfect knowledge of pathology, surgical bacteriology, and anatomy'.[22] Two years later Gibney's student and associate, Royal Whitman, became the first to define the field as 'that division of surgery which treats of disabilities and diseases of the locomotor apparatus and of the prevention and treatment of deformities of the framework of the body'.[23] While this definition did not specify the inclusion of acute injuries into orthopaedics, it set out the basis for the modern specialism by identifying the locomotor system as its therapeutic territory. It also established the basis for professionalization within orthopaedics by identifying the field as a special branch of general surgery.

Not until the First World War did Jones define orthopaedics as 'the treatment by manipulation, by operation, and re-education, of disabilities of the locomotor system, whether arising from disease or injury'.[24] Long before then, though, he was aware of how the field was coming to be reconceived. From the 1880s he was visited by various of the charter members of the AOA and, in 1890, he was elected an honorary member of the Association. Between 1892 and 1893 he also collaborated in various publications with John Ridlon, who was then the honorary secretary of the AOA.

Jones was not the only Briton to be made aware of the professionalization of orthopaedics in America. Several other general surgeons also became honorary or corresponding members of the AOA at this time.[25] Among them was Alfred Tubby (1863–1930), since 1891 the honorary surgeon to both the National Orthopaedic Hospital and the Evelina Hospital for Sick Children. Three years after being elected assistant surgeon to the Westminster Hospital in 1895, Tubby became Surgeon-in-Charge of its new orthopaedic outpatient department,[26] a career step which exemplifies the late nineteenth-century pattern for consultant surgeons. Although his experience in 'orthopaedics' was more limited than Jones's, he shared the generalist approach to the specialism. Both men also shared the conviction that their specialist work was becoming better-known outside Britain than within it. Thus driven together, they became close friends, collaborated on various works in the field,[27] and, in 1894, established the British Orthopaedic Society (BOS), of which Tubby was to be the secretary.

THE BRITISH ORTHOPAEDIC SOCIETY

Information on the BOS is sparse, but the idea for it seems to have come mainly from Jones and to have been largely inspired by the AOA. Like the AOA, it took inspiration from the contemporaneous organization of other bodies 'devoted exclusively to a special branch of practice'. Telling in this connection is the fact that at the inaugural meeting of the BOS, held in Bristol in August 1894 during the annual meeting of the British Medical Association, it was 'decided not to elect a President, but rather to ask one of the members present to act as Chairman at each meeting, *thus following a precedent of some of the younger Societies*' (my emphasis).[28] Though precisely which societies were being imitated was not specified, the suggested egalitarian preference is worthy of note. Subsequent meetings were held in London and provincial centres, with programmes consisting of clinical demonstrations and papers. In May 1895 the Society visited Liverpool, taking in work at the Royal Infirmary and Southern Hospital, and a demonstration by Jones at the Medical Institution of the treatment of intractable cases of talipes.

In all, some 44 persons joined the BOS, of whom 27 were from the provinces and 17 from London. Only five of the members did not hold an MRCS qualification, while 27 were Fellows either of the Royal College of Surgeons of London or Edinburgh. The Society lasted only four years, however, and the slim volumes of its biannually published *Transactions* (begun in 1896) ceased in 1899. At any one time the BOS had no more than 33 members and its meetings were thinly attended from the start. Twenty-four of the 44 who joined the Society made contributions to orthopaedic literature, but only four members were deemed to have sufficient reputations to be elected to the AOA (Jones, Tubby, E. Noble Smith and W. J. Walsham); and only four were to become members of the postwar British Orthopaedic Association (Jones, Noble Smith, Tubby, and an assistant of Jones, the radiographer, C. Thurston Holland).

The Society's lack of success can be attributed to many factors. Not unimportant was that general practitioners were 'a little apt to put aside the question of deformities as dry and abstruse', as Jones was informed in 1894.[29] Given what has already been said about the status of conservative surgery in Britain, somewhat less weight should probably be accorded to the retrospective reckoning of some, that the Society was eclipsed by the high noon of Lister's prestige and the dazzle of radical surgery.[30] Nevertheless, relative to the attractions of other areas of contemporary surgery, orthopaedics could scarcely hold a candle. The corrective work with children, when not carried out in low-status orthopaedic hospitals, was conducted in equally lack-lustre outpatient departments in general hospitals. Although promising young surgeons like Tubby might be appointed to such departments, it was usually only because more prestigious hospital openings could not be found for them. For such men, the appointment to an orthopaedic department (and/or an orthopaedic hospital) was usually regarded as a toe-hold on the bottom rung of the hospital hierarchy.[31] It was seldom an acknowledged 'first step' to a career in orthopaedics. Typically, the 25-year-old Frederick Treves secured an appointment as clinical assistant at the ROH upon his arrival in London from Dorset in 1878, but quickly moved on to and up the career ladder of the London Hospital.[32] Not only were few inpatient beds attached to appointments in orthopaedics (even in the orthopaedic hospitals), but neither was much teaching connected with them, except on an *ad hoc* basis. As the report of the managing committee of the City Orthopaedic Hospital noted in 1905, the lack of integration into medical curricula further served to render 'the nature of Orthopaedic work . . . but little understood, even in the Medical profession'.[33] Hand in hand with the poor status of orthopaedics went its low remuneration. Because it concentrated on childhood diseases and deformities which were disproportionately among the poor, it provided little scope for private practice among the middle and upper classes. What room there was for the latter – for example, in the treatment of curvature of the spine – continued to be jealously guarded by senior general surgeons based in the voluntary hospitals.

Yet neither the popular and professional image of orthopaedics nor the realities of its practice wholly account for the failure of the BOS. After all, in America the status problems for the specialism were not dissimilar, yet the AOA survived. Nor should too much importance be attached to the fact that there were differences between members of the BOS in their therapeutic orientations. The AOA, no less than the BOS, had its share of dogmatic conservative surgeons as well as those, like Sayre, who leaned to the radical end of the spectrum. Newton Shaffer, for instance, one of the founders of the AOA, bitterly opposed the introduction of operative surgery into orthopaedics, believing that to mix it with mechanics was to endanger both.[34] Likewise, E. Noble Smith, of the City Orthopaedic Hospital in London, distinguished himself in the 1880s as an opponent of osteotomy and an advocate of purely manipulative techniques in the cure of deformity.[35] But most of

the members of both societies were more like Jones and Gibney in that
they relied on different therapies according to the type of case. Jones
followed in the conservative footsteps of Thomas, but he did not feel
compelled to 'insist, as Thomas would have done, upon surgical rest as
a principle of treatment in every case'.[36] William Mayo was to praise
Jones for applying 'real surgical principles to the treatment of
deformities', by which he seems to have meant operative surgical
techniques, as opposed to the use that Jones made in after-treatment of
'mechanical contrivances . . . as an adjunct to surgery'.[37] But Jones was
seldom regarded as a wizard with the scalpel;[38] more often he was seen,
as by Lord Dawson in the 1920s, as the world's greatest manipulative
surgeon – in other words, as a highly qualified bonesetter.[39]

At root, the different fates of the BOS and AOA lay in differences
between the respective contexts for specialization. While orthopaedic
practitioners in both countries in the late nineteenth century needed to
stress that orthopaedics was a branch of modern general surgery in
order to distinguish themselves from the older type of medically
qualified or unqualified specialist, the economic and intellectual
underpinnings of this view differed. In Britain, where general surgeons
in voluntary hospitals (especially the London teaching hospitals) stood
at the apex of the medical hierarchy, it was appropriate to articulate
specialist interests wholly in terms of serving the scientific interests of
generalists. Otherwise, claims for specialization might appear as
threatening not only to the confidence claims of general surgeons,
and to their physical territory (hospital beds), but also to their
privileged access to private patients. So long as securing an appointment as a general surgeon in a voluntary hospital was the pathway to
private practice, it was financially counter-productive for would-be
specializers to couch arguments for specialization in any other way.
This was reflected in the report in the *Westminster Review* in 1881 on
the progress 'in the various "special" departments of surgery', where it
was maintained that 'with this increasing cultivation of specialism, the
integrity of medicine is maintained by the greater recognition of broad
scientific principles as the only true basis in every speciality. [The
advancements in the special departments] . . . are refinements of
surgery, and not in any way independent of it.'[40] Such integrative
notions, buttressing the political status quo in hospital medicine, were
not contested in Britain before the 1900s, and then only obliquely.

In late nineteenth-century America, however, the idea of the
'essential unity' of surgery was coming to be superseded by more
Spencerian notions of 'natural' divisions of labour and of struggles

between groups. Such rhetoric was more appropriate to that country's less regulated, less formalized structure of medicine in which general surgeons exercised less authority, where private pay-bed hospitals – accessible to general practitioners – were more prevalent, and where medicine was increasingly perceived in commercial business terms and conducted from private 'offices'.[41] To be identified as a properly qualified specialist surgeon in this open-market context was to gain a competitive edge, whereas in Britain it was to run the risk of losing access to elite private practice. Unlike AOA members, few in the BOS were in a position to run such a risk. Jones was exceptional, but he shared the interconnected professional, financial and intellectual reasons for endorsing generalism.

It is not surprising, therefore, that the *Transactions of the British Orthopaedic Society* never extended beyond clinical discussion of techniques and procedures. Although members of the BOS adopted the American reconceptualization of orthopaedics, they felt the need to stress that it was '*merely* a special development of the conservative methods of surgery applied to abnormal conditions of the organs of locomotion'.[42] Far from hinting at any battles for turf, this statement barely suggested a separate therapeutic territory, and it was hardly an incentive to recruitment.

There were also other practical and conceptual problems obstructing specialist identity and specialty formation in British orthopaedics. Not least important was the difficulty of putting into practice a specialty embracing the whole of the locomotor system irrespective of the type of injury or the age of patients. Even in America, where there was more room for manoeuvre, only in a few places was successful entry made into the adult market – and then usually only for the treatment of chronic deformities as opposed to acute injuries. In 1895 the Boston orthopaedist, Joel Goldthwait, organized America's first clinic for adult cripples (at the outpatient department of Boston's Carney Hospital), and in 1904 he was among the orthopaedists involved in directing the medico-mechanical department founded at the Massachusetts General Hospital. Over the next six years the latter department was to treat nearly 5000 patients for non-united fractures, dislocations, stiff joints, sprains, arthritis and deformities.[43] In 1905, Goldthwait's orthopaedic colleague in the department, Elliot Brackett, remarked: 'It is within the remembrance of even most of the younger men when the orthopaedic work was almost entirely confined to children. Now a large, if not the larger, part is devoted to adults.'[44] In the same year, Gibney reported on over 154 cases that had passed through his adults' ward in the New York Hospital for the Ruptured and Crippled. His accompanying plea 'for hospitals in which the adult cripple, male and female, may receive the same scientific treatment that has so long been meted out to the children' was, however, little realized outside these major centres before the war. As in Britain – always excepting Jones's work in Liverpool – orthopaedics was compelled to remain a branch of practice focused on the chronic deformities of children. But from the perspective of potential specialty formation, this focus in itself raised problems for it overlapped with the more exclusive medical interest in children of other specialist groups and institutions.

CHILDREN'S MEDICINE AND SURGERY

The demarcation between work conducted in childrens' hospitals and that in orthopaedic institutions was seldom clear either in theory or in practice. Of the 44 members of the BOS referred to earlier, no fewer than 23 held appointments in children's hospitals: 11 worked in children's hospitals only; 12 held appointments at both children's and orthopaedic hospitals; and only six held appointments solely at orthopaedic hospitals. It is unclear whether such overlap itself partly accounts for the failure of the BOS. What is certain is that at the level of institutions there was increasing rivalry. The managing board of the City Orthopaedic Hospital, for example, remarked in their report for 1905:

> It is not generally known that Orthopaedic Hospitals were the first to deal with the special surgical afflictions of childhood and that later they embraced the treatment of deformities and disabilities in general. Hospitals for Sick Children were more recent in origin than Orthopaedic Hospitals, they have been more lavishly supported, partly because their name can be understood by the average person without the use of a dictionary. In spite of the invaluable work that some of the Children's hospitals have done in improving the general medical and surgical treatment of the disorders of childhood they have never taken the place of Orthopaedic Hospitals.[45]

Such defensive rhetoric suggests that by this date children's hospitals had outpaced and outclassed orthopaedic ones. This was especially so in relation to surgery, and in relation to bone and joint surgery in particular.[46] Although it was only with the introduction of osteotomy in the 1870s that children's hospitals became 'surgical' to any significant extent, thereafter it was within them that much of the new antiseptic bone and joint surgery was taken up and developed. (Osteotomy, after all, was mostly applicable to children in the correction of the knock knees and bow legs resulting from rickets.) At the Glasgow Hospital for Sick Children, for instance, Macewen conducted 833 antiseptic osteotomies between 1883 and 1892.[47]

Before the 1870s children's hospitals were primarily for medical cases. Of the 29 in Britain by 1889 (19 in the provinces, 10 in London), most had been established by physicians, and in some of them, as late as the 1890s, the wisdom of practising any kind of surgery had been openly questioned.[48] In Edinburgh, for example, the children's hospital had no surgical department until 1887. More characteristic was the Birmingham Children's Hospital where, out of a total of 1063 inpatients in 1862, 45 (a mere 4.2 per cent) were surgical cases.[49] By the end of the century, nearly two thirds of inpatients in children's hospitals were surgical, the great majority for the treatment of tuberculosis of the bones and joints. Thus children's hospitals – far more so than the few orthopaedic establishments in Britain[50] – became important sites for the practice of surgery in general and the development of expertise in bone and joint surgery in particular. Largely free from the vested interests of the senior surgeons in major voluntary hospitals, children's hospitals came to provide professional havens for aspiring junior surgeons anxious to try out new ideas and techniques, and to gain the experience necessary to move up hospital career ladders.

The epitome of such sites was the Children's Hospital in Boston, Massachusetts, established in 1869. By the 1880s approximately 65 per cent of all its surgery involved cases of tuberculous bones and joints, and, as one historian has remarked, it was this 'orthopaedic surgery' that was largely responsible for the transformation of the image of the Hospital from that of an orphanage steeped in medical fatalism to a beacon of curative medical activism attracting the attention of some of the most promising and progressive surgeons.[51] It was there in the 1890s that Harvey Cushing, later renowned as a brain surgeon, studied orthopaedics under E. H. Bradford, and thus came to interest William Halsted in establishing an orthopaedic department at the Johns Hopkins University Medical School.[52] It was at the Boston Children's Hospital, too, that such future leaders of the orthopaedic specialism in America as Brackett, Goldthwait, Robert Lovett and Robert Osgood cut their teeth and, in the process, established orthopaedics as a special branch of general surgery. Here too, the young Mancunian, Harry

Platt, would serve apprenticeship in 1913, the consequences of which were to be felt in the development of British orthopaedics in the postwar period.[53]

It is not clear to what extent the 'orphanage' image applied to children's hospitals in Britain before the 1880s. Most of them (as distinct from children's homes and outpatient dispensaries for children), largely dealt with short-stay cases, such as for acute abscesses and cleft palates, and served in part as institutions for childhood fever cases. It may be that the subsequent building of separate infectious disease hospitals (which were largely occupied by children) partly accounts for the number of beds made available for surgical cases thereafter. Whatever the exact sequence of events, it is clear that, just as in Boston, it was largely on the basis of surgery for bones and joints that the image of children's hospitals was recast and surgical reputations and careers established. For Manchester's Thomas Jones (1848–1900), for instance, an appointment as surgeon to the Pendlebury Children's Hospital from 1874 to 1880 was the pathway to an appointment as consultant surgeon at the Royal Infirmary and to a lectureship in Practical Surgery at Manchester's Owens College. His book of 1887 on the conditions and diseases of bones helped win him the Chair of Surgery.[54]

Thomas Jones's successor at Pendlebury from 1882, G. A. Wright (1851–1920), was another who distinguished himself internationally through his work in bone and joint surgery before being promoted to the Royal Infirmary and to the Chair of Systemic Surgery in the Manchester Medical School. A corresponding member of the AOA and a contributor to their *Transactions*, Wright made important contributions to the study of hip diseases in children, and co-authored with the senior physician at Pendlebury, Henry Ashby, what was to become the main textbook in the field, *The Diseases of Children, Medical and Surgical* (1889; 5th edition 1905).[55] A third figure to emerge out of apprenticeship at Pendlebury, E. D. Telford (1876–1961), was to become consultant surgeon to the Manchester Royal Infirmary in 1908 where he was responsible mainly for orthopaedic cases. In 1922 he was appointed Professor of Systematic Surgery at the University Medical School.[56]

The Edinburgh surgeon, Harold Stiles (1863–1946), is another who made his name largely on the basis of the bone and joint surgery he conducted in children's hospitals. His reputation in this area was such that, by 1911, when the Mayo brothers were seeking to establish an orthopaedic department at their Clinic, it was to him and to Robert

Jones in Liverpool that they dispatched the general surgeon Melvin Henderson for training.[57] During the war, Jones appointed Stiles the Deputy Director of Military Orthopaedics for Scotland; he was made an honorary member of the BOA; and in the 1920s and 1930s he played an active part in founding Scotland's first orthopaedic hospital for crippled children, the Princess Margaret Rose, in Edinburgh, and in founding the Chair of Orthopaedic Surgery at the University of Edinburgh.[58] The historian of the Mayo Clinic who referred to Stiles as the noted *orthopaedic* surgeon of Edinburgh may well be forgiven.[59]

The career of Stiles was in one respect significantly different from that of Thomas Jones, Wright or Telford: instead of moving from children's surgery to work with adults in a general hospital and in private practice, he did the reverse. To the consternation of his contemporaries he sacrificed his position as a full surgeon at the Edinburgh Royal Infirmary in 1892 to take up an appointment at the Edinburgh Children's Hospital.[60] His reputation in children's surgery was such, however, that, like Robert Jones in bone and joint work, he was able to survive in specialist private practice.

For most others in children's surgery this was not an option; for children's medicine and surgery, as for orthopaedics on its own, the restricted nature of the private market compelled most to remain in general surgical practice. Outside the private sector in Britain it was impossible to survive as specialists in these fields. Unlike in America, there were no salaried posts in university medical schools, nor indeed many teaching beds for these subjects in general hospitals. By contrast, at the Massachusetts General Hospital the orthopaedic outpatient department was staffed by 16 salaried surgeons, and in 1911 (through efforts by Goldthwait) America's first inpatient ward for orthopaedics in a general teaching hospital was created (with 22 beds by 1913).[61] Comparable relations between the surgery of children, university-controlled hospitals and orthopaedic specialization existed in Baltimore and New York.[62] Although most orthopaedics in prewar America was still conducted in outpatient clinics, an AOA survey of 1910 revealed that in 41 medical schools orthopaedics was taught as a separate branch of surgery by 35 full- or part-time salaried professors, and that in 51 schools it was combined with general surgery and was taught by 13 professors.[63] In Britain it was not until after the Second World War that equivalent means for the reproduction of career opportunities in orthopaedics became available.[64] Here, again, we have an indication of the reason why the intellectual rationalization of generalism remained intact.

ORTHOPAEDICS VERSUS 'PAEDIATRICS'

Although 'paediatrics' as such did not exist in nineteenth-century Britain, insofar as a distinction can be drawn between children's medicine and orthopaedics, this distinction was becoming less marked in the prewar period. In the treatment of bone and joint tuberculosis, rickets and infantile paralysis, 'orthopaedists' and 'paediatricians' alike were increasingly laying stress on the importance of hygiene, diet and fresh air. Infantile paralysis (poliomyelitis for the most part), though an unlikely concern for orthopaedists from a modern surgical point of view, had in fact long been associated with orthopaedics because of the use of braces and splints in its treatment.[65] After 1893 it became more closely wed as a result of the pioneering work on tendon transplant of Jones and Tubby. Significantly, Jones's early writings on tendon transplant in cases of poliomyelitis appeared in the *American Journal of Pediatrics*.[66]

Both potentially specialist fields of endeavour also shared the same occupational outlook in that they cut across the more usual organ-specific basis for specialization. On the one hand this meant that both encountered problems of conceptualization within the structure of medicine – at the Johns Hopkins Medical School, for example, paediatrics was at first a part of internal medicine.[67] On the other hand, though, aspirants to either field could proclaim as a virtue their generalist orientation to whole-person medicine – the one, on the basis of patient age, the other (increasingly), on the basis of concern with the entire locomotor system.[68]

It is less surprising therefore than it might at first appear, that in 1900 Alfred Tubby became the co-founder (with the ophthalmic surgeon Sydney Stephenson) of the Society for the Study of Diseases in Children (SSDC), and that Robert Jones became an active member on its Council and a contributor to its journal. Among others on the SSDC Council were Stiles, Telford, Henry Ashby, Charles Macalister (who joined Jones at the Southern Hospital after training under Ashby at Pendlebury), and Dawson Williams, the editor of the *British Medical Journal* from 1881 to 1928, who held an appointment at the East London Hospital for Children.[69]

Like the founders of the BOS, the Council of the SSDC looked encouragingly to technical and professional developments elsewhere in children's medicine and surgery. In particular, it turned to the American and Philadelphia paediatric societies and to the 'father' of American paediatrics, Abraham Jacobi. With high approval British paediatricians noted that a chair in paediatrics had been established in New York as early as 1860 and that a second chair had been created at Harvard in 1888.[70] Personal links with members of the American societies and publications in their specialist journals encouraged progressive postures as well as anti-establishment disdain for British 'backwardness'. An important provincial component to the membership of the SSDC doubtless contributed to this outlook.[71]

But the SSDC had as difficult a time as the BOS in convincing the rest of the medical profession of its worth. Interestingly, its apogee appears to have been reached in 1901 when Jones and Macalister hosted the provincial meeting in Liverpool[72] – thus coinciding with the demise of the BOS. But thereafter the SSDC fell into disarray. In addition to being confronted by anti-specialist general surgeons, especially in London,[73] the SSDC had problems over the recognition of general practitioners within its ranks. The latter's fears of a specialism poised to encroach on one of their main sources of livelihood seems also to have been behind the BMA's threatened closure of its section on Diseases of Children in 1896. Significantly, on that occasion, it was the members of the BOS, recognizing that orthopaedics 'had a very special reference to diseases peculiar to children', who successfully lobbied against the move.[74] The BMA's Children's Section survived into the interwar period, but only just – the professional precariousness of this whole area of medicine being widely evident.[75] It is not therefore difficult to understand why the SSDC should have decided in 1907 to dissolve itself in exchange for specialty shelter in the Royal Society of Medicine (RSM) as a section on Diseases of Children. The move was bitterly opposed by Jones for reasons which remain unclear, but which probably had to do with the perceived surrender of children's medicine and surgery to metropolitan and establishment interests. Preceded by a bitter internal struggle within the SSDC, the move led to Jones's withdrawal from the Society and to his permanent estrangement from Tubby.[76]

That in 1913 orthopaedics itself became a subsection of the RSM does not vitiate this probable explanation for the rupture between Jones and Tubby. However, two other circumstances outside Jones's control require mention. The first is that underlying the Royal Society of Medicine's incorporation both of the SSDC and orthopaedics were its own political interests. From around the turn of the century the RSM had been earnestly seeking to improve its position, especially in relation to the BMA, and it was in the context of this struggle that it actively encouraged the affiliation of budding specialist groups. In the same year that the RSM absorbed the SSDC, it absorbed the British Electro-Therapeutic Society, and in 1912, the year before orthopaedics was embraced, it also took in urology and proctology (including the recently formed British Proctology Society).[77]

Precipitating the inclusion of orthopaedics into the RSM was, second, the advent of the International Medical Congress at which orthopaedics was to have its own section. Since 1890, when the Congress had been held in Berlin, members of the AOA had lobbied for a separate section, and these efforts were crowned in 1909 when Jones was chosen by the Americans to act as the section's president.[78] Thus, after this date, Tubby and the London orthopaedists were in a strong position to lobby the RSM for a separate orthopaedic section, wielding the international status of orthopaedics as a stick, and using the advent of the meeting in London as an excuse. Whether Tubby was also hoping to upstage Jones is not known; it is worth noting, however, that upon the formation of the orthopaedic section in the RSM, Tubby nominated as its first president Ernest Muirhead Little, the son of William Little.[79]

Despite appearances, then, professional coherence in British orthopaedics on the eve of the Great War was largely illusory. So far as the orthopaedic sub-section in the RSM can be taken to represent London neo-traditional orthopaedics, and so far as Robert Jones and the place of orthopaedics at the International Medical Congress of 1913 can be taken to signify other sets of interests, 1913 might better be regarded as the moment when the different worlds of orthopaedics came into collision. Before we can add to this, though, something more must be said about these other interests as they developed in the period from around 1900 to the eve of the Great War.

THE POLITICS OF PROVINCIAL SURGERY

William Mayo, in commenting in 1907 on the 'force of circumstances' that compelled Jones to concentrate on orthopaedics while remaining a general surgeon, was isolating one of several characteristics that identified Jones within a significant cohort of British surgeons. Among others in this group (also visited and lauded by Mayo on his tour of British surgery in 1907), were Stiles of Edinburgh, Victor Horsley of the National Hospital for Paralysis and Epilepsy in London, Rutherford Morison of Newcastle, and Berkeley Moynihan of Leeds.[80] Like Jones, these were all talented and highly motivated general surgeons who, in the course of seeking to extend their reputations, increasingly restricted themselves to specific areas of surgery: Stiles and Morison concentrated on the surgery of children, Horsley on neurosurgery, Moynihan on abdominal surgery.[81] Like Jones, too, none of them were self-interested advocates of specialization; although they had no time for 'reactionary' opponents of specialism, they remained strongly committed to the idea of the essential unity of surgery. Marking them off from the older generation of surgical generalists was their further and overriding commitment to expertise and efficiency, and their desire to be judged on the grounds of efficiency and productivity rather than, say, on the quality of doctor/patient relations. As was frequently said of the work of Jones, so it was to be remarked of Stiles, for instance:

> he has the largest amount of surgical work of any man in Edinburgh. He is the chief surgeon of the Children's Hospital and Chalmers Hospital [for children], and has a private infirmary, giving him, in all, not far from 100 beds. He is 45 years old, an indefatigable worker of the American type, and is well known to us as the translator of Kocher's 'Operative Surgery'. He is rapid, exceedingly accurate, and has no frills of any kind.[82]

Or, as Ernest Jones, Freud's biographer, was to say of Horsley (after having served as his house-surgeon in 1902): '[he needed] to fill the unforgiving minute with sixty seconds worth of distance run. He got through more work in a day than anyone else I have known, making use of each moment ... yet with all his swiftness he was never hurried.'[83] 'Efficiency' in this context, as Susan Reverby has noted, 'signified a hardworking, disciplined, unemotional person, willing to break with traditions to become an effective expert'.[84] 'Fatigue' on the other hand, as Anson Rabinbach has shown, symbolized the 'subversion of modernity'.[85]

Such preoccupations are not readily separable from another of the characteristics of these surgeons: their involvement in and identification with the basic sciences and with physiology in particular. Before the turn of the century, as the eminent physiologist, Edward Sharpey-Schäfer reflected in a lecture of 1923 on 'The Relations of Surgery and Physiology,' 'it might have been admitted that there was a certain amount of interdependence between Medicine and Physiology, but no one would have thought of connecting Surgery with anything but Anatomy.'[86] What the American surgeon, George Crile, described in relation to his work on the body's reaction to shock (begun under

Horsley in 1895) as 'physiological surgery',[87] was highly conspicuous in the 'new neurology' of Horsley and in the 'new cardiology' developed by Thomas Lewis, James Mackenzie and others.[88] But it was no less evident – at least as a style of thought or an orientation – in the orthopaedics of Jones and his American colleagues. Indeed, a senior figure in British medicine would later maintain that '[modern] orthopaedic surgery stands not so much for the correction of deformity as for *a vision dominated by the study of physiology and the respect of function*' (my emphasis).[89]

Central to this vision was a holistic comprehension of the body in terms of an integrated dynamic system, the efficiency and economy of which was seen to derive precisely from this organization. (Would-be modern orthopaedics, which was predisposed to such elaborations as a consequence of the commitment to conservative surgery, was quintessentially physiological in this sense, since its primary concern was with the restoration of parts of the locomotor *system* for the sake of their maximum use-efficiency or *function* – the restoration of such function being seen as dependent on the condition of the patient's body and mind as a whole.) The value attached to efficiency and industry by these physiological surgeons was thus constitutive of their perception of the body – or rather, of their functionalist idealization of the body. As a model for organization, this physiology was also part and parcel of the cohort's concern with new ways of managing patients, as exemplified by the division of labour and other rationalizations of clinical procedures effected by Jones in Liverpool. In fact no separation is possible between these physiologically informed rationalizing concerns and the application of 'scientific management' to business and industry at this time – the latter itself being derived from physiological idealizations of organization and function. We shall return to this in subsequent chapters; suffice to say here that the idea of efficiency inherent in the physiological outlook bore both a moral and a managerial load.

In spite of the fact that the cohort did not advocate specialization as such, then, their celebration of surgical expertise was not without threat to others in medicine. Especially to the defenders of surgical generalism in the metropolitan medical establishment who claimed mastery of surgery as a whole, the meritocratic assault of the 'young turks' on the traditional enclaves of power, status and authority was only too apparent – even if the abstract physiological mediation of the meritocratic values was not. After all, much of the cohort's growing international fame derived not from work conducted within the major teaching hospitals, but from that done in small specialist hospitals, outpatient departments, and/or private clinics such as Jones's in Liverpool. Not unlike the centre of excellence that William and Charles Mayo built up in remote Rochester, Minnesota, most of these clinics, with the exception of Horsley's, were geographically distant from the traditional metropolitan centres of medical excellence and authority.

The potential of such havens of expertise to challenge metropolitan authority had, in fact, been demonstrated some years earlier. In 1883, a year after the American Surgical Association was formed by certain specialist general surgeons,[90] some 200 Fellows of the Royal College of Surgeons, mostly from the provinces, rallied behind the call of the Birmingham gynaecologist, Lawson Tait, to organize a British Association of Surgeons. Tait, who was one of the first consultant specialists in Britain, was then a vice-president of the Royal College of Surgeons, and his threat to form a surgical association allied to the BMA forced the Royal College to change its rules and allow provincial surgeons representation on its council.[91] The positive response of the Royal College of Surgeons to the threatened Association of Surgeons effectively nipped the latter proposal in the bud, but in a context in which the civic pride and political strength of major provincial cities were growing as a result of new sources of commercial wealth, and in which local educational and health systems were expanding, the concession did little to dampen the underlying spirit of dissent from metropolitan medical power, privilege and exclusion. Mayo echoed some of this when he commented in his report of 1907:

> Under the present conditions it is nearly impossible for a man from the provinces, no matter how great his ability, to obtain a position on the staff of any of the London hospitals . . . This does not mean that the general average of surgical work in London is poor, but it does mean that some of the most modern surgery is to be seen in the provincial towns.[92]

By the 1900s the strength of this provincial anti-traditionalism was compounded through identification with and emulation of all that was internationally best in surgery. Among the cohort, the strongest links, both personal and professional, were with preeminent American surgeons such as Crile (with whom Jones stayed during one of his visits to the USA),[93] Cushing and Halsted of the Johns Hopkins University Medical School, the Mayo brothers, and J. B. Murphy of Chicago. Most of these surgeons, it should also be noted, had more than a passing interest in bone and joint surgery, and several ran flourishing private clinics where they routinely dealt with such cases.[94] Along with others, these doyens of progressivism in American surgery came together in 1903 to form the Society of Clinical Surgery, which was masterminded by the 'prince of organizers', Franklin Martin, the earnest disciple of Murphy.[95] Several were also involved in the founding in 1905 of the journal *Surgery, Gynecology and Obstetrics* (edited by Martin), and most of them were later to constitute the inner core of the American College of Surgeons, established in 1913, and also masterminded by Martin.[96] The foundations of the American College of Surgeons were in fact laid in Britain in 1910 when the Society of Clinical Surgery made its first trip abroad in order, especially, to visit the clinics of Horsley, Moynihan and Jones.[97] Ironically, it was after a gala dinner in London at the Royal College of Surgeons that the idea was mooted of a College in the USA, the object of which would be to standardize in American hospitals the kind of surgery that was best represented in Britain by the elite of dissident 'provincial' surgeons.

The Society of Clinical Surgery's visit to Britain was at the invitation of Moynihan who, a year previously, had organized the Provincial Surgeons' Club on the model of the American Society of Clinical Surgery. Among the members of the Provincial Club, in addition to the specialist general surgeons already mentioned, were such later eminent figures as Sir William de Courcy Wheeler of Dublin, John Lynn-Thomas of Cardiff, A. M. Henry Gray of London (then a student of Horsley), and Ernest Hey Groves of Bristol, all of whom were to become involved with Jones in his wartime organization of orthopaedics.[98] The object of the Provincial Surgeons' Club, like its American counterpart, was to bring like-minded men together in order to advance their surgical skills through visits to important clinics and, simultaneously, cement professional and political ideals. Hey Groves later confessed, for instance, that it was the Provincial Club that gave him the ideas on hospital reform that he deployed in the 1930s. It was also the pathway by which he came under the 'gentle but compelling genius' of Jones, and hence to be among the select group that met in November 1917 to plan the BOA, of which Hey Groves was to be the president in 1928.[99]

In 1913, four years after the founding of the Provincial Surgeons' Club, Moynihan took another leaf from the book of his American confrères and launched the *British Journal of Surgery*. Hey Groves was the co-founder and editor, and among those on the editorial committee were Jones, Morison, Stiles, and Lynn-Thomas.[100] Outstanding British surgeons, especially provincial ones who were not a part of Moynihan's network, were soon brought in; among them, secured by outright flattery, was Macewen of Glasgow, regarded by many at the time – himself especially – as the greatest surgeon in Europe,[101] and Sir Rickman Godlee, the nephew of Lister, who was then the President of the Royal College of Surgeons.[102] Moynihan also cultivated the support, or at least stroked the feathers, of the presidential successors to Godlee at the Royal College of Surgeons, Sir W. Watson Cheyne and Sir George Makins, though neither was to be closely identified with the provincial circle.[103]

The existence of the Royal College of Surgeons and, to a lesser extent, the Royal Society of Medicine, continued to present formidable obstacles to the formation of new national associations of surgeons in Britain before the war.[104] Nevertheless, it is apparent that in Britain, as in America, similar efforts were going into the political restructuring of surgical organizations so as to give representation to the interests of the new generation of surgeons with their claims to industry, efficiency and expertise. The conferring of honorary fellowships on Crile, Murphy, Mayo and other American representatives of the new surgery by the Royal College of Surgeons during the International Medical Congress of 1913 was acknowledgment of the new generation's ideals and aspirations.

Whether Jones's place in the politics of provincial surgery permits us to claim that by 1913 his orthopaedics was a world apart from that conducted in London, at least three things should now be clear. First, the 'new orthopaedics', as exemplified by Jones's work in Liverpool, did not mainly descend from the old orthopaedics, but rather, was in large part a branch of the new surgery. Second, the new surgery, and above all the new orthopaedics, was developed in Britain mostly outside the major hospitals, either in private clinics, such as Jones's, or in 'junior' institutions, such as the children's hospitals. Finally, the new surgery, in challenging the traditional dominance of London in medical education, practice and political authority, can be seen as partly an expression of the expansion of provincial wealth and influence.

To perceive these social and political developments as the essential background to modern orthopaedics is not to imply that scientific and technological factors can be ignored. As the technique of tenotomy was basic to William Little's orthopaedic practice (as possibly 'Enlightenment science' was to Andry's coining of the word), so osteotomy, X-rays, aseptic procedures, and other prewar technical procedures, such as arthoplasty (joint reconstruction), tendon transplants, bone-grafting, and open surgery for the reduction and fixation of fractures, were all important to the orthopaedics of Jones and his colleagues. So too, were new bodies of scientific knowledge, such as that on the structure and function of nerves, the histology of bone cysts, bacteriology, the laws of heredity, endocrinology, and the biochemistry of nutrition.[105] While most of these innovations did not emerge from those who might have styled themselves 'orthopaedists',[106] and few were seriously pursued in orthopaedics before the war, identification with them could be an important symbolic expression of modernity. This chapter has not dwelt on the history of these innovations, mainly because they constitute the familiar stuff of most 'internalist' accounts of orthopaedics and of medicine in general. But it would be wrong to conclude that they can therefore be juxtaposed to political agency in the understanding of specialty formation. On the contrary, the scientific and the technological need to be seen as part of the social and political processes that opened the space for the new orthopaedics. Not to acknowledge this would give grounds for reinforcing notions of technological determinism and/or models of 'natural growth' in the development of medical specialties. Indeed, it would strengthen the idea of the historical 'inevitability' of medical specialization against which this chapter has argued.[107]

4

THE CAUSE OF THE CRIPPLED CHILD

Unlike the politics of surgery between the 1880s and the First World War, the history of the child has been extensively pursued. The demographic, socioeconomic and ideological conditions that contributed to the late-Victorian and Edwardian preoccupation with, and revaluation of, childhood are now familiar. Known too, in large part, are the shifting philanthropic and state political responses to childhood poverty, illness, deprivation and abuse, to say nothing of 'deviancy', delinquency and 'degeneracy'.[1] It is also widely acknowledged that as education became compulsory after 1870, so the schoolroom took on the character of a laboratory for the medicopsychological surveillance, control and testing of children. Very largely, this was the site where childhood was reconstructed in predominantly psychological and medical terms.

Little attention, however, has been paid to the medicalization process itself. Historians of such specialisms as paediatrics and obstetrics, which were broadly dependent on the 'child health and welfare movement', have tended to concentrate on the internal politics of professionalization, and have sometimes been criticized for privileging the role of technologies.[2] Historians of child psychology, child guidance and mental testing, on the other hand – although attentive to the rise of medical 'experts' – have produced mostly intellectualist accounts of the development of psychological thought, or have contributed to debates over state education and social policy.[3]

This chapter's concern with how the social 'discovery' of the 'crippled child' contributed to the shaping of modern orthopaedics prior to the First World War facilitates an exploration of the medicalization of the child which cuts across many of these historical compartments. The growth of services and institutions for these children by would-be orthopaedic surgeons illuminates a context in which socioeconomic, ideological, political, technical and professional factors were increasingly intertwined and interdependent. Within this context modern notions of 'child health' became linked to major tendencies in welfare and social reconstruction.

DISCOVERING CRIPPLED CHILDREN

Like the growth of interest in children in general towards the end of the nineteenth century, the preoccupation with physically handicapped children was common to the whole of the Western world. If Britain was different in any respect, it was perhaps only in the diversity and multiplicity of its voluntary agencies for such children – over 40 of which emerged between 1870 and 1914.[4] Hitherto, handicapped children had been the subject of little attention. True, the early nineteenth-century reform movement had spotlighted children crippled through accidents in unsafe factories, mines and workshops, but that concern had faded with the passage of legislation restricting the employment of children.[5] In any case, these were not the 'crippled children' of the late nineteenth century. The latter, personified by Bob Cratchet's son in Dickens's *A Christmas Carol*, were typically the victims of disease rather than of dark satanic mills. In particular, they were sufferers from rickets, poliomyelitis and, above all, tuberculosis of the bones and joints. Until the late nineteenth century emotional and financial outlay on such children had been regarded as wasteful – in part, because they were 'thought unlikely to outlive childhood'.[6] They had been shunned, too, for supposedly reflecting the sexual sins of their parents – syphilis being among the major predisposing causes of crippling.[7] Thus a small Cripples' Home and Industrial School for Girls, established in 1851 in Marylebone, London, was one of the few special institutions for these children. It was to be another 14 years before the National Industrial Home for Crippled Boys was established in Kensington.

Medical care, as distinct from purely custodial provision, was also rare. In London, a few privileged cases of bone and joint tuberculosis might find their way to the 30-bed Hospital and Home for Incurable Children, at Maida Vale, or to the only other voluntary hospital for this purpose before the 1880s, the Alexandra Hip Hospital (established in 1867 and extended to 68 beds in 1899).[8] But most were destined to languish in workhouses along with the blind, epileptic and insane.[9] In a society whose dominant ideology measured individual worth by the ability to be productive, indigent cripples were simply 'in-valid' as humans.

The American sociologist, Viviana Zelizer, has convincingly argued that the introduction of child labour laws and compulsory education towards the end of the nineteenth century were a part of a general process whereby children became emotionally priceless to parents and to the community at large.[10] The wage-earning 'non-child' of the labouring poor was transformed into the economically worthless 'child-scholar' and appropriated into a romantic ideal of childhood. The trend was 'unmistakable', Zelizer writes:

> In the first three decades of the twentieth century, the economically useful child became both numerically and culturally an exception. Although during this period the most dramatic changes took place among the working class, the sentimentalization of child life intensified even among the already 'useless' middle-class children. (p. 6)

If, as seems the case within this context, crippled children of the labouring poor came to be all the more precious for being all the more economically 'useless', their end-of-the-century 'discovery' might be described as doubly indebted to the introduction of compulsory education. For, in Britain at least, it was the failure of the Elementary Education Acts of 1870 and 1880 to make any provision for mentally and physically handicapped children which rendered these groups more than ever conspicuous.

Philanthropic endeavour drew further attention to handicapped children. Initially, much of this philanthropy was local, missionary in tone, and directed at highly specific targets. An early example is John Groom's Crippleage and Christian Mission for Watercress and Flower Girls, established in London in 1863. More typical by the 1880s was the Manchester and Salford Crippled Children's Help Society, which emerged from a Band of Kindness intended to inculcate in children Christ's kindness to animals.[11] It was further characteristic of this philanthropy that much of it was undertaken by middle-class women, many of whom were themselves seeking more emancipated lives. Their charity, supposedly enshrining selfless female qualities, was often deeply sentimental, as reflected in such titles as the Crutch and Kindness League, Guild of Good Samaritans, League of Hearts and Hands, and the Guild of the Brave Poor Things. Increasingly, however, such charity was superseded by, or came itself to aspire to, a more 'scientific' form. Reflecting and influencing this shift was the Charity Organization Society (COS), established in 1867 by urban professionals and businessmen to ward off mendicity and charity 'abuse', and to deal with 'problems of character' by placing charity on a 'rational' or 'scientific' footing. 'Scientific' in this context meant impersonal, business-like management, in contrast to the allegedly 'soft' and indiscriminate charity of evangelical bodies like Dr Barnardo's.[12] Increasingly, the COS was also to perceive its mission as defending the socioeconomic order against socialist threats.[13]

Whether 'sentimental' or 'scientific', however, the charity extended to crippled children served an ideological purpose. At a time when confidence in the social philosophy of individualism was becoming shaky, it offered powerful reinforcement to the ethic of self-help. Although some of the voluntary organizations only provided boots, braces and crutches, and others offered no more than fresh-air holidays, behind them all lay the object of rendering crippled children capable of literally standing without crutches. Ironically, in light of the economic devaluation of children of the labouring poor in the late nineteenth century, crippled children were to be taught 'the dignity of labour'. Despite handicap, they were to be made 'useful and independent' – able 'to stand up to the emery wheel of competition'.[14] The acclaimed cripples' home at Alton, in Hampshire, established in 1906 by the carpet manufacturer, Sir William Purdie Treloar, was founded expressly as a 'labour colony'.[15] Similar was the training school for cripples founded in 1903 at Chailey, Sussex, where the motto 'Men Are Made Here' was engraved over the boys' entrance.[16] Cases deemed unimprovable were excluded from these homes.

It might be argued, then, that what was different in late Victorian society was not the abolition of the ideology behind the former invalidation of cripples as humans, but rather, the greater interest in capitalizing on cripples as reinforcement for that ideology. The new, ostensibly more sympathetic attitude to cripples was often expressed as if it were radically at odds with the 'brutal doctrine' of the survival of the fittest. Yet, in reality, it was a means of bringing cripples and other actual or potential 'deviants' into an individualist world of social and economic struggle. As Ruskin summed it up: 'This is the help beyond all others; find out how to make useless people useful and let them earn their money instead of begging it'.[17]

Such intentions did not necessarily conflict with the more obvious change at the end of the century: the movement of the state into child welfare. Although it was not until 1899 that permissive legislation was passed giving local authorities the statutory power to make arrangements for ascertaining the numbers of children suffering from physical defects and to provide for their education, in the 1880s and 1890s, obstacles in the way of such legislation were steadily overcome. By the 1880s, strict adherents to the doctrine of governmental non-intervention were hard to find and, in relation to physical and mental health in particular, it was clear long before the spectre of 'physical deterioration' became a national-cum-nationalist obsession during the Boer War, that sweeping state action was called for.[18]

There was no necessary conflict between the actions of the state and liberal individualism. For contemporaries to appreciate that *some* social problems might in certain respects be beyond the scope of voluntary effort was not to submit either to Fabian state centralism or to socialist collectivism. As the Education Acts themselves reflect, the role of the state as perceived by Gladstonian liberals was that of a junior partner in the social and moral undertakings of voluntary organizations. Far from stultifying voluntary effort, drying up the 'fountains of charity', or undermining individualist philosophy, the state was seen as husbandman to voluntarism.

By the turn of the century there were several areas of social reform where the boundaries between state and voluntary action had become indistinct. Many of those who became involved in voluntary activities for crippled children were by then also serving on School Boards, local councils and Boards of Guardians – the election of women and working men to which became possible after 1894. The Settlement Movement of the 1890s also aimed, specifically, 'to foster a spirit of co-operation between public and private agencies'.[19] Shunning the labels 'philanthropy' and 'charity' and encouraging 'civic responsibility', settlement workers – most of whom were women from the educated middle class – tended to be less concerned with deterring the idle, than with stimulating welfare schemes. The Guilds of Help were similar in character, some even tending to municipal socialism. The first was begun in Bradford in 1904, and by 1910 there were over 60, mostly in the north of England.[20] Progressive MOsH, as well as persons employed by the Board of Education and the Local Government Board, also participated in voluntary work on behalf of the unhealthy and deprived, and encouraged others to join with them. That others did so is, of course, partly a reflection of the fact that notions of 'unhealthy and deprived' – the language of poverty, environment and social amelioration – had begun to displace the hereditarian casting of crippledom and the association with degeneracy.[21] Meanwhile, the COS was steadily consolidating its authority through rationalizations and coordinations of public and private welfare services in the name of efficiency. In a similar spirit, around 1900, various of the organizations involved with crippled children in Birmingham, Manchester and Liverpool formed federations to avoid wasteful duplication of effort and to encourage greater public expenditure.[22]

The largest voluntary organization specifically concerned with the crippled children of the poor was the Invalid Children's Aid Association (ICAA), established in 1888 as an offshoot of the COS. Its primary concern was with providing adequate home care for crippled children – hence with inspecting domestic conditions and 'strengthening family ties'. But the ICAA also arranged for care in convalescent homes: of 221 cases investigated during the first year of its operation, 106 received treatment away from home.[23] By 1900 the ICAA was operating in a quasi-official capacity as a central organizing, advising and coordinating body, dealing with cases referred by public health authorities, MOsH, School Medical Officers, district nurses, midwives, GPs, and hospital outpatient departments and dispensaries. By 1919, when its army of women volunteers in branches across London had over 33 000 crippled children on the books, it was an agency much consulted by the Board of Education which regarded it as a collaborator and source of expertise.[24]

With the development of these more formalized functions around the turn of the century, leading figures in the crippled children's movement came to be regarded less as philanthropists or agents of charity than as voluntary social workers. Margaret Beavan (1877–1931) of Liverpool was typical. Inspired by the work of the Liverpool Victoria Settlement in the 1890s, she became involved with the local Kyrle Society for Children; in 1907 she became the secretary of the city's Invalid Children's Association, and in 1910 organized the after-care services for the children attending Robert Jones's clinic at the Royal Southern Hospital. In 1913, with the assistance of Jones and George Newman, the Chief Medical Officer to the Board of Education, she established the Leasowe Open-Air Hospital for children suffering from tuberculosis of the bones and joints. On the strength of these achievements, in 1927, in Liverpool, she became Britain's first woman Lord Mayor, though two years later she disgraced herself as 'Maggie Mussolini' when she stood as the Tory candidate for the depressed constituency of Everton.[25]

Mrs (later Dame) Grace Kimmins (1870–1954), who was married to the chief educational inspector for the London County Council (LCC), was another who moved from settlement work in the 1890s to more ambitious projects for cripples. Inspired by her success at founding the Guild of the Brave Poor Things, she went on to develop the training school at Chailey, in Sussex, on funds raised in a public appeal by the Duchess of Argyll, but later supported by the Board of Education.[26]

By far the most notable of this generation of women activists on behalf of cripples was the novelist, Christian revivalist, settlement worker, COS activist and granddaughter of Thomas Arnold, Mrs Humphry Ward (1851–1920). In 1899, at the Passmore–Edwards Settlement in London, she opened Britain's first rate-supported school for invalid and crippled children. She also led the campaign that secured the passage of the Elementary Education (Defective and Epileptic Children) Act of 1899. Mrs Ward's husband was the art correspondent for *The Times* and it was through a series of letters that she published in that paper in 1900 that the resolution was passed by the London School Board in October 1900 to take up the permissive statutory powers contained in the 1899 Act enabling special provision for crippled children. As Chairman of the Education Sub-Committee of the Joint Parliamentary Advisory Committee to the Local Government Board, Mrs Ward was also largely responsible for securing 'the Crippled Children's Charter' – the clause in the Fisher Education Act of 1918 that made it *compulsory* for local education authorities to make special provision for crippled children within ten years. Unlike Margaret Beavan, Mrs Ward never aspired to political power, nor did she approve of women claiming it. In 1908 she founded the Women's National Anti-Suffrage League. However, with a son in Parliament from 1910 to 1918 (who was also active on the Joint Advisory Committee), and with a daughter married to George Macaulay Trevelyan, Mrs Ward was never far from the corridors of power, nor without influence within them.[27]

MEDICAL INTERESTS AND APPROPRIATIONS

Few of the voluntary organizations with which these women were associated were primarily concerned with meeting the medical and surgical needs of crippled children. The COS, in an influential publication of 1893 on *The Epileptic and Crippled Child and Adult*, scarcely invoked medical authority, and nowhere projected medical science as a potential solution. In general, helping handicapped children had more to do with humanitarian movements and concerns over education, than with medical aspects of child health.[28] Not until the late 1920s did the ICAA change its object from that of the 'supervision and assistance of invalid and crippled children' to 'obtaining for them the best medical treatment'.[29] Although voluntary organizations were not unaware of developments in children's medicine and surgery, and frequently claimed the medical treatment of cripples as their first consideration, their meetings and publications reflect greater interest in inculcating in cripples 'the love of God and their neighbours', and teaching their parents to avoid spoiling and/or exploiting them commercially.[30]

Nor was much attention directed to curative procedures in the homes that some of the charities provided. Retrospective accounts of these institutions by medical authors understandably emphasize that treatment 'was either non-existent or so notoriously bad that the term "cripples' home" was associated with pity and contempt'.[31] But this is to misread the purpose of these homes. Most were intended first and foremost as places for education and training. Chailey, for instance, had no medical officer until 1927,[32] and the medical officer appointed to Lord Treloar's Cripples' Hospital at Alton in 1908 – the later-to-be-famous, Sir Henry Gauvain (1878–1945) – was an inexperienced 30-year-old who had only just completed his house appointment at St Bartholomew's Hospital.[33] Since 'the problem' of the crippled child was articulated primarily in moral and educational terms, and was underpinned by socioeconomic legitimation, medical men were largely irrelevant to its solution – at least as healers.

Some of those who worked for the ICAA at the turn of the century *did* propose that crippled children 'should be put forward for surgical treatment in hospitals, and their defects and deformities rectified to such an extent that they could safely attend schools for ordinary and healthy children'.[34] But aside from being essentially an educationalist's perspective, this was also a minority view. Behind it was the assumption that there would always be cripples who could never be helped by surgery. Surgeons themselves shared this view, many of them admitting to a large class of 'stationary cripples' who, as a result of congenital malformations, accidents of childbirth, infantile paralysis or long-standing rickets, were beyond the surgical pale and were capable of help only through special *educational* and training facilities.[35] Surgery was regarded, at best, as merely a means of stabilizing the condition of the physically handicapped; at worse it contributed to the manufacture of cripples. Not until well into the interwar period was it generally believed that crippling diseases were both amenable to treatment and largely preventable.

Before 1900, medical men and medical institutions had little interest in the social/educational movement catering for the handicapped, and few had any involvement with the voluntary agencies.[36] The West London Society for Sick Children, established in 1902, was novel in having a management committee composed of 'physicians and surgeons and persons experienced in medical and social work among the poor'. Original, too, was the Greengate Dispensary in Salford, where, within a specialized medical setting for the treatment of paralyzed and disabled children – itself unique – efforts were made in 1903 to accommodate a day nursery run by the Manchester and Salford Sanitary Association. The Dispensary also had close connections with the Manchester and Salford Crippled Children's Help Society, and in 1906 it was certified as a 'residential school' by the Board of Education. It is perhaps significant, however, that the combined educational and medical work at the Greengate Dispensary was only begun after the death of its emigré founder, the American Dr Theodore Grimké, when his medically unqualified widow took full charge. It is also significant that this innovative institution was founded by an emigré.[37]

The wider medical interest in the social problem of crippled children that began to emerge around 1900 was closely connected with the growth of medical interest in two other groups of children, the sub-acute chronically ill, and the mentally handicapped and epileptic. The sub-acute chronically ill – mostly, but not exclusively, those suffering from forms of non-pulmonary tuberculosis – came into prominence in the 1880s. Ironically, their plight became more apparent partly as a consequence of the growth of surgery in children's hospitals, for this left many of the new surgical wards swamped with such cases. At the Pendlebury Children's Hospital, for instance, where 123 cases of rickets and 507 cases of bone and joint diseases were treated in 1882, it was claimed to be necessary to restrict the number of surgical beds available for such cases.[38] At Edinburgh's Royal Hospital for Sick Children, the senior surgeon reported in 1893:

The deformities of childhood – harelip, spina bifida, cleft palate, phimosis, and clubfoot – come to us in numbers, and hydrocele is very common. But the two great classes of disease which throng the waiting-room, and almost choke the beds, are joint diseases and glandular swelling and suppuration. Had I four times the number of beds at my disposal that I have, I could fill them all in a week with cases of spinal diseases, and hip-joint disease in its late manifestation. . . . Only the very worst can be admitted, where psoas abscesses have to be opened, or hip abscesses to be drained, or joints to be excised.[39]

The problem was one not just of hospital beds; it was also that most of these cases required long-term post-operative care. If allowed into children's hospitals, they reduced the patient turn-over rate – the 'success' record upon which the funding of most of these charity hospitals depended. Consequently, such cases were only allowed in when their conditions became acute; once relieved, they were discharged as outpatients. But few were able to attend outpatient departments on a regular basis, and those that could were subjected to the widely varying mercies of the different (usually junior) medical officers in charge. Without continuity of treatment and prolonged convalescent care, these patients entered a vicious downward spiral of ill-health, drifting in and out of hospitals as acute short-stay cases, rarely achieving lasting cure. It was commonplace to see such children much improved under hospital conditions, only to witness their return some months or years later in more wretched states.[40]

Some specialists in children's medicine in London – Timothy Holmes and Howard Marsh among them – signed a letter to *The Times* in November 1899 drawing attention to this 'serious difficulty'.[41] But the problem was first tackled by Jones and Macalister in Liverpool in the late 1890s, of which the institutional outcome was the Liverpool Country Hospital for Chronic Diseases of Children (Heswall), established in 1898.[42] Macalister, then 38, had only recently joined Jones on the staff of the Royal Southern after his training at Pendlebury. He recalled that it was around 1887 as a 'consequence of experience culled from a Children's Clinic opened under the auspices of the Medical Missionary Society at the south end of [Liverpool]' that he came to the realization that the traditional outpatient and short-stay inpatient treatment of chronic cases was therapeutically counter-productive and wasteful of time and money.[43] But there was more than merely enlightened humanity and economics behind the idea for more sophisticated institutional provision; at least in part, inspiration came from the model for professionalization provided by a colleague of Macalister and Jones at the Royal Southern, William Alexander (1844–1919), a general surgeon who had a large practice in gynaecology. In the early 1880s Alexander had become interested in the surgical treatment of epilepsy, partly as a result of his experience in operating on the displaced uteri of women who suffered from convulsions. In 1888, after visiting the epileptic colony in Bielefeld, Germany, he established Britain's first epileptic colony, at Maghull, near Liverpool. By carefully selecting his cases and making much of his surgical interventions, he cultivated a spirit of curative optimism in this hitherto bereft area, and thereby obtained considerable private financial backing.[44]

Jones and Macalister were following in Alexander's footsteps when, in the face of similar pessimism and resignation over chronically ill children, they proclaimed that few were so incurable that they could not recover if placed in a proper healing environment. To provide such an environment involved soliciting the support of the locally well-off and influential, as well as the support of the Ladies' Committee of the Liverpool Home for Incurables.[45] Unlike Alexander, however, Jones and Macalister had little interest in establishing a separate institution or 'colony'.[46] Rather, they hoped to obtain space for their work within the Liverpool Infirmary for Children which had close connections with the premier medical institution in Liverpool, the Royal Infirmary. But the Children's Infirmary resisted the approach, perhaps perceiving it as self-interested. In the end it was the far less prestigious Children's Convalescent Home at West Kirby (established in 1881 for children mainly suffering from tuberculosis of the bones and joints)[47] that allowed the campaign committee behind Jones and Macalister to rent 22 beds. In 1909 they opened the separate open-air country hospital at Heswall.[48]

In more ways than one, therefore, 'something new was being urged in an original way' by Jones and Macalister at West Kirby – the medical treatment of the chronically ill. Such was the claim at the opening ceremony by one of its backers, the editor and proprietor of the *Liverpool Daily Post and Mercury*, Sir Edward Russell. 'The hospital', Russell hoped, 'might be the beginning of a national undertaking, and . . . the country at large would realise the importance of so treating the rising race that nothing of which science could make a certainty should be left to chance.'[49] Certainly, Jones and Macalister selected their cases as carefully as Alexander had selected his in order to prove their curative claims.[50] But the success of the project probably owed as much to the role of Russell and that of two other men – the influential William Carter (1870–1907), senior honorary physician to the Royal Southern, who was also behind the establishment of the Liverpool School of Tropical Medicine in 1898, and Andrew Gibson, the only son of a recently deceased local shipping magnate, who was to invest over £45 000 in the project. The strength of such patronage made it difficult for the Liverpool Infirmary for Children to hold out against the pressure of Jones and Macalister to link Heswall to the Infirmary by making it the official country branch – the negotiations for amalgamation eventually coming to fruition in 1919.[51] Long before the war, however – indeed, before the institution at Heswall had been built – the gospel of dealing medically with chronically ill children had begun to be preached. As taken up by Tubby in an address to the SSDC in 1903, it was shown to be intimately bound to the need for rationalizing both hospital services and hospital funding.[52] As echoed by the LCC in a report of 1907, 'the usual Hospital method of treating such cases' could no longer be regarded as efficient.[53]

At the same time that the plight of the chronically ill was beginning to be taken up by doctors, the fate of mentally handicapped children was also attracting attention. Again, specifically medical interests tended to trail behind the growth of social welfare initiatives. And in most places it was only after medical beach-heads had been established in relation to the mentally handicapped child that medical interest in the social problem of crippled children burgeoned. For example, John Thomson, the Scottish paediatrician and colleague of Harold Stiles at the Edinburgh Children's Hospital, became interested in the relationship of mental handicap in children to impoverished social environments in the 1890s and then extended this interest to the plight of physically handicapped children. In 1902 he helped found the Edinburgh Cripples' Visitor Society which, in 1904, became the Edinburgh Cripples' and Invalid Children's Aid Society. He also organized the Edinburgh Play Centres, modelled on those established by Mrs Ward in London.[54]

Another who became involved with physically handicapped children through involvement with the mentally handicapped was Dr William A. Potts of Birmingham. Trained like Thomson at the Children's Hospital in Edinburgh, Potts became honorary physician to the National Association for the Feeble-Minded (established in 1896) before being invited in 1909 to chair (and be the sole medical person on) the Birmingham Education Committee's Special Sub-Committee of Enquiry on Physically-Defective Adults and Children.[55]

A more significant figure in children's medicine at the turn of the century who became involved with the social problem of the crippled child via the problem of the mentally handicapped was the senior physician at the Pendlebury Children's Hospital, Henry Ashby. In 1897 he was approached by Mary Dendy, a member of the Manchester School Board and the founder of the Lancashire and Cheshire Society for the Permanent Care of the Feeble-Minded, to help establish the distinction between 'backwardness' and 'feeblemindedness'.[56] Ashby's report to the School Board, published in the same year, led to plans for building two special schools in Manchester, as well as to his involvement in the politics of special education. In 1903 he was solicited to investigate for the Manchester Educational Committee the needs of the physically handicapped, the outcome of which was the establishment in 1905 of Swinton House, the first residential school for crippled children in Britain financed from local taxes.[57]

THE SCHOOL MEDICAL SERVICE

As Ashby's involvements reveal, schools were the places where medical and social interests in the handicapped joined forces. As early as 1888, Dr Francis Warner, an expert in child psychology, undertook an inquiry on behalf of a joint committee of the BMA and the COS into the physical and mental condition of 50 000 children in 106 schools.[58] Such sites became more open, and medical involvements deeper, with the introduction of the School Medical Service (SMS) into England in 1907 and Scotland in 1909.[59] Although historians have come to question the success of the service in improving the health of children,[60] no such doubts surround the encouragement it gave to those with specialist interests in the child. Orthopaedists recognized that the abandonment of the policy of leaving the discovery of crippled children to 'non-medical lay officers and teachers' considerably extended their interests and authority.[61] In 1907, at the International Congress on School Hygiene, Reginald Elmslie, newly appointed to the orthopaedic department at St Bartholomew's Hospital, argued the need for special schools where 'the services of a skilled orthopaedic surgeon' could be obtained. He became an active member of the Medical Officers of Schools Association, and in 1908 was appointed to the BMA's special subcommittee (appointed by the LCC's Education Committee) to consider the medical treatment to be provided for London's school-children under the 1907 Act. Not surprisingly, the subcommittee was to urge the immediate establishment of special school clinics throughout London and the appointment of specialists.[62] It was around this time, too, that the medical press began to criticize those, like Lord Treloar, who were involved with crippled children but who were not medical 'experts'.[63]

Professional medical interests thus came to be seen as directly linked to the social/educational advance of the crippled child in general, and to the disbursements from local educational and public health authorities in particular. Indeed, setting up state-approved schools and hospital schools for physically 'defective' children ('PD' schools) was increasingly recognized by those establishing them as a means to funding long-stay institutions for the medical care of crippled children. The first to exploit this opportunity was the West Kirby Home in 1902;[64] the Royal Orthopaedic Hospital, influenced apparently by schooling in children's hospitals in America, arranged for the LCC to fund its hospital school in 1903.[65] Heswall, perhaps because of its private sources of income, was only certified by the Board of Education in 1913, at which time it had a teaching staff of six.[66]

The Education Act of 1907 which introduced the School Medical Service did not require that 'physically defective' children be referred to specialists, but it made grants available for medical treatment.[67] The Act also gave encouragement to gathering statistics on the causes of physical handicap. (Francis Warner's surveys into 'defects in development' scarcely mentioned the physically handicapped, except in relation to mental degeneracy.) Early surveys were to reveal that for London there were as many as 13.9 such 'cripples' per 1000 of the population, for Birmingham, 5.8, and for the county of Lancashire, 5.9. However, as the pioneer school medical officer James Kerr pointed out in 1926 in citing these statistics, 'all were probably about the same figure in reality',[68] for it was difficult to determine what actually constituted 'physical defectiveness' or, more particularly, 'a cripple'. Mrs Ward's definition of a cripple as 'a child who by reason of physical infirmity was unable to attend an ordinary elementary school' was accused of being so broad as to apply to 30 to 40 per cent of schoolchildren.[69]

The problem of definition, along with negligence in the collecting of statistics on the part of less 'progressive' local authorities, meant that there never was – and perhaps never could be – anything more than a rough estimate of the size of the problem of the physically handicapped. Newman in his annual reports to the Board of Education found it impossible to offer absolute figures: in 1919 his statistics for crippled children ranged from 3.5 per 1000 in country areas to 22.9 in 'an area outside Manchester'. 'Such wide differences', he commented, suggests the adopting of different standards by the investigators, with an outbreak of poliomyelitis in one area or a wide definition of rickets in another, or unrecognised tuberculosis in a third. The average of the returns . . . is 8.6 per 1,000, but some of the large returns are clearly exceptional, and perhaps it would be fair to assume that in England and Wales as a whole somewhere between half and one per cent of the children of school age require treatment and education as cripples. This yields, say, between 30,000 and 60,000 children in total.[70]

Statistics on the *causes* of crippling were somewhat easier to arrive at. In 1909 Elmslie reported on 2141 children admitted to the LCC's 'PD' Day Special Schools, and comparable figures on 3631 children were supplied by Dr W. H. Hamer for the LCC in 1914 (Table 4.1). The 1911 Birmingham survey conducted by Potts (Table 4.2) looked specifically at 'crippling' among both children and adults and was unique in its comprehensiveness. All the surveys revealed that the major source of crippling among children – between 30 and 40 per cent – was tuberculosis of the bones and joints.

Table 4.1   Causes of crippling in London, 1909, 1914

		Elmslie (1909)	%	Hamer (1914)	%
1.	Tuberculous disease:				
	Spine	323 ⎫			
	Hip	376 ⎬ 863	40.2	1145	31.5
	Knee	137			
	Various	27 ⎭			
2.	Infantile paralysis	252	11.7	625	17.2
3.	Spastic paralysis	131	6.2	170	4.7
4.	Progressive paralysis	15	0.7	54	1.5
5.	Various deformities	335	15.6	399	11.0
6.	Heart disease	268	12.6	705	19.4
7.	Various chronic diseases (chorea, 41; phthisis, 18)	142	6.6	285	7.8
8.	Various slight defects (delicate/nervous children)	135	6.4	28	0.8
9.	Congenital deformities	—	—	220	6.1
Totals		2141	100	3631	100

Sources: Board of Education, *Annual Report by the Chief Medical Officer for 1910*, p. 202; idem., *Annual Report by the Chief Medical Officer for 1914*, p. 171.

Table 4.2   Major causes of crippling in Birmingham, 1911

	Under 16 Years		Over 16 Years	
Causes	Cases	%	Cases	%
Tuberculosis	285	39.5	206	24.9
Infantile paralysis	175	24.3	73	8.8
Rickets	73	10.1	—	—
Congenital deformities	71	9.8	47	5.7
Apoplexy (or birth palsy)	58	8.1	116	14.0
Accidents	25	3.5	133	16.1
Scoliosis	13	1.8	8	0.9
Pseudo-hypertrophic muscular paralysis	5	0.7	—	—
Necrosis of bone	5	0.7	—	—
Other	18	1.5	55	6.9
Total	728	100.0		
Rheumatoid Arthritis			29	3.5
Rheumatism			33	3.9
Rheumatoid Fever			10	1.2
Venereal Disease			39	4.7
Spinal Myelitis			11	1.3
Amputation for blood poisoning			8	0.9
Unclassified			60	7.2
Total			828	100.0

Source: Birmingham Education Committee, *Report of a Special Sub-Committee of Enquiry concerning Physically-Defective Adults and Children* (1911).

Such facts and figures multiplied before the war in proportion to the applications for central government grants for 'PD' schools and medical facilities. By 1911, after the passage of the Finance Act granting £1 500 000 for sanatorium work – a move said to be much encouraged by Lloyd George's visit to Treloar's Alton in that year[71] – some 62 day and residential schools for the education of crippled children had come into being, accommodating 4617 children and certified as 'special schools' by the Board of Education. Fifty-five of these schools – the great majority of them in London – were receiving annual grants from the Board.[72] Few of the residential schools called on specialized medical provision, however, and it was not until after the war that the hospital and hospital school came to be seen by the Board of Education as having a prior claim over the purely educational needs of cripples. Nevertheless, by 1913, Newman was paying particular attention in his annual reports to the medical provisions at Alton, Heswall, Leasowe and three or four other 'residential sanatorium schools for children suffering (mainly) from surgical tuberculosis'. All of these institutions were then receiving substantial sums from the Local Government Board, through the TB clause – Section 16 (1) – of the National Insurance Act of 1911. The 'orthopaedic' hospitals at Leasowe and at Coleshill, Warwickshire, were launched through this provision and were largely sustained by it.[73] At Heswall, this source, together with Board of Education funds, amounted to £2700 per annum before the war.[74] Local Poor Law authorities were also an important source of funding. Newman noted in his *Annual Report* for 1913 that whereas the overwhelming majority of children at the special schools for the deaf, blind, mentally defective and epileptic were sent by local education authorities, the largest proportion of physically defective children were sent by Boards of Guardians – presumably in an effort to reduce overcrowding in workhouse infirmaries.[75]

The medicalization of physically handicapped children thus came to be heavily reliant upon several branches of government. As it did so, the careers of those involved in the care and cure of such children were increasingly bound to legislation in public health and welfare. Thus specializers in orthopaedics were to have much in common with MOsH, obstetricians, and those with special interests in infectious disease hospitals and in the treatment of tuberculosis and venereal disease.[76] But it is not clear to what extent would-be orthopaedic specialists in Britain calculated their professional advancement in these terms, or would have wanted to. Before the war there was no well-defined career track in orthopaedics, nor – subsequent to the collapse of the BOS in 1898 – a forum for the deliberation of such strategies. Efforts to exploit the social/educational movement on behalf of crippled children thus tended to be local and individual, rather than professionally stategic for 'orthopaedics'. 'Exploit', moreover, is hardly the right word; according to Mrs Ward, even after the war, when Jones and 'his lieutenants' sought to make a general movement on behalf of crippled children, they knew 'very little about the existence of the physically defective schools! – and were both delighted and surprised to hear that there was so much knowledge and sympathy on the subject already in existence'.[77] Newman expressed a similar view in November 1919, noting that Jones and his colleagues 'appear to be wholly ignorant of the fact that a scheme of this nature has been working *for ten years!* . . . is steadily going forward, . . . is receiving grants from the Board of Education, [and] has been described in our Board of Education medical reports for years'.[78]

Nor is it possible to do more than suggest that the involvement of general surgeons in the care of handicapped children was linked to a process of collective image-making. It may be that such involvement was felt to counter the image of being merely a mechanic or, worse still, a blood-stained craftsman. Like dinner-socials and surgeons' travelling clubs, public work of this sort may have lent surgeons a certain social cachet, equivalent to the intellectual status conferred by Lister and germ theory. But in the absence of appropriate collective biographical information, this must remain speculative.

FOREIGN MODELS

More certain is that Jones and his colleagues before the war were becoming aware of how specialization in orthopaedics in other countries was being advanced through the involvement of orthopaedic surgeons in the crippled children's movement. Developments in America and Germany in particular were important. As early as 1893 the Boston orthopaedist, Edward Bradford, after visiting an institution in Italy for the education and training of children suffering from rickets, opened the Boston Industrial School for Crippled and Deformed Children – the first of its kind in America.[79] Five years later, directly as a result of the interventions of the St Paul orthopaedist, Arthur J. Gillette, Minnesota became the first state to accept at public expense the treatment, care and education of needy cripples.[80] New York followed suit in 1900, Massachusetts in 1905, and many other states before 1914.[81] Although charity organizations were involved in the pursuit of this legislation, in all cases it was government that undertook the building of specifically *curative* centres in which orthopaedic surgeons (or surgeons with a particular interest in orthopaedics) were normally engaged as surgeons-in-charge.[82]

Hand in hand with professionalization in American orthopaedics went the specialism's articulation of the 'social problem' of the crippled child. In 1909, for example, AOA leaders staged a major symposium on 'What Shall We Do With Our Cripples?' within the section on orthopaedic surgery of the New York Academy of Medicine.[83] Although as late as 1916 the paediatrician Abraham Jacobi maintained that in child welfare as a whole, 'Laymen [more precisely, lay-women] have advanced ahead of the medical profession',[84] in relation to the crippled child, orthopaedists were well-ensconced. And more medical initiatives followed after the establishment of the Children's Bureau in 1912, signifying the acceptance by the federal government of responsibilities for promoting the health and welfare of the young. One reflection of such initiatives was the extensive *Bibliography of the Education and Care of Crippled Children* (New York, 1913) compiled by Douglas McMurtrie;[85] another was the *American Journal for the Care of Cripples*, founded in New York in 1914.

In Germany before the war medical involvement with the movement for crippled children was more developed than in America. Indeed, much of the American orthopaedic interest in 'the problem' stemmed directly from discussions at the Congress of Orthopaedic Surgery held in Berlin in 1908, which was attended by leading American orthopaedists.[86] In McMurtrie's *Bibliography*, about half of the 724 recent books and articles listed were written by German medical men, most of them specializing in children's orthopaedics.

Particularly noteworthy at the time, as well as illustrative of the medicalization of the social problem of the crippled child, was the work of Konrad Biesalski (1868–1930). Trained in the orthopaedic clinic of Albert Hoffa in Berlin, and from 1908 to 1923 the editor of the *Zeitschrift für Krüppelfürsorge*, Biesalski claimed to be the first person 'to distinguish a cripple as a sick person'.[87] Between 1903 and 1909, at a time when tuberculin tests were revealing that some 90 per cent of working-class children were infected, Biesalski orchestrated the various religious and secular agencies involved with cripples in Germany to undertake a national census – the only such country-wide enumeration before the late 1930s.[88] Defining a cripple as 'a person who as a result of a bone, joint or muscle disease, a nervous disease or the absence of an important limb or part of a limb is permanently disabled in such a way that . . . earning capacity is affected', Biesalski's census located some 110 000 such persons. Of these, he considered 55 000 to be in need of institutionalization under the care of a qualified medical expert.[89]

The importance of Biesalski's work to orthopaedists in other countries extended far beyond his valuation of the cripple as primarily a medical problem (in particular as an orthopaedic problem); equally important was his success at harnessing the considerable social, political and economic power of the various German educational and welfare agencies for crippled children. It was largely as a result of his efforts that in 1909 a number of these organizations joined together to form an association for the care of cripples, the purpose of which was to 'spread information about the many possibilities of help which are offered through the development of modern orthopaedic surgery and pedagogy' and, in general, to redirect public opinion away from fatalistic views of cripples to medically optimistic ones.[90]

OSWESTRY

It was not until after the war that Jones and 'his lieutenants' endeavoured to organize and orchestrate a national association for the care and cure of crippled children similar to Biesalski's in Germany. Nevertheless, the foundations for such a movement were laid before the war. Indeed, the regional development that came to be centred on what today is the Robert Jones and Agnes Hunt Orthopaedic Hospital at Oswestry, Shropshire, was itself influential internationally before the war. In many ways the history of children's orthopaedics at Oswestry recapitulates the intersection of social and medical interests outlined in this chapter; additionally, it draws attention to the major therapeutic innovation to be associated with orthopaedics in relation to crippled children in the interwar period – open-air therapy.

The origins of orthopaedics in Shropshire go back to the turn-of-the-century founding by the nurse Agnes Hunt (1867–1948) of the Salop Convalescent Home for Women and Children at Baschurch, a hamlet near Oswestry. Herself partly crippled from youth by osteomyelitis, Hunt underwent nurse training in the 1880s, first at the Sea Side Hospital and Convalescent Home for Children at Rhyl, then at the West London Hospital in Hammersmith, and finally at the Salop Infirmary. In 1892 she qualified as a Queen's Jubilee District Nurse. After eight years of district nursing she turned to the convalescence business, partly because of her health, and partly to placate her indomitable widowed mother.[91]

Both Hunt's training and her mother bore upon the subsequent course of events. The children's hospital at Rhyl had Hugh Owen Thomas as its visiting surgeon, and it was one of the first hospitals in the country 'to advocate fresh air as an integral part of the treatment of cripples'.[92] From her training there Hunt claimed to have learned 'the paramount importance of two things in the treatment of long-standing chronic cases – fresh air and happiness'. Equally important, she learned that 'no nurse was worth her salt if she had not the joy of life in her, and the power of sharing it with her patients'.[93] Such caring wisdom was driven home by Hunt's experience at the West London Hospital, which at that time had not, as she put it, 'been touched by the finger of Florence Nightingale'.[94] As these remarks also indicate, Hunt's orientation was curative, rather than educational; she was later to reflect in reference to the contrasting ambitions of her contemporary, Mrs Kimmins at Chailey, 'I did not mind much whether they [crippled children] were educated to take their place in the world or not.'[95] As for Hunt's mother, she had previously been a Poor Law Guardian in West Kensington as well as a force to be reckoned with on the Committee of the COS at its London offices in Hammersmith Broadway. Hunt was thus well-schooled in the ways of late-Victorian charity before she embarked on what was to be her life's work at Oswestry. From her mother she also acquired not only some of the funding for the convalescent home, but the manner of bullying the public authorities who were to maintain what was ultimately to be a major curative institution. Doubtless it also helped that her brother Rowland was a local Justice of the Peace and the Unionist MP for Ludlow from 1903 to 1918, and that her brother-in-law, Sir Frederic Kenyon, was the well-connected Director of the British Museum.

Within the 'story' of the transformation of Baschurch from a small convalescent home for delicate children who might benefit from good diet, nursing and fresh air, to an entirely open-air hospital catering for hundreds of mainly tuberculous cripples, much was to be attributed to the teachings of Thomas as received by Hunt through her training at Rhyl. However, the immediate stimulus for the open-air system adopted at Baschurch was slightly more mundane. According to Sister Teresa Fraser, who arrived at the home in May 1901 just before the therapeutic transformation began, open-air treatment was started because of the expected arrival of a patient suspected of early phthisis, who was deemed in need of isolation. The home was then already overfull with about 20 patients (only six of whom were 'cripples'), and its liquid assets amounted to no more than £5. Efforts were therefore made to render a disused attic habitable for the phthisical patient.

Suddenly the laziest member of the [attic clearing] party had an inspiration: – 'Surely open air is the latest treatment for consumption; what is the matter with the cart shed?'. . . . [T]here was a great deal the matter with that cart shed. . . . But we were exhausted, and anything was better than further effort in the garret. . . . Thus, with the immediate object of evading too strenuous labour, open-air treatment was inaugurated at Baschurch Convalescent Home.[96]

As the quotation indicates, the idea of open-air therapy was already 'in the air'. Indeed, by 1905 it was only the *lack* of controversy surrounding it that drew medical comment.[97] Certainly, Thomas's advocacy of open air for the treatment of tuberculous and other crippled children in the slums of Liverpool and, institutionally, at Rhyl, was not unique. Thomas, for all his individualism was, as we have seen, well within the conservative surgical tradition of which the tonic virtues of fresh air formed an essential part. Within British medicine, the advocacy of open air in the treatment of consumption goes back a good deal further than George Bodington's essay of 1840 *On the Treatment and Cure of Pulmonary Consumption, Rational and Successful*.[98] But for much the same reason that it is mistaken to attribute the 'discovery' of the crippled child to medical interests, so it would be erroneous to restrict the advocacy of open-air to a medical tradition. By the time Bodington's work was rediscovered and republished by the Sydenham Society in 1901 an essentially non-medical fresh-air charity movement had for some time been gathering force. By 1897, this movement, which endeavoured to render air into a commodity of bourgeois philanthropy, had already become the subject of at least one doctoral dissertation.[99] Of course, there were other more medical inputs to the advocacy of open air in Britain, among them the late-nineteenth century development of open-air TB sanatoria in Germany.[100] But in this medical venture, as in others involving open air, the 'medicalization' and 'scientization' of air[101] either followed upon, or were contemporaneous with, social interests in the management, control and prevention of threatening disease and behaviour – a fact which was to be made further manifest in the history of the extensive open-air school movement of the early twentieth century.[102]

In 1901 it had yet to be demonstrated that good nursing under uninterrupted open-air conditions was beneficial for all types of crippled children. However, the sociomedical context for such an experiment was clearly as favourable as the economic incentives were strong. The success of open-air therapy at Baschurch was thus assured well before Jones became involved with the home in 1904. By then, the economic advantages of open air at Baschurch had been doubly proven, for not only were open-air sheds cheap, but it was shown that

Source to Plates 4.1 and 4.2: The Heritage of Oswestry (Robert Jones and Agnes Hunt Orthopaedic and District Hospital, 1975), p. 46. (By kind permission of the Robert Jones and Agnes Hunt Orthopaedic and District Hospital.)

Plate 4.1   An open-air 'ward' at Baschurch, *c.* 1910

Plate 4.2   Child patients at Baschurch, *c.* 1910

the practice, in conjunction with good food and good nursing, hastened the curative process. Patient numbers were now averaging over a hundred annually, and there was increasing demand for the admission of more cases of surgical tuberculosis. Unlike those admitted to Treloar's Home, these were severe cases, requiring from six months to two years for recovery. The need for such an institution, together with the curative optimism surrounding it, brought funding from branches of the COS in Hammersmith and Shrewsbury, the ICAA in London, the Birmingham Cripples' Union, the Women's University Mission, the Kyrle Society of Liverpool, as well as from Poor Law unions. As a 'result of working on Charity Organization Principles', it was claimed, efficient coordination was everywhere apparent.[103]

Jones became involved when Hunt began transporting some of her cripples to the Nelson Street Clinic for treatment, and after she had sought his expertise for her own condition at the Royal Southern Hospital. Impressed by the strength of her character, as much as by the results she was obtaining in her open-air sheds, Jones consented in 1904 to become the consulting orthopaedic surgeon to Baschurch. Thereafter, until the outbreak of the war, he regularly performed Sunday surgeries once a month.[104] Significantly, the description 'convalescent home' was dropped from the title page of the annual reports in 1905, and the word 'orthopaedic hospital' was placed in its stead shortly after an operating theatre was installed in 1907. Thurston Holland, Jones's technical apprentice at Nelson Street, was brought in as the radiologist after X-ray equipment was installed in 1909, and other of Jones's assistants from Liverpool were recruited as visiting consultants. Thus before the war Oswestry had become, as it was to to remain until long after the Second World War, the spiritual home and show-piece of the self-consciously styled Liverpool school of orthopaedics – a reference both to the place, and to the commitment to Thomas's and Jones's brand of conservative surgery.[105] Locally, nationally and internationally its fame spread: in 1907, the young Shropshire GP, Gathorne Girdlestone – later the first professor of orthopaedics in Britain – came into Jones's circle;[106] and it was here that Jones ushered many of those who came to observe his clinics in Liverpool. As with the latter, so with Oswestry, the surviving impressions are only of those who came to marvel and praise. Among visitors in 1908 were Stiles, Forbes Mackenzie (a Canadian orthopaedist), and Dawson Williams, who, as editor of the *BMJ*, would later be helpful in the preparation and publication of Jones's and Girdlestone's postwar propaganda on a national scheme for the care of cripples.[107] Another of the visitors was

Arthur Gillette, then president of the AOA, who, after inspecting the hospital in 1908, returned to Minnesota to found America's foremost open-air children's hospital.[108] Directly and indirectly, the developments at Oswestry influenced other homes for crippled children in Britain, including that at Heswall where, in 1909, Hunt was recruited to organize the open-air wards, and where, in 1910, Sister Teresa Fraser became the Lady Superintendent.[109]

More was to come out of Oswestry than just its open-air therapeutics for crippled children. Beginning around 1910 the hospital pioneered the development of orthopaedic clinics, 'thrown out like great tentacles from the Central Hospital'.[110] To some extent this was a logical consequence of the relative isolation of Oswestry from the major urban centres. Without the provision of outlying clinics it was difficult to control the problem of post-hospital patient relapse.[111]

But from the very beginning the Shropshire clinics were intended as more than mere extended outpatient/after-care units. Besides facilitating shorter periods of institutional care and allowing for maximum use of inpatient facilities (which by 1913 had risen to 314 patients, whose average stay on the ward was down to three months),[112] the clinics were also intended to act as diagnostic and preventive medical centres. GPs and welfare workers were encouraged to refer not just evident cases of deformity, but also suspected cases. Remedial treatment could thus be carried out early and 'crippling' nipped in the bud. The clinics could therefore effect startling economies: whereas the cost for an inpatient was well over £100 per year, that for outpatient treatment was only a few pounds.[113] At the same time, the clinics educated social and health workers, as well as the public at large, to the value of orthopaedics. In this sense, as well as in their preventive and follow-up functions, the Oswestry orthopaedic clinics (of which there were 13 in operation by 1918 serving some 600 patients) were apiece with the clinics for TB that some local authorities introduced after the National Insurance Act of 1911 and which were organized on a national basis after the Public Health (TB) Act of 1921. Similar were the clinics for women and children which were organized nationally after the Maternity and Child Welfare Act of 1918. As with the local authority clinics, the Oswestry orthopaedic ones, with their combined preventive and curative functions, anticipated one of the important strands in postwar medical planning.

By 1917 the Shropshire County Council was sending all of its cases of surgical TB to Oswestry, and in 1918 the County Council became the first public body in Britain to take full financial responsibility for the care of all crippled children under 14 years of age within its boundaries. By this date, too, 13 other Poor Law unions were sending patients to Oswestry, thus bearing the bulk of the institution's costs, then estimated at 12*s* 8½*d* per child per week. After 1917, through negotiations with the public health authorities of Staffordshire and Shropshire, the hospital also secured the financial involvement of the Local Government Board in the hospital's school and in a nurses' training centre.

Before the Great War, then, Jones and his colleagues had begun to cultivate a site which was significant in being separate both from the work of general surgeons in general hospitals and from the narrow 'orthopaedics' conducted in the old orthopaedic hospitals. It was also increasingly important socially and politically in being linked to broadly-based GP and local authority welfare and educational services. But Oswestry was not simply the outcome of would-be specialists capitalizing on social initiatives in the name of the crippled child. In truth, the site was opened because of a process of mutually defining social and professional interests. Orthopaedic specializers did indeed come partly to determine their interests in the course of capitalizing on interrelated social, ideological, political and medical developments, but that very process was inextricably bound to one in which those involved with social and welfare issues were defining *their* interests. The same confluence of professional and broader social and political interests also shaped the particular logic of medical organization that can be seen emerging around Oswestry, in which would-be hospital-based experts sought to oversee 'efficient', regionally integrated services. To better understand this, though, we need to move beyond the crippling diseases of childhood to the other site for the making of modern orthopaedics – accidents and trauma.

5

HAPPENINGS BY ACCIDENT

Like crippled children, accidents had to be brought into focus before professionalizing strategies could be articulated in their name. Unlike crippled children, however, the medical 'discovery' of accidents, which was roughly contemporaneous, was unrelated to any widespread humanitarian movement. While fatal and non-fatal accidents were prevalent in the Victorian period, and were perhaps on the increase,[1] they were not much considered before the last decades of the century. Concern over factory reform in the 1830s and 1840s had generated the potent image of the child factory cripple, which helped pave the way for the regulation of child labour in factories and mines. But that agitation not only failed to direct attention to the adults who comprised the majority of accident victims, it served to deflect attention from them.[2] During the mid-Victorian 'age of equipoise', injured workers were as out of mind as out of the sight of the 'respectable' middle classes. To the Manchester medical practitioner, John Roberton, writing in 1860, it appeared that 'if "plague, pestilence, and famine" used to be the evils most feared in bygone times, contusions, fractures, dislocations, burns, and other kinds of injury may well be regarded as the dread of the present', but with this crucial difference, that 'whereas the plague and the pestilence were the terror of all ranks, the rich as well as the poor, the bodily injuries now so frequent are the lot mainly of the labouring classes'.[3] Because of this class specificity, accidents appeared not to threaten the splendour of the mid-Victorian city in the way that issues of public health manifestly did. Significantly, Roberton found it almost impossible to track down any worthwhile statistics on industrial injury.

For the most part, it was only when a major disaster struck that public attention turned to matters of accident prevention and provision. A horrible railway collision, a coal pit collapse, or a boiler explosion in a factory, might trigger the setting up of a relief fund for the victims and their families, along with a commission of inquiry and, sometimes, the building of a hospital. But such responses probably did as much to divert attention from the routine of fatal and serious non-fatal industrial injuries; usually acknowledged as 'acts of god' and

79

attributed to negligence, such events and the responses to them sustained the convenient fiction of accidents as arbitrary, individual happenings, whether occurring in the home, on the street, or in industry. Not until the 1880s and 1890s was this view significantly modified.

The medical profession did not play a major role in rendering accidents socially and politically visible. Even in their most visionary prewar plans for the 'Dawn of the Health Age'; provision for accidents seldom figured.[4] While important late nineteenth-century legislation was to provide financial compensation for the victims of accidents, it did not address medical provision, nor was there medical pressure that it should. Workmen's compensation legislation did not serve – as the Education Acts had for crippled children – to problematize accidents as an issue open to specialist interests. Prior to the First World War, trauma and fracture cases remained among the routine work of GPs and hospital general surgeons. Not until 1913, in the wake of a BMA report on the treatment of simple fractures, were there a few calls for 'skilled surgical experts' or 'fracture specialists' to work in 'specially equipped wards'.[5] And not until well into the interwar period were such calls seriously acted upon.

Traditional orthopaedics had few claims to this area. As prewar textbooks on surgery confirm, 'orthopaedics' continued to be associated with the manipulative and mechanical treatment of congenital deformities, dislocations, paralysis, contractures, knock knee, and bow leg.[6] Even in the 1899 'Supplement' to the American *Cyclopaedia of the Diseases of Children*, the subject of fractures was assigned to a general surgeon, while the only entry by an orthopaedic surgeon was that on 'Swimming, Dancing, Bicycling' – then perceived as common causes of deformity.[7] From around 1900 a few hospitals in the USA began to have fracture cases referred to orthopaedic departments,[8] but in most places, on both sides of the Atlantic, it was usually only fractures that had refused to mend and had become 'deformities', that were thus assigned.[9] Although in 1896 Jones encouraged the BOS to 'assert their right to widen the domain of their specialty' by incorporating fracture treatment,[10] the prewar medical history of accidents was to have little to do with bids for fracture cases by aggrandizing would-be orthopaedic specialists. Jones's interests were hardly typical. Moreover, in the debate over the proper treatment of simple fractures, which led to the BMA's report of 1912, Jones declined to participate;[11] behind the subsequent call for 'fracture specialists' stood members of the new surgical cohort, but not specifically orthopaedic lobbyists.[12] Even in America, where the AOA was in a better position to stake the claim for fracture treatment (and did so as early as 1894), there was little interest in seriously pursuing it before the war. As an AOA President reflected in 1948, 'it is singular that fracture problems were so seldom discussed at the early meetings of this Association'.[13]

This chapter explores how the organizational basis for the treatment of traumatic injury was brought into focus towards the end of the nineteenth century. As with the emergence of medical provision for crippled children, so for accidents, neither simple humanitarianism nor 'the impact of modern medicine' (as some extraneous factor) provide sufficient explanation. The process of the medical discovery of accidents was not inevitable, but nor was it 'accidental'. Rather, it was intimately bound to the reconstruction of the late Victorian social and economic context.

PROFESSIONAL INDIFFERENCE

The livelihoods of many Victorian practitioners depended on the treatment of workplace injuries, and from around mid-century many hospitals became preoccupied with accident cases. But the focus of the medical profession for most of the nineteenth century remained more on the care of individual patients than on any special organization which 'accident cases' as a class might require. Before the 1880s, medical interest in accident provision was at best sporadic. Few surgical texts paid the slightest attention to 'first-aid' for accident victims,[14] and there were few, if any, nineteenth-century sequels to such medically-authored self-help texts as Stephen Bradwell's *Helps for Suddain Accidents Endangering Life* (1633) and S. A. D. Tissot's *Advice to the People . . . upon any inward or outward accident* (1767). Medical texts for practitioners did not appear much before the last quarter of the century; W. P. Swain's *Surgical Emergencies . . . a manual for the use of general practitioners*, for instance (5th edition 1896), being issued in 1874. Even the therapeutics of traumatic injury were of little general interest. While the treatment of trauma cases featured in the debates over conservative surgery, those debates did not extend beyond questions of clinical outcomes. The only types of traumatic injury other than fractures to be specifically discussed in medical texts before the end of the century were railway and street accidents. These were significant because they frequently involvd members of the middle class, raised sensitive matters of insurance and medical jurisprudence,[15] and invoked concern over the pace of modern life.

Brief notices on the causes and locations of accidents were sometimes given in the profession's main journals. Rarely, however, were they discussed in terms of institutional provisions or medical treatment. In 1868, the *British Medical Journal* was 'startled' to learn from the Accident Assurance Company

> that 2,000,000 of persons, or one in every ten of the population, receive injury every year from accidents of a more or less serious character. Of these 10,000 are killed, or die from the direct effects of the accident. The total amount paid in claims by this company is £250,000 since 1849.[16]

But the conclusion the journal drew from these figures did not touch on professional interests; rather, they were seen, in the light of the accepted belief that accidents were a matter of *private* concern, as providing 'striking evidence of the advantage reaped by the public in making provision against those casualties to which all are more or less liable'. Not until 1885 did the *BMJ* comment on accidents in factories. Informed by the Factory Inspector's Report that there had been 8904 serious accidents in factories and workshops during 1884, the journal again expressed no medical interest. It simply concluded that 'some legislative action must be taken for bringing factory-owners more to account', especially with regard to extending and enforcing the legislation on the fencing of machinery.[17] At most, such references reflect how the profession came to share some of the general public's growing awareness of the economic importance of accidents.

Virtually alone among mid-Victorian practitioners in having a medicopolitical interest in the subject was John Roberton. His paper of 1860, quoted above, focused on the 'high ratio of severe casualties' received at the Manchester Royal Infirmary, and was intended to elaborate a package of reform measures. In particular, Roberton called for better ambulance transport, the provision of first-aid services in populous communities with surgeons equipped to answer emergency calls, increased hospital beds for surgery, improved organization of surgical wards, and the relocation of those wards to more salubrious locations (preferably outside city centres). Roberton also advocated the use of case records to enable uniformity and continuity in patient-care.[18] In almost all these respects, his ideas were well in advance of the time, as indeed was the idea put forward in the local press in response to his paper – that the Manchester Infirmary be sold and the profits used for the erection of four outlying hospitals connected to four or five 'accident rooms' in the various industrial quarters of the city.[19] In effect, this idea, which was similar to the French plans of the 1780s for the Hôtel-Dieu, was that which came to be implemented during the Great War with the establishment of Casualty Clearing Stations linked to larger base hospitals. A version of it also came to exist in Manchester in 1914 after the Infirmary was relocated to the outskirts of the city.[20]

Roberton's interest in accidents was largely bound to his critique of the Manchester Infirmary and to his ideas for hospital reform in general. Similarly, in some of the smaller industrial towns some surgeons focused on the growth of accident cases primarily in order to persuade the local wealthy to build hospitals. In some cases, as in Huddersfield in the 1840s, they were successful.[21] But this was not always the case, nor necessarily a shared professional strategy. In Oldham, where the matter was raised in 1868, at least one surgeon was of the opinion that no such special provision was required. '[N]umerous workshops and cotton factories have supplied a fertile field for the treatment of acute surgical cases,' he confessed, but

> these, with the exception of a comparatively small number sent to the Manchester Infirmary, have been treated, through the liberality of the employers of labour, as private patients at their own homes. The comfortable circumstances and, as a rule, the excellent conditions of the dwellings of the working population, have enabled this to be done, not only without disadvantage, but . . . with positive benefit to those unfortunate enough to suffer accidents by machinery.[22]

In this case, a commitment to the status quo hinged on contemporary concern with cross-infection in hospitals, and partly, it would seem, from a desire not to lose patient fees from paternalistic employers. (Perhaps, too, there was the ambition to prevent workers from agitating for compensation from employers or, as bad in the medical profession's book, to prevent workers from agitating for hospitals under their own control.) The quotation also reminds us that the removal of accident cases to hospital was by no means universal even where hospitals existed. In some colliery districts, as late as the 1920s, an anti-hospital 'home instinct' was said to be still 'very great'.[23]

Even if removed to hospital, accident cases were not an obvious means to professional advancement. Unless they occurred continuously and in significant numbers, as during war, they were not easy to organize around or to exploit within professionalizing rhetorics. As with crippled children, however, the main deterrent to professionalizing investment was economic: since accidents were primarily the lot of the working class, they provided no entrée to lucrative private practice. That Jones contemplated abandoning the Nelson Street Clinic once he had obtained his appointment at the Southern Hospital and had begun private practice among the monied is partly illustrative of the fact that such work was not considered a pathway to fame or fortune.

Only with regard to fractures was the situation slightly different. These, after all, were not uncommon among horse-riding ladies and gentlemen. Fractures, moreover, had obtained a certain lustre in hospital medicine as a result of serving as the surgical basis for the therapeutic contest between Listerians and the anti-Listerians from the late 1860s. The high mortality rate from amputations in cases of compound fractures – cases in which the bone breaks the skin – was much diminished as both sides endeavoured to justify their claims.[24] In the process, there also resulted better organization of surgical wards and better quantification and standardization of surgical cases. Because of this, the mid-century complaint that the powers of British surgery 'have been more sparingly dealt out in the cause of a broken bone, than of any other structure' became less common.[25]

But like the response to major disasters, the special attention that came to be directed to certain serious fracture cases tended to deflect attention from the circumstances surrounding *most* such cases. These continued to be dealt with, not in well-managed inpatient wards by reputation-seeking (or defending) senior surgeons, but in the low-status outpatient or casualty departments where they were often dealt with haphazardly by dressers and junior staff.[26] Hence the accusation of a contributor to the *BMJ* in 1874, who, after quoting from Macaulay's *History of England* that

'every bricklayer who falls from a scaffold, every sweeper of a crossing who is run over by a carriage, may now have his wounds dressed, and his limbs set, with a skill such as, one hundred and sixty years ago, all the wealth of a great lord ... could not have purchased'

went on to say:

It is true [that every bricklayer and sweeper] may have the benefit of such skill if, perchance, his injury has been sufficiently severe to cause him to be admitted as an in-patient; but otherwise he may, unfortunately ... become the victim of such carelessness and unskilfulness as, we may reasonably hope, was not common even two hundred years ago.[27]

Forty years later similar criticisms would be heard in the course of the debate over the treatment of simple fractures.[28] After the war, it was to be picked up by orthopaedists and woven into their rhetoric for the building and control of hospital fracture clinics. At this point, though, in the mid-1870s and thereabouts, the criticism was emerging within a much broader debate over the function and funding of charity hospitals and, in particular, over the proper use of outpatient departments – a subject to which we will return.

VICTIMS

To the majority of potential victims, accidents mattered dearly. For labourers, even a relatively minor injury could bring untold financial and emotional hardship. A simple fracture of the arm, for instance – among the most common of the injuries taken to GPs and hospital casualty departments[29] – would keep a person from work for at least four weeks even if it was treated under the best conditions by a skilful surgeon.[30] Restoring good function after a simple fracture of the leg could take up to six months.[31] During such time, if one were not a member of a friendly society, the only recourse was to friends and relatives, or to the dreaded Poor Law. (In London, an 'Accident Relief Society' was established in 1838, to assist 'the families of the suffering poor when plunged into distress by accident'; but this was only for those who were hospitalized, and for precious few at that, to judge by the fact that by the time the Society faded away in the 1870s it had relieved only 103 such families.)[32]

Often fractures were carelessly set, or were not allowed to set properly because of a worker's desperation to return to employment. Then as now, fractures of the leg generally involved some degree of shortening and hence limp, however they were set. Thus movement and dexterity were often impaired, and in a labour market in which these were crucial for almost every kind of skilled and semi-skilled occupation, this frequently meant the loss of employment. A worker might be kept on – in a factory, say – as one of the many unskilled 'to-ers and fro-ers', but this was entirely at the discretion of the employer, who might or might not be concerned with his reputation. In any case, it would mean a cut in wages, perhaps by as much as half for a skilled worker. And these were the lucky ones: for many skilled and semi-skilled workers, such as those in shipyards, mines, quarries and the building trades, there were few 'odd jobs' or 'light work' to fall back on. Moreover, since much labour was hired out on a piece-rate basis, many workers were outside the scope of such gestures from employers. Handicapped, they would join the pool of unskilled labour and dodge the degradations of the poor-house as best they could.[33]

Not until the Workmen's Compensation Act of 1897 was a major step taken by the state towards the elimination of these causes of pauperization. Fought for by the unions and proposed within the same social context as reform of the Poor Law,[34] the Act made it easier for some workers to claim financial redress for injuries sustained at the workplace. The Act also greatly increased the reporting of accidents. But it had little direct bearing on the medical provision for accidents or, for that matter, on their prevention.[35] As far as most workers were concerned, these matters were still left in their own hands.

In certain trades and industries workers had long taken it upon themselves to insure against times of illness.[36] Within that self-help tradition, towards the end of the nineteenth century, many of them also began to take positive steps for their medical care in the event of industrial injury, though initially the incentives were middle-class. One such step was the organization of branches of the St John Ambulance Brigade. Constituted in 1887, a decade after the St John Ambulance Association was established, the brigades were intended to 'carry out medically approved first aid and assist in the transport of the sick and injured'.[37] Before the war there were over 23 000 Brigade volunteers, most of them working-class. Much of their first aid also appears to have gone beyond that which was 'medically approved' at head-quarters. For example, around Sunderland and Middlesbrough, where there were over 1000 volunteers by 1899, Brigade members are reported to have carried out bonesetting among various trades as well as for football clubs and friendly societies.[38] Similar work also seems to have been undertaken at collieries, few of which were without at least one Brigade member by the turn of the century.

Another positive measure undertaken by workers at this time was the organization of work-place collections for hospitals. The most notable example of this was the Hospital Saturday Funds, which were modelled on the church collections for the Hospital Sunday Funds. The Saturday Funds spread rapidly from around 1870 and were eventually federated into a national structure.[39] Their primary organization remained local, however, and it was to hospitals serving mainly industrial constituencies that most of their money was channelled. By the end of the century many such hospitals were heavily dependent on these workplace collections and there is evidence that in some locations members of the workers' committees, realizing their economic power, took up places on the boards of management – for example, at the Poplar Hospital in 1891.[40]

It would be wrong, however, to overestimate the importance of either of these developments on their own for bringing accident provision into medical focus. Not only were they limited in their extent, but they were not autonomous. Both were a part of wider socioeconomic developments affecting hospitals – developments that, in the first instance at least, might simply be seen as predicated upon the increases in the number of accident cases arriving there.

HOSPITALS

While it was customary in the nineteenth century for patients to enter voluntary hospitals by means of a letter of recommendation from a subscriber to the charity, serious accident cases were usually granted an automatic right of entry. As such, the number of hospital beds taken up by accident victims could not easily be controlled. Since Poor Law infirmaries did not admit accident cases, this was a problem peculiar to the voluntary hospitals, and it was one that intensified as the century advanced. At the London Hospital, for instance, the number of accident cases received between 1842 and 1908 rose from 5503 to 18 501 – an average increase of 500 per annum – with as many as 28 532 cases being received in 1901.[41] In some heavily industrialized areas, the trend was noticeable earlier, as at the Manchester Royal Infirmary where about 20 per cent of outpatients and 50 per cent of inpatients were accident cases from the mid-1840s – mostly, as at the London Hospital, the victims of contusions, 'wounds' and fractures. As Roberton had noted, many of the accident cases arriving at the MRI were of such a serious nature that the hospital's mortality statistics were significantly increased; in the 1850s over half the mortality 'was directly or indirectly the result of these cases'.[42]

Even hospitals with largely rural catchments, such as the Norfolk and Norwich General, were receiving increasing numbers of accident victims from around the mid-century.[43] The greatest pressures, though, were felt by hospitals near docks. In Liverpool, both the Northern (1834) and the Southern (1842) hospitals were established expressly to cater to the injuries of stevedores. The Stanley Hospital in Liverpool, where Jones obtained his first appointment, had originally been built for the treatment of Diseases of the Chest and Diseases of Women and Children, but by the 1870s it was dealing mostly with accidents sustained by dock workers.[44] Coping with the needs of this particular high-risk group was also the *raison d'être* of the Poplar Hospital, established down-river from the London Hospital in 1855, largely for the employees of the East India Company Docks. By 1890 it was officially titled the 'Poplar Hospital for Accidents', advertised itself as 'open all hours for the treatment of accidental injury or emergency', and was treating, annually, some 750 inpatients and 12000 out-patients.[45] Slightly further down the Thames, the Royal Victoria Albert Dock Hospital (a 14-bed branch of the Dreadnought Seamen's Hospital, Greenwich) was functioning similarly. The Miller General Hospital, still further down-river, at Greenwich, was founded in 1885 largely for the reception of riverside accident cases, the numbers of which rose from 46 in 1885 to 3267 in 1893.[46]

Another generous supplier of mangled bodies was the iron and steel industry. It was specifically for this constituency, in south Stafford-shire, that a four-bed cottage hospital was opened in Walsall in 1863.[47] Unlike similar small hospitals intended for accident cases (such as that established in Ashton-under-Lyne after a colliery explosion in 1858, or that set up beside the foundry and shipyards of Barrow-in-Furness in 1866),[48] the Walsall Cottage Hospital became famous as a result of the nursing, budgeting and propaganda skills of the Anglican nun, Dorothy Pattison ('Sister Dora'). By 1872, when the hospital was treating over 2000 outpatients and 200 inpatients, it was drawing patients from as far afield as Glasgow, Liverpool and London and had become a model for accident hospitals elsewhere. In 1878, Queen Victoria's daughter, the Princess Imperial of Prussia, wrote for its rules in order to establish a similar institution for industrial casualties in Berlin, and in 1890, Lawson Tait praised it as a shining example of what a cottage hospital could accomplish in the treatment of serious accidents.[49]

Whether built for the purpose of receiving such cases or simply inundated by them, hospitals had their meaning transformed. Traditionally, charity hospitals had been places for the 'deserving poor'; the 'undeserving' were relegated to Poor Law infirmaries, while those who were in employment preferred care at home. Accident cases were not, of course, wholly responsible for the shift towards the end of the century to increasing numbers of employed workers as patients in the voluntary hospitals. To a degree, this was merely a further symptom of the economic, social and cultural changes that were rendering hospitals, rather than homes, the main resort for medical care. In tandem with the changes in the immediate context of labour out of which accidents arose were the changes in urban demography consequent upon both the immigration of ethnic groups and the emigration to the suburbs of certain sections of the middle and working classes: Furthermore, while changes in the size, structure and function of households affected the potential for home care, changes in the labour market meant that there were many more urban 'strangers' without the homes in which to receive such care. These changes, as much as real and perceived ones in the quality of hospital nursing and medical care, contributed to new patterns of hospital use.[50] In general, the reception of accident cases in hospitals contributed to and confirmed the trend.

Part and parcel of this transformation were changes in hospital funding. Formerly to the 'deserving poor' had gone the fruits of 'sacred charity'. But once employed persons began to be treated in charity hospitals, the hospitals became the objects of a self-serving philanthropy on the part of industrialists and workers. Many hospitals, such as that at Walsall, were created by the 'gift' of local industrialists; others, as we have noted, were coming to rely on worker contributions; and everywhere both general and specialist hospitals were establishing formal and informal types of pre-payment schemes with employers' and workers' friendly societies. While one of the effects of these developments was to tie voluntary hospitals closer to local economies, another was to call into question their function as 'charities'. The latter issue was rendered particularly acute in connection with the use of outpatient departments, especially in the major voluntary hospitals in London. The problem of so-called outpatient 'charity abuse' partly involved accident cases, but for our purposes here, it is as important for illuminating the wider context of rationalization into which accidents were being brought.

OUTPATIENT 'CASUALTIES' AND THE CONTEXT OF RATIONALIZATION

The increase in the number of attendances at outpatient departments during the second half of the nineteenth century has been described as 'something totally new in medicine'.[51] Numbers began to rise between

1835 and 1850, but it was between 1850 and 1911 that they reached 'crisis' proportions, with a million and a half attendances by 1887 in London alone.[52] Increases as great as twelve-fold were not unknown in both metropolitan and provincial hospitals.[53]

Although familiarly described as the problem of 'outpatients', in fact, as the *Lancet* pointed out with regard to St Bartholomew's Hospital in 1865, it was 'in the casualty division that the increase of numbers has been most marked'.[54] This remained the case through to the Great War: in 1910, among the ten teaching hospitals in London, casualty cases comprised between 38 and 82 per cent of all outpatients (the average being 64 per cent).[55] 'Casualty' patients did not necessarily refer to serious or emergency cases however. While it is significant that the word had no fixed definition common to all hospitals, it usually included the most trivial cases as well as the most urgent. 'Outpatients', on the other hand, normally referred to those 'who, after receiving a regular letter of admission, are entitled to the advice of the assistant surgeons and physicians'.[56]

How many 'casualties' were serious accident cases cannot readily be determined from the statistics on 'outpatients', and the proportion obviously varied between hospitals. From different sets of figures for the London Hospital, it appears that of 133 000 'casualty' cases received at the outpatient department in 1910, around 19 000, or 14 per cent, were serious accident cases (presumably admitted and re-classified as inpatients).[57] This figure is in accord with that given in an inquiry of 1870 in which it was concluded that three-fifths to nine-tenths of outpatient cases were of a trivial nature.[58]

In seeking to account for the startling rise in casualty attendances at hospitals contemporaries were not unmindful of the demographic changes around them. For the most part, however, they chose to focus on inadequacies in Poor Law medical relief and on the erosion of the traditional system of subscribers' letters for entrance into hospital. Above all, they loaded blame on to those workers and their dependents who availed themselves of outpatient services yet were deemed financially able to contribute something to their medical care. Loud in railing against such 'abusers' were general practitioners, who saw their livelihoods seriously threatened by outpatient attendances. Representing them in part was the Medical Attendance Organization Committee which sought to have 'casualty departments ... strictly limited to accidents and street emergencies'.[59] Equally strident was the Charity Organization Society which wanted to see the development of Provident Dispensaries where workers would pay a few pence per week against the event of medical need, and in the process learn the virtues of providential behaviour.[60]

Neither GPs nor the COS were interested in accident provision as such, and it would be wrong to suggest that accident cases played any causal role in enabling the COS to involve the profession as a whole in the campaign to 'repress the demoralising tendency of gratuitous aid so freely lavished'.[61] Those in hospital medicine converged on the issue largely because they saw in it one of the central problems facing their future – the fragmentation of functions and resources consequent upon the rise of specialty interests. Outpatient departments, it needs recalling, were staffed not only by some of the most junior doctors, but also by some of the most ambitious aspirants to specialization. Here, doctors not only acquired the power, usually denied them, to admit patients to hospitals, but they were able to cultivate specialty interests under relatively unconstrained conditions.[62] Progressives such as Edward Sieveking promoted medicine's turn towards specialization and criticized the practice in medical education of merely 'walking the wards'. At the same time, however, they worried about any shift towards the use of specialist hospitals in medical training, and sought to resolve this anxiety by transforming the outpatient departments of teaching hospitals into 'polyclinics'. Within these clinics, Sieveking argued in 1868, students would not only be able to pursue the range of developing specialisms, but they would be encouraged to follow through the care of patients from the clinic to the home, thus overcoming the 'habit of looking upon the patient simply as an object of observation, a nosological curiosity'.[63] By these means, neither medical education nor patient-care would be fragmented; the integrative pursuit of the one would be homologous with the pursuit of the other.

Although the idea of the polyclinic was never fully taken up in Britain, and the training of doctors in the de-objectification of the patient not at all, by the 1890s most outpatient departments in general hospitals were well supplied with specialty clinics. It was there that the majority of the rising generation in surgery were based, and it was from them, increasingly, that there came the demand for the better utilization of outpatient departments in medical education.[64] Such calls, which aimed essentially to shift the balance of power in hospital medicine, were often accompanied by others for improvements in the segregation of outpatient cases and for greater standardization and uniformity in medical and surgical procedures – for more 'precise methodological plans and steadfast business-like habits [in surgery]'.[65]

Almost all such calls were cast within and legitimized through the rhetoric of charity abuse – in terms, as Sieveking had put it, of 'acquir[ing] more certain means of sending the real pauper to the workhouse infirmary; and . . . of frightening away the well-to-do mechanic and tradesman, whose means and position do not justify their becoming recipients of charity'. Behind such statements, increasingly, stood the spectre of the bankruptcy of the voluntary hospitals, and, hence, if not their disappearance, their coming under the control of the state.[66] Arguments for rationalizing charity and for rationalizing the clinical space for therapeutic procedures and professional advancement thus proceeded hand in hand in a common political and economic context.

However, the case for applying system, uniformity and expertise (or division of labour) in the management of medicine and welfare did not derive simply in response to the outpatient problem. Throughout society such concerns gained ground wherever greater complexity, waste and inefficiency appeared to threaten. In the administration of the Poor Law, in particular, where preoccupations with preventing improper access to 'free' services had always been paramount, such concerns extended at least as far back as Chadwick's vision of the workhouse system 'as a set of specialized schools, hospitals, asylums, almshouses . . . [in Unions] as units in a rational local government system'.[67] In 1866, the Association for the Improvement of London Workhouse Infirmaries was founded with the aim of reclassifying and rehousing the workhouse population into more manageable, differentiated, groups. In the following year a central health authority – the Metropolitan Asylums Board (MAB) – was created to carry this out.[68] Apiece with these developments, in the voluntary sector, was the founding of the COS itself in 1867. As suggested in the previous chapter, basic to the COS's 'science of charity' was a distancing of individual experiences of poverty and illness; 'efficiency' in welfare was to emerge from the operation of the same abstract principles of rationalization that served to maximize profits in business and industry – in particular, by differentiating, systematizing and coordinating the individual units or agencies involved in order to eliminate wasteful duplication of effort, that is, 'scientific management'.

The last decades of the nineteenth century witnessed an acceleration in the application of these principles. Faster, cheaper and larger means of production, as well as faster and cheaper means of communication and transport enabled vast increases in the size and scale of enterprises.[69] In tandem with these changes, the 'Second Industrial

Revolution' called forth sciences of administration which were bent on ever greater system, uniformity, and efficiency. Management 'science', in turn, increasingly granted moral authority to the administrative process. Time itself was standardized around the world in 1884 (a move promoted by the interests of American railroad companies), whilst the space between peoples and places was, as Marx put it, 'annihilated' as a consequence of the faster transport and communications.[70] Even the spoken English of the English middle class became standardized around this time, shedding its regional differences.[71] Indeed, increasingly, individualism, autonomy and independence were eroded as everything and everyone became construed functionally as a 'unit' of larger *inter*-dependent wholes. Municipally administered networks for urban sewerage, water supplies, transport, electricity, gas, telecommunications, and police and fire services were all conceived in this period.[72] In part to facilitate such services, many cities turned themselves into 'corporations', while for the metropolis, the London County Council was created in 1888 to assume responsibility for education, sewage disposal, housing and the Poor Law hospitals.[73] Such developments were pursued morally in the interest of social order and control, whilst being driven by, and functioning primarily for, economic efficiency (making money) or eliminating inefficiency (wasting money). Metaphors of the body were created in which these complex interactive systems were perceived as natural. The new physiology of the 'young turks' in medicine and surgery, in which the functional efficiency of the body as a whole was seen to derive from the complex interaction of the parts, was at one with this economic ideal. In tandem with the new physiology – in part drawing upon it – was the version of scientific management elaborated in the 1890s by the engineer Frederick Taylor, in which the virtues of standardization, systematization and specialization were celebrated for their effect in maximizing production.[74]

This group of ideas – known simply as 'rationalization' in Britain[75] – was adopted in the voluntary hospitals towards the end of the century in the course of the gradual displacing of lay governors by business managers. Although little research has been conducted into this transition in British medicine (in contrast to American medical history),[76] its salient features are discernible through such exemplars of the new breed of managers as Sydney Holland – the second Viscount Knutsford – and Henry Burdett. Knutsford (1855–1931), was one of the directors of the largest group of shareholders of the London docks, in which capacity he played a central part in the rationalizations of capital and labour that took place on the docks during the last decades of the century. (In the famous dockers' strike of 1889, which was intimately linked to those reorganizations, Knutsford was the leading exponent of the employers' case.) In 1891 he 'drifted into hospital work', restructuring the management of the Poplar Accident Hospital (which served his own workforce), before moving on, in 1896, to do the same for the London Hospital. Knutsford remained the Chairman of the London until his death, by which time he had fully succeeded in his aim of 'reorganizing, rebuilding and refinancing' it.[77] During this time he also played key roles in virtually every organization and committee for the more efficient organization of medical services in London, including those aimed at ambulance services for the victims of accidents.[78]

Henry Burdett similarly combined a career in hospital management with one in banking, the stock exchange and corporate finance. Unlike Knutsford, however, who also enjoyed an aristocratic lifestyle, Burdett (1847–1920) never relaxed from applying business principles to medical charity. As a result of his various publishing ventures in the field (especially his *Hospitals and Charities Annual* and his journal *The Hospital*), by the 1880s he had emerged as the foremost British exponent of rationalization in medicine.[79] As early as 1869, whilst managing the Queen's Hospital, Birmingham, he had devised and put into practice a system of uniform hospital accounting which was later widely adopted.[80] In Birmingham he also became involved with the Hospital Sunday movement, and he was later to write on the need for placing the Sunday and Saturday Funds under a single management body and establishing a uniform system of collection.[81] Unsurprisingly, among Burdett's many other concerns was charity 'abuse' in outpatient departments. For the curbing of it he advocated the use of almoners to means-test patients, and the imposition of patient fees.[82] The issue of outpatient abuse was in fact to be rendered by Burdett one of the two main concerns of the Hospitals' Association (subsequently the British Hospitals' Association), a management body which Burdett helped bring into existence in 1884 after delivering a paper on hospital administration to the British Social Science Association. The other main concern of the Hospitals' Association – to which we shall turn in a moment – was the coordination of ambulance services.

Both Knutsford and Burdett were also involved in the organization of the Prince of Wales' Hospital Fund for London (subsequently the King's Fund). Established in 1897 to help bail out the impoverished voluntary hospitals, and thus stave off dreaded interference by the

state, the King's Fund was to be far more successful than either the
COS or the Hospitals' Association in its role as a rationalizing body.
As the historian of the Fund has put it, 'Taking its cue from Burdett
and other like-minded officials, the Fund sought to create a new
generation of capable, well-paid professionals who would revamp
hospital management.'[83] By wielding the power of the purse over
institutions that were increasingly desperate for funds, it was able to
eliminate many duplicated small-scale medical services. The amalgamation in 1907 of the three small orthopaedic hospitals in London, for
instance, was directly the result of such pressure from the King's
Fund.[84] But it was to prove more difficult to exert this kind of pressure
on the major voluntary hospitals, and virtually impossible to enforce
coordinations of services between them. The history of the efforts to
organize an ambulance service for London are revealing in this respect;
while further illuminating the context of rationalization in which
accident victims were literally brought to hospital, this history also
points to the forces of resistance to cooperative action.

THE ARRIVAL OF THE AMBULANCE

Calls for ambulance services for the victims of accidents began to be
heard in most British cities in the early 1880s. In part this related to
impressions of the urban landscape being disfigured by increasing
numbers of street accidents.[85] A Society for Preventing Street
Accidents and Dangerous Driving was formed in London in 1879.[86]
But the immediate inspiration for what was soon to blossom into an
'ambulance movement' was an article in the *BMJ* in July 1881
describing the New York ambulance system.[87] The author, Dr
Benjamin Howard of New York (then resident in London), who was
the promoter of the horse-drawn Howard Ambulance, was keen to see
a similar system established in London. Howard was to find a powerful
ally in Ernest Hart, the editor of the *BMJ*, and in an editorial in the
same issue of the *BMJ* that carried Howard's article, Hart passionately
endorsed the need for such a system for London. 'Out of eighty-five
hospitals in this metropolis,' Hart pointed out, 'five of them, last year
alone, received over twenty-two thousand patients, nearly half of the
number in some of them being emergency cases.' Yet in not one of the
85 hospitals, nor in the entire police and fire department was there a
single ambulance wagon. 'Ask . . . how any one . . . managed to get
from the scene of the accident . . . to the hospital which received them,

– and the answer has to be supplied by our own imagination.'[88] Most accident victims in cities were in fact left to the police, who either carted the victim to his or her home (if it was near), or else called upon fee-charging stretcher-bearers from the voluntary hospitals to perform this function.

The time for change was ripe, however, and a number of initiatives in accident provision were then just beginning. In 1880, at the same time that the metropolitan police ambulance service was being partly over-hauled, 35 voluntary first-aid posts in various parts of the metropolis came to be manned by the St John Ambulance Association.[89] The Metropolitan Asylums Board also started an ambulance service in 1881, although this was only for the transport of infectious patients and was not extended to include surgical, medical and psychiatric cases until the 1900s.[90] Far from having any links to curative hospital treatment, the MAB's service was more a measure in preventive medicine, or a part of an isolation policy to protect other people.

The exclusion of accident cases from the MAB's ambulance service serves as a reminder of the fact that, unlike fire or police services which were funded from the rates, ambulance services for accident cases were not easily legitimized in terms of protection to private property and the social order. Partly because of this, and partly because only the voluntary hospitals received accident cases, plans for the organization of a metropolitan ambulance service were conceived around the voluntary hospitals. Benjamin Howard, armed with letters of introduction from Hart, played an active role in these proceedings. But no longer did he stress, as he had in his original communication in the *BMJ*, the harmony of the voluntary *and* the municipal elements in the New York system.[91] Remaining mute on that point, he gained a partial success when, in December 1881, the London Hospital purchased one of his ambulances and set about to organize a metropolitan ambulance system.[92] The London Ambulance Service soon failed, however, for want of support from the other major voluntary hospitals.[93] Proud of their independence and in fierce competition with one another for charity funds, the major voluntary hospitals were islands of local autonomy, unwilling to cooperate on most projects, let alone on those that might involve municipal or local authority controls. Moreover, the voluntaries had neither the financial nor the professional incentive to develop a public service for accident victims. Such services promised only to compound problems concerning casualty cases. Whereas outside London the ambulance movement flourished – notably in Glasgow, Liverpool and Manchester, where there were fewer hospitals to coordinate and the schemes were closely connected with the police and other services on the rates[94] – in the metropolis the movement fell on stony ground.

The failure of the voluntary hospitals in London to coordinate an ambulance system for accident cases drew the attention of Burdett and the Hospitals' Association; by the turn of the century their publications on the subject were only rivalled in number by those on outpatient 'abuse'. Precisely how this concern was sparked is unclear. Burdett's interest seems to have emerged while he was the secretary of the Dreadnought Seamen's Hospital, Greenwich, in the late 1870s and early 1880s, when there was considerable discussion on the use of ambulance ships on the Thames.[95] Some of his and the Hospitals' Association's thinking on the subject derived from Howard, some from the setting up of the MAB's ambulance service, and some from French examples.[96] More fundamental, however – to Burdett and his colleagues as to everyone else – was military influence, as is well indicated through the titles of early articles on the subject, such as Mayo Robson's 'First Aid to the Wounded in Civil Life', or popular texts, such as Reginald Harrison's *The Ambulance in Civil Practice* (1881; 5th edition 1902).[97] Like the word itself, most 'ambulance' technology derived from the battlefield; the Howard Ambulance, for instance, having been designed for the Union Army in the American Civil War. It is not incidental, of course, that towards the end of the century the military – especially in its idealized 'Prussianized' form – was increasingly looked upon as a model of organization, achieving efficiency and social control through tight hierarchical structure.[98]

For Burdett and his colleagues, military ideas on ambulance services were received partly through the St John Ambulance Association. One of the three founders of the Association was the frustrated military careerist, John Furley (who was also the founder in 1884 of the civilian Invalid Transport Corps). Furley organized the British voluntary ambulance services during the Franco-Prussian War of 1870–1 and the Russo-Turkish War of 1876, and was the director of the *ambulances volantes* of the French Army during the Commune War of 1871. It is hardly surprising therefore that his model for civilian ambulance services was wholly militaristic with 'corps', 'divisions' and 'brigades' (designed to be 'uniform or interchangeable') ascending up a hierarchic chain of command to a central authority.[99] Another important influence (as much on Furley as on Burdett and his colleagues) was Surgeon-General Sir Thomas Longmore, the professor of military surgery at Netley. Longmore's *Treatise on the Transport of Sick and*

Wounded Troops (1869) was the first work devoted exclusively to the subject. Written in the shadow of the Crimean and the American Civil Wars, the book was directed to perfecting military needs, but the principles of organization, communication, and intercommunication that it outlined, as well as the apparatuses that it discussed – including the Howard Ambulance – readily translated to civilian use. Longmore himself had some interest in civilian ambulance services and (like Knutsford and Sieveking, too), he played a part in the St John Ambulance Association.[100]

Typical of the works advertised in Burdett's *Hospitals and Charities Annual* was that by the Army Medical Officer, G.J.H. Evatt, *Ambulance Organisation, Equipment, and Transport* (1884).[101] Evatt's work was unashamedly drawn from Longmore's, and like it focused largely on military arrangements. The chapter on civil ambulance arrangements, however, exemplifies the translation of military to civilian organization. Perceiving this translation as already partly effected in the municipal ambulance arrangements of New York, Chicago and Boston, Evatt looked forward to the day when London and other British cities would be similarly mapped out into 'battle zones' with 'lines of communication' ('telephonic') running from the streets to the hospitals and police departments, where ambulance wagons would stand at the ready along with 'trained medical officials'. He went on:

> The existing hospitals, and the many other municipal hospitals needed, should be distributed with system over our great city. A chain of outposts in the shape of municipal dispensaries should bring medical relief within a quarter of a mile of every citizen. Here first aid should be ever ready, and here the outpatients now swarming and crowding at our great hospitals should be dealt with in detail and by districts. At certain hours in the morning, midday, and evening, the sick-transport waggons from the great central hospitals should call at these outlying dispensaries, and carry in comfort the cases chosen for admission to the district central hospital. But far more than this is needed, for a ring of great hospitals, combining in the same extensive grounds both convalescent and treating sections, should surround London at a distance far removed from the smoke and overcrowding of our great city. (pp. 25–6)

Evatt's ideas, including that for 'a strong municipal government' to carry them out, were reminiscent of Roberton's plans for Manchester in 1860, and in general terms they anticipated much postwar thinking on the organization of hospitals and health services.[102] Except for his radical suggestion that 'the London Hospitals should come under a central Board, and their funds be "pooled" in a common fund, having in reserve the municipal rates to fall back upon', Evatt's ideas were essentially those extolled by Burdett in his various publications. In 1893, for example, the same year in which Longmore's book was reissued as *A Manual of Ambulance Transport*,[103] Burdett commented favourably on the suggestion

> that outpost hospitals, or receiving wards in connection with the principal general hospitals should be opened in the out-lying districts [of London], where the majority of the patients reside, so as to secure the minimum of suffering and risk to poor persons meeting with accidents.[104]

Like Evatt, Burdett was looking across the Atlantic, in particular to New York and Boston, where experimental schemes of this type had recently been tried in order partly to accommodate the medical needs of the new middle-class suburbanites. In Boston, fully-staffed two- and three-bed district hospital substations were established and linked to the City Hospital via an ambulance service.[105] Effective ambulance services thus rationalized the space between hospitals and undermined the need to build more large expensive ones. Furthermore, by transporting patients to the most appropriate places for treatment, wasteful duplication of hospital services could be avoided. Enthusiastically, Burdett concluded that 'some such system as this must sooner or later be applied to London and all large towns'.[106]

But like Roberton's plans for the victims of accidents in Manchester, those relayed by Burdett and his colleagues for London died still-born. Although the climate for such rationalizing schemes had ripened considerably (not least because of widespread militarism), it proved impossible to compel the voluntary hospitals to act in the coordinated manner required. Not until 1913, four years after the Home Office had deemed it 'efficient and economic' to have an ambulance service for street accidents in London, was a scheme worked out with the London County Council. When it finally came into operation in February 1915 it was not overseen by the voluntary hospitals, but by the London Fire Brigade.[107] For the most part, the voluntary hospitals remained insular and uncooperative. For Burdett, this failure of 'intercommunication' was second only to the failure of the voluntaries to curb outpatient abuse – 'the most remarkable proof of the impossibility of inducing those responsible to act together and enforce the necessary reforms'.[108]

THE MANCHESTER SHIP CANAL PROJECT (1888–93)

In contrast to the circumstances that worked against the introduction
of a coordinated ambulance service in London were those surrounding
Robert Jones's organization of the accident service along the 35-mile
work-site of the Manchester-to-Liverpool Ship Canal. The Canal,
which was to turn land-locked Manchester into the fourth largest port
in the UK by 1903, was regarded as one of the engineering wonders of
the world, if not, perhaps, 'the greatest non-martial triumph that has
ever been witnessed'.[109] Historically, it stands as an emblem of
modernity. Although begun on private capital, it soon emerged as a
bold experiment in municipal enterprise, with the Manchester
Corporation becoming the largest shareholder in the Ship Canal
Company. The Canal not only helped to 'network' the 'Greater
Manchester' area as a whole, and directly link it to the outside world,
but it also set a precedent for the subsequent involvement of the
Corporation in electrification schemes, sewerage, tramways, housing,
and the development of the country's first major industrial estate at
Trafford Park.

The casualty service was spiritually apiece with the project as a
whole. Intended to meet the needs of the 10 000–20 000 navvies
employed in the diggings, the service in its final form consisted of three
casualty hospitals spaced between, and connected to, the many
separate work-sites by a network of over 250 miles of railway. Each
of the hospitals was staffed by a resident house-surgeon (persons who
had previously worked for Jones as dressers), a qualified nurse as a
matron, a ward nurse under the matron, a cook and a handyman. An
external medical service, manned by local GPs, was also connected to
each hospital to cater for the general health of the navvies and their
families and to supervise public health facilities. A clear division of
labour was thus established to deal efficiently with every contingency,
and unity of control and continuity in the treatment of injuries under
the house-surgeons seems to have been routine. Jones himself was
called out only to perform more serious operations, an effective
communications system being established for this purpose. In all these
respects, the service was a model of a uniform, integrated accident and
medical service, the hallmark of which was maximum efficiency in the
reception and treatment of cases – in all, some 3000 of which (mostly
fractures) being dealt with during the Canal's construction.[110]

Unlike the circumstances prevailing in Roberton's Manchester or,
later, in Burdett's London, those surrounding the treatment of the

Plate 5.1 Constructing the Manchester Ship Canal, 1888–93

Source: By kind permission of the Greater Manchester County Record Office and the Manchester Ship Canal Company.

casualties on the Ship Canal project were unencumbered by a diversity of administrative and political interests. Financed entirely by the Ship Canal Company, the accident service was essentially a one-man show, centrally administered by Jones and requiring neither negotiations with nor coordination among different hospitals. Indeed, it existed partly because the voluntary hospitals at either end of the Canal were too far removed from most of the construction sites to be effectively utilized.[111] Even had the voluntary hospitals in Manchester and Liverpool been able to cope with the accident cases from such a large working population, it is unlikely that they would have wanted to. Their resources were already stretched, and many of them were reluctant to treat gangs of transient workers. Bitter memories still lingered of costly experiences with navvies injured during the earlier booms of railway construction.[112] The circumstances behind the accident and medical service on the Ship Canal were thus akin to those pertaining at various mining operations and railway workings in America, where company hospitals were established to deal with similarly large working populations isolated from existing medical facilities.[113] Jones's accident service was not, therefore, an achievement in the intercommunication of independent voluntary hospitals, but rather, in the organization of a self-contained service.

Whereas Roberton's interest in accident provision was primarily linked to medical reform, public health and philanthropy, and Burdett's to rationalizations of charity and ideals of sociomedical management, Jones's interests were tied to the immediate needs of capital and labour. The Ship Canal project was in fact a continuation of Jones's existing involvements: not only was he, at this time, the medical officer to several insurance companies, but he was also routinely involved with employers and with workers' organizations at both the Nelson Street Clinic and at the Stanley Hospital (and at the Southern after his appointment there in 1889). Among the supporters behind his first (unsuccessful) application for the post of honorary surgeon to the Southern in 1882 were the secretaries of no less than nine major trade unions in the region.[114] It was these contacts that helped secure his appointment as consulting surgeon to the Canal project. Although the Ship Canal Company had sought to hire him as surgeon to the works (at a reputed £3000 per annum), Jones turned the offer down and accepted instead an honorarium of £1000 per annum in order to be able to continue with his private practice.[115]

More importantly, Jones's involvement in the project and its medical service emerged in a context in which it was increasingly difficult for employers to be indifferent about accidents. As a result of debates in the 1870s and 1880s over employers' liability, employers and unions alike were sensitive to the problem. Although it is unclear whether the Ship Canal Company took advantage of the clause in the Employers' Liability Act of 1880 that enabled and encouraged employers to bribe or force their workers to 'contract out' of the Act – unclear, because of the extensive involvement of sub-contractors – the Canal medical service can be seen as a part of the welfare bargaining with unions that often accompanied 'contracting out'.[116] Certainly the service was not simply a product of enlightened paternalism; there is little evidence that the Company had any interest in the long-term welfare of its employees, some 1700 of whom were left permanently incapacitated and another 250 temporarily so as a result of injuries sustained during the Canal's construction (according to an official of the Navvies'

Union in 1892).[117] The Company's main interest was in short-term intensive labour, and it is known to have imported cut-price American labour whilst trying to keep the wages of labourers at 4½*d* per hour. Moreover, the main contractor of the Canal had agreed to pay £100 per day for every day that the construction extended beyond the scheduled completion date.[118] Thus, quite apart from other considerations, it was expedient to have a service on hand that would assist in avoiding the loss of work-time through injuries, especially minor injuries. Since the days of the railway booms, accidents had been known to be expensive, for they entailed not only the loss of the injured man's labour, but also that of the mate who accompanied him to home or hospital.[119] But prior to the end of the century employers seldom expressed concern about this. Now, however, in a context of greater interest in labour power and labour wastage, keeping a 'fit' labour force proceeded hand in hand with the discovery by employers of 'genuinely efficient ways of utilizing their workers' labour time ("scientific management")'.[120]

The Manchester Ship Canal has been seen as marking the beginning of a revolution in local business organization. Manchester became one of the great 'tentacular cities' of the modern age, as much through its revived commercial and financial functions as through its new electric tramways and expanding cultural institutions – the latter culminating in its own university, and in the 'new journalism' of its own newspaper, the *Manchester Guardian*. 'The city made communications more than ever the basis of its economic empire.' The Canal was thus both a symbol of the erosion of 'the individualism of traditional "Manchesterdom"', and the realization of 'corporatism'.[121]

Similarly, it might be said that the Ship Canal project symbolized and realized the transformation of accidents from individualized happenings to a type of medical concern requiring its own organization and professional politics. Such a transformation was neither accidental nor reliant upon medical advances. Largely, it depended upon a new valuation of labour, just as the drawing of crippled children into medical organization depended upon a revaluation of childhood. In both cases, medical interests were not central to the revaluation process itself, though they were structured and shaped by it. In short, expertise and efficiency in the medical management of accidents became constitutive of preoccupations with the elimination of waste in industry, business and welfare. Given the extent to which the metaphors and apparatus of this efficiency drew upon military experience, it is perhaps not surprising that it was to be in war itself that the greatest intensification of this 'efficiency' in medicine was to occur.

6

THE GREAT WAR

Much has been written about the First World War in the making of the modern world – of the paradox of that awesome dance of death giving birth to modernist vision.[1] But the significance of 1914–18 for modern medicine and its image has been little explored. Instead, on the basis of an abundant medical literature, it has largely been accepted that, for all its horrors, the 'Great War' was not only great for medicine, but was good for humanity in general, encouraging medical innovations, stimulating new therapies, drugs, surgical techniques, and so on.[2] This chapter is not concerned with challenging this overtly positivist, implicitly militarist, and profoundly simplistic message so much as with pointing to its irrelevance in interpreting one of the assumed-to-be most significant and enduring of the medical benefits of the war: specialization.[3] Through the wartime history of orthopaedics, we shall show, the perception of specialization as a straightforward beneficiary of war – comparable, say, to the munitions industry – is at best superficial, at worst wrong.

The history of orthopaedics lends itself to this purpose, for its supposed transformation during the war to a 'sophisticated branch of surgery' has often been taken as paradigmatic of the positive relationship between war and specialization.[4] Certainly, to the first generation of modern orthopaedists, this relationship seemed indubitable. As one of their number recollected in 1959:

> It is a chastening thought that but for the impact of the First World War on the surgery of injuries of the locomotor system, the emergence of orthopaedics as an independent surgical speciality covering a wide field, and playing an important role in under-graduate and post-graduate teaching, might have been long delayed.[5]

Since around 65 per cent of all the casualties of the war involved impairment of locomotor functions (principally as a result of fractures caused by bullets and shrapnel),[6] the war was indeed 'great' in providing an opportunity for the kind of orthopaedics practised by Jones in Liverpool. By 1918, 20 special Military Orthopaedic Centres had been established in Britain under Jones's direction, commanding some 20 000 beds.[7] At the hub of this military orthopaedic empire was the 800-bed hospital at Shepherd's Bush, London, a former Poor Law infirmary requisitioned by the War Office in March 1916, which was soon to become a national showpiece of military medicine. Its success was matched by Jones's *Notes on Military Orthopaedics* (1917), a compilation of his articles in the *British Medical Journal* in 1916, for which a large reprinting was called in 1918. The formation of the British Orthopaedic Association at the end of the war symbolized the success of modern orthopaedics still further.

Through 'the halo' that the war cast on the treatment of traumatic injury,[8] the repute of orthopaedics was radically transformed from the days of the struggling BOS in the 1890s. No longer was the specialism the butt of medical jokes – a pretentiously labelled medical backwater primarily for the mechanical treatment of crippled children.[9] And undermined almost entirely was the mistaken etymological association of 'orthopaedics' with disorders of the feet. What Jones chose to call 'military orthopaedics' encompassed not only the treatment of all types of fractures, and all diseases and injuries of joints, but also cases of gunshot wounds, nerve lesions, and plastic surgery of the neck, face and jaw. By ensuring that the commanding officers of military hospitals were armed with this knowledge of the scope of orthopaedics,[10] Jones did more than merely educate those army generals who had snorted about 'orthopaedics' being 'a damned silly name to give a military hospital'.[11] By the end of the war, few were unaware that 'orthopaedics' applied to the treatment of the injuries and diseases of the locomotor system as a whole by surgery and other means. By then, too, the imagery of orthopaedics had penetrated even the most ethereal reaches of society: the socialite, Lady Diana Manners, for instance, after falling through a skylight on Victory Night, is said to have declined the services of Sir Arbuthnot Lane for the open surgical treatment of her fractured femur, because 'she had learnt that such fractures had been treated very successfully in the recent war on what was called a "Balkan beam"'.[12] Meanwhile, the troops at the front came to experience first-hand the basic principles and benefits of modern orthopaedics. In 1915 Jones persuaded the Army Medical Service (AMS) to introduce the various splints that he and Hugh Owen Thomas had devised, and in January 1916 he embarked on a tour of the base hospitals in France to demonstrate their use.[13] By 1917 the splints were not only in wide use at the base hospitals, but they had become standard issue at the front and a matter for regular drills at the regimental aid posts. In particular, Thomas's light and versatile leg splint for treating fractured femurs (technically his 'knee bed splint'), came to be regarded by officers and men alike as virtually a sacred object ('St Thomas splints').[14] Such familiarity with the equipment of orthopaedics makes explicable the fact that immediately after the war when support was mustered for 'Labour's Own Hospital' – the Manor House Hospital at Golders Green in London – it was under the respected title of the Industrial Orthopaedic Society that thousands of workers subscribed.[15]

In America, the contrast between prewar and postwar orthopaedics was no less striking. 'In our country, previous to the War,' reflected Joel Goldthwait, one of the leaders of the specialism during the war, 'the orthopaedic surgeon was being tolerated more or less, but was not cordially welcomed by the general surgeon; it was not many years removed from the old so-called "strap-and-buckle days"'.[16] But by the war's end, the kudos of orthopaedics was high. The AOA mushroomed from a small assembly of around 35 mostly East Coast surgeons to a fully national body of over 300 members; no longer could 'practically all the subjects pertaining to Orthopaedic Surgery . . . be found on the programme of a single meeting', as had been the case before the war.[17] In Europe, too, especially in Italy and in Germany (where Konrad Biesalski organized orthopaedic clinics and educated the public to the care of war cripples),[18] the outlook and confidence of orthopaedic practitioners was much enhanced. A major surgical speciality was seen to have emerged.[19]

Precisely because this wartime 'making' was not inevitable, however, its history serves to expose the fallacy of a simple causal relation between war and medical specialization. Contrary to the claims of the official wartime medical historians and others since, the sheer volume of locomotor injuries during the war did not compel the specialism's rise.[20] Even were it to be supposed – as it often is in writings on medical specialization – that the medical demands of the war were somehow capable in and of themselves of calling forth supplies of medical experts, the experts in orthopaedics in 1914 were hardly the most obvious group to be beckoned. As we have seen, just prior to the war a space for orthopaedics had opened up in general hospitals, but it was a space for a specialism focused on the congenital deformities of children. Although a handful of orthopaedists before the war, besides Jones, had some experience in handling trauma – including two who served as medical officers in the war in South Africa[21] – only Jones's practice in Liverpool transcended the conventional parameters of the specialism.

But even to the limited extent to which Jones can be said simply to have 'exploited' the wartime situation, this exploitation was far from being a straightforward consequence of military medical need. In reality, rendering the war into an 'orthopaedists' war' depended on certain fortuitous political and economic circumstances in which the worth of Jones's expertise was able successfully to be demonstrated. More broadly, it relied on the emergence of a context in which the managerial skills that were fundamentally a part of that expertise could be effectively exercised. The wartime enabling of Jones's 'new orthopaedics' thus required more than merely the assemblage of those wartime conditions that in retrospect can be seen as advantaging specialty development: the aggregation of large numbers of patients suffering from the same types of problems; the authoritarian structures in which the supply and maintenance of specialist facilities could be prioritized; and the salaried employment of doctors that eliminated the peacetime financial constraints on full-time specialty work. Undeniably, these were important handmaids to specialty advance, but they cannot account for the formation of military orthopaedics, or account for how and why fractures, for instance, came to be seen as the 'problem' for which 'orthopaedics' was the solution.

Crucial to the wartime making of modern orthopaedics was the negotiation and occupation of a political space in medicine for reorganizing medical work and power relations generally – in effect, the opening of the space that had been a part of the agenda of the 'modernists' in surgery since the 1880s. What was *not* involved was the promotion of specialist interests as such on the part of a self-consciously orthopaedic lobby. The professional body, the BOA, was only established at the end of the war. Significantly, its inspiration and initial planning came not from Jones and his British colleagues, but from the American orthopaedist, Robert Osgood.[22] It is also among the many ironies in the history of orthopaedics in Britain that the BOA was to be one of the few tangible benefits of the war for orthopaedists.

To discuss orthopaedics and the Great War is to do more, therefore, than merely raise questions about measuring the 'goodness' of war for specialization. Above all, it is to challenge the assumption that war simply imposes its will upon passive medicine – as if war were purely a military event in a world disconnected from other social and economic processes. Suffice it to say here, that in confronting that view, the wartime history of orthopaedics makes it difficult to perceive medical specialization as merely 'standing alongside' the munitions industry, as it were, as a major beneficiary of the organized violence of the war.

Rather, its history compels us to see both medicine and the military as standing in the same socioeconomic context as the munitions industry – as part and parcel of the same historical process. War and specialization as abstract forces in the history of medicine can thus be de-mythologized.

In the path of this project, however, lies the myth of the making of modern orthopaedics itself and, in particular, the mythic role of Robert Jones. That Jones has to be set apart from other historical actors by virtue of his unique prewar experience in the large-scale organization of trauma too readily allows the wartime achievement of the specialism to be cast entirely as his handiwork. In the hagiography of Jones this interpretation is strengthened by the projection of his war work against a backdrop of military medical ignorance, disorder and incompetence. In addition to upstaging other key figures and marginalizing the political context of both medicine and the military, this projection also involves being lax with the chronological sequence of events. Before we begin to discuss the wartime making of orthopaedics, then, attention needs directing to the myths of its making.

MYTHS OF MAKING

It is true, of course, that at the outbreak of the war few persons had any idea of either the nature or the extent of the casualties that would be received. Previous military experience, especially that in the Boer War of 1899–1902, reinforced the importance of disease-preventive hygienic measures over those for the treatment of wounds. This emphasis, along with the fact that most of the gunshot wounds in the veldt were surgically 'clean', meant that few doctors in 1914 were prepared for the lethal septic wounds produced by the machine-gun fire, shrapnel bombs and high-explosive shells on the richly manured and bacteria-infected soil of the Western Front.[23] In a context in which as many as 300 000 casualties could result from a single encounter,[24] where dressings and antiseptics often seemed powerless against suppurating wounds, where septicaemia, pyaemia, gas gangrene and other forms of blood poisoning were rife, and where compound fractures could be as fatal as those of a century before, prewar surgical knowledge and experience were poor comfort.[25] Pressed for time, space, and essential supplies, surgeons confronting infected compound comminuted fractures (the most common type of injury in which the soft tissue was damaged and the bone was both broken and shattered) saw little alternative but to amputate. During the first months of the war, a staggering 80 per cent of cases of gunshot wounds involving fractures, and 60 per cent of uncomplicated knee-joint cases, resulted in amputations, mostly as a consequence of infection to the wound rather than of the wound itself.[26]

In many ways the surgeons were in a far worse position than their predecessors in or out of war, for their training had been predominantly in the slow-moving 'no-touch' rituals of aseptic surgery. By 1914, few surgeons were operating on scrubbed-down kitchen tables; most were now working in meticulously clean hospital operating theatres. Moreover, as a result of the growth of conservative surgery over the previous forty years, experience in the amputation of limbs was limited. Out of 5483 major operations performed at St Thomas's Hospital in 1913, for example, only 34 had been amputations.[27] Not only among civilian surgeons, but among military surgeons too, the practice of amputating was far less familiar than a half-a-century before. Since the Franco-Prussian War of 1870–1, when the amputation rate was brought down from its usual 30–40 per cent of the wounded in battle to around 16 per cent, the rate had continued to fall – to as little as 0.5 per cent among the wounded Japanese in the war with Russia in 1904–5.[28] It is hardly surprising, therefore, that those soldiers who survived amputations during the early months of the war often arrived back in Britain with stumps that resembled the necks of axed chickens.

As late as 1918 the limb-fitting hospitals in Britain were still receiving cases in which the nerves and blood vessels had been inadequately sutured, or where no allowance had been made for the post-operative retraction of muscles and skin.[29] But, for the most part, the situation was well in hand long before then. By 1915 substantial improvements had come about simply as a result of surgeons becoming accustomed to the medical conditions of war and modifying their practices accordingly. By 1916 the amputation rate at the base hospitals was down to a modest 25 per cent of all gunshot joint wounds and fractures, and by 1917 it had fallen to 7 per cent. The death rate among troops from amputations fell from 33 per cent to 15 per cent in 1916 and to 8 per cent in 1917.[30]

Orthopaedics had less to do with these downward curves in the statistics than orthopaedists would later suggest. According to one set of figures (apparently gathered in 1916 but not published until 1919), the mortality rate for compound fractures before the intervention of Jones and his colleagues was as high as 80 per cent.[31] But such statistics – not unchallenged at the time[32] – should be treated with caution;

issued after the war they were designed to legitimize the place of modern orthopaedics in civilian hospital medicine. Although Jones can be credited with helping to reduce the need for many amputations by vastly improving the primary treatment of fractures through the introduction of Thomas's splints in 1915, the main reason for the reduction in amputations and in the mortality from this cause was the introduction (officially in the summer of 1916) of Alexis Carrel's method of carrying an effective germicide (Dakin's solution) to the recesses of an infected wound.[33] Even the practice of segregating cases of fractured femurs in special hospitals appears not to have come from Jones, but rather from the observation by consultant surgeons in France of the impressive results obtained by the French through such practice.[34] Most of Jones's work after his appointment as Director of Military Orthopaedics in March 1916 was conducted in Britain rather than at the front, the latter consisting mainly of measures in organization to improve upon what by then had already been substantially improved.

In the wake of the Boer War it was also assumed that most of the surgery in future wars would be conducted in hospitals set well back from the scenes of battle. More efficient communications and transport would enable most of the sick and injured to be treated expertly at base hospitals. Indeed, in the event of a European war, it was expected that much of the major medical and surgical work would be conducted in Britain. For this purpose, 23 home Territorial Force General Hospitals were organized as early as 1908. Connected to the major London and provincial teaching hospitals, the 'territorials' were to be run during war by consultants drawn largely from the honorary staffs of the teaching hospitals who would be paid on a part-time salaried basis.[35]

Until the Great War was well under way it was anyone's guess how many beds in the territorial hospitals should be allocated to surgery and how many to medicine.[36] More historically significant, however, is the fact of the prewar organization itself. Conducted with a view to rendering military medicine more efficient, the planning was detailed. For the handling of casualties in the field of battle, for instance, it extended from the regimental stretcher-bearers and aid posts, through the various dressing and clearing stations, on down to the base hospitals and convalescent camps.[37] What is more, this structure remained largely intact for the duration of the war. Changes occurred only to the intended functions of some of the parts. Thus, far from being 'absorbed in chaos' or 'blighted by mismanagement' before the arrival of Jones and his colleagues, the organization could be regarded as a model of rationalization. Knutsford, who inspected the arrangements at Boulogne and Calais early in November 1914 in response to criticisms in the press about the maltreatment of the wounded in transit, returned home 'deeply impressed with the general efficiency of the whole organization'.[38] Such efficiency, moreover, was to be considerably improved well before Jones came on the scene, as a result of the two major wartime changes to the prewar organization.

The first of these changes occurred at the Casualty Clearing Stations – destined to be the best-known of all the medical units at the front. Sited well beyond the expected range of artillery fire, the CCSs were planned as temporary, portable structures where patients would be checked over and disposed of, or, if seriously wounded, 'cleared' for ambulances and transported either to the permanent base hospitals or back home to the territorial hospitals. But the CCSs soon came to undertake definitive surgery and to retain patients for a part of their recovery period. The amputation of limbs, the traction and fixation of fractures, debridement (the removal of shattered bones and shrapnel from wounds), and other means of treating wounds all came to be practised at the CCSs, some of which became semi-permanent structures with as many as a thousand beds. This change was partly a result of the unexpected volume of serious casualties during battles, the transport for which proved difficult. But more important was the widespread realization among medical officers of the fatal consequences of delaying the treatment of septic wounds. Thus this change in function was also central to the reduction in amputations and the mortality resulting from that cause. Moreover, it too was largely complete by 1915.[39]

The other major organizational innovation in medicine during the war was the development of special hospitals, such as those for head and facial injuries, neurasthenia (mainly shellshock), epilepsy, cardiac disorders and tuberculosis. Again, this development took place relatively early in the war. In the case of neurasthenia, the President of the Psycho-Medical Society wrote to the War Office in the first month of the war – before the problem of shellshock had attracted much attention – proposing an organization of hospitals for such cases. The first of these hospitals was opened in January 1915.[40] Although what can be seen in retrospect as the first of the special orthopaedic hospitals was also established early in 1915, the others were not established until 1917 and thereafter, or until after the raising of what

Jones was to refer to as 'an orthopaedic conscience' – akin to the 'aseptic conscience'.[41]

RAISING AN ORTHOPAEDIC CONSCIENCE

According to Joel Goldthwait, writing in the 1930s, orthopaedics developed as an important specialism during the war because 'very early [it] became obvious' that one of the basic principles of Jones's orthopaedic treatment of crippled children – continuity of treatment, or 'to see from the very beginning an end result' – could be constructively applied to the casualties of war.[42] Though true in part, such a statement begs more questions than it answers: exactly *when* this realization occurred is left as vague as the question of *to whom* it became obvious, and *how*. Answers to the latter can be approached by first addressing the problem of chronology.

While Jones may well have been aware of the applicability of his work to disabled soldiers from the start of the war, he had in fact little opportunity for demonstration before 1916. For the first twenty months of the war he was attached to the Liverpool-based 1st Western General Territorial Hospital with the rank of Major. In this capacity, according to his nephew and biographer, he was only too aware that he had no power to 'coerce the war machine'. Apparently:

> watchful and active – he prepared his ground with great pains and discretion and then, at the proper moment, submitted his proposals in such succinct and convincing shape that they could usually be granted with acclamation and with credit to the authorities. However much he wanted, he rarely asked for more than he was likely to get at any particular moment. Thus, he never embarrassed the authorities, and kept the path open for further demands as current opinion advanced.[43]

Jones's first opportunity for advancing his interests came towards the end of 1914, when the potential manpower needs of the war became strikingly apparent through the devastating battles of Mons, the Marne, the Aisne and Ypres. After investigating the hospitals in the Western Command as a part of the normal course of his duties, he submitted a damning report on the wasteful consequences of evacuating patients too soon from hospitals in order simply to relieve pressure on beds. Pointing to the need for prolonged treatment in many cases, and the economic benefits that could be expected, he suggested experimenting with a special hospital for reconstructive surgery.[44]

The report quickly reached the War Office where it made its way to the Director-General of the AMS, Sir Alfred Keogh. It could not have fallen into better hands. As the Director-General of the AMS between 1905 and 1910, Keogh had been responsible for the major reforms in military medicine designed to prevent a repetition of the 'wastage' of manpower in the Boer War. As the Rector of Imperial College of Science and Technology from its founding in 1908, Keogh also occupied a central place among Edwardian progressives in the organization of scientific and medical education. R. B. Haldane, who was also involved in those reforms and who, as Secretary of State for War, had encouraged Keogh's military reforms in the 1900s, assured Field Marshall Kitchener upon Keogh's recall to his old post, that he was 'one of the best organisers I ever knew'.[45] Keogh had his enemies, as might be expected, but 'especially among the young men', as an American correspondent in London reported in September 1914, he was felt to be 'exactly the man for the place at the time'.[46]

It is unsurprising, therefore, that Jones's submission of detailed plans to eliminate military 'wastage' were favourably received. In January 1915 permission was granted to secure as many beds as necessary (up to 400), to be set apart at the requisitioned Poor Law Infirmary at Alder Hey, outside Liverpool, for the accommodation of those cases in military hospitals which would likely benefit from the treatment Jones proposed.[47] The military hospitals were circulated on the subject and Jones was put in charge of the Surgical Division. Although there are no statistics to determine the number of patients who were returned to military service from Alder Hey in 1915, by 1918, when the bed space had grown to over 1000, claims were made of 75 per cent of patients returned to active duty.[48] 'The experiment proved so successful', Jones recalled, 'that I was practically given a free hand to increase our beds in Liverpool and start similar establishments in other centres.'[49]

In fact, the road to expansion was not so simple. In retrospect, Alder Hey may have signalled the economic legitimacy of the principles practised by Jones upon the disabled – of patient segregation and continuity in their treatment – but it was to take more than merely example to secure a military orthopaedic empire, especially one centred in London. After all, implicit to Jones's work was the declaration that ordinary surgeons in general hospitals were not competent to handle a large percentage of the wounded. From this perspective it could be argued – contrary to Goldthwait – that the more 'self-evident' Jones's principles in the treatment of the disabled became, the less inevitable was their application.

Even the support of the Director-General of the AMS was not enough. Although by 1917 Keogh was personally involved with Jones's schemes and did tours of inspection with him,[50] before the Spring of 1916, his hands were tied. Partly as a result of the administrative reforms that he had been involved with before the war, he was barred from making independent policy decisions and appointments; whereas before 1904 the position of Director-General included being a member of the Army Council administering the War Office, thereafter it was a post that came under the control of the Adjutant-General's office.[51] As a consequence of this restructuring, Keogh was further circumscribed in his actions and appointments by the civilian medical establishment, upon whose support and cooperation he had largely to depend. But senior members of the Royal Colleges, in particular, were by no means as enamoured of Keogh as younger members of the profession, and the more that those outside the metropolitan medical establishment gained influence in military medicine, the greater became their anxieties. Hence the rumour that when Keogh came to appoint Jones Military Director of Orthopaedics in March 1916, senior members of the Royal College of Surgeons made a formal protest to their president in the hope that this would lead the Army Medical Council to rescind the appointment.[52] Keogh is alleged to have predicted such a reaction: 'If I [appoint Jones] the London surgeons will have my head on a charger', he is reported as saying.[53] Moynihan, who prompted this remark after threatening Keogh that a scandal would break out over his head if he did not appoint Jones, is claimed to have retorted, 'Very well, Sir Alfred, if you will not have Robert Jones, you cannot have me.' According to Moynihan, Keogh therefore appointed Jones 'as he could not do without me!'

Moynihan's account of how Jones came to be installed at AMS headquarters (with an office just down the corridor from Keogh's) doubtless got taller in the telling, but it is significant here as much for what it imparts of the medicopolitical fears surrounding Jones's appointment as for what it conceals of the context in which Moynihan could pretend to command such authority. Obscured is the fact that during the first few months of 1916 the fortunes of both the military and the AMS were at an all-time low. The war then seemed set to drag on for years; manpower shortages led to the introduction of conscription in January 1916; and increasingly there were doubts about the ability of the AMS to cope with the number of casualties. To many, innovation seemed wanting. 'There are thousands of ... problems', complained Dr Wilfred Grenfell in February 1916, writing from France. 'But except in laboratories, where excellent work is being done, everyone is so busy & everyone hates any new departures, till it is proven or is conventionalized, & so no advance is made.'[54]

More specifically, Jones's appointment coincided with the break-down of army medical services in Mesopotamia. As this news broke, there also emerged allegations of incompetence against certain surgeon-generals in France – allegations which reflected badly on Keogh's own competence. At this same juncture, Keogh was criticized for never having called together his Advisory Board – a Board stipulated by the Committee on the Reorganisation of the Royal Army Medical Corps in 1901 on which he had so prominently served. All of these issues were raised in the House of Commons on 15 March 1916, and they were sufficient to cause Keogh to tender his resignation.[55] It is not clear if he calculated that this action would provoke a rallying in his support, but that was its effect, the upshot of which was a strengthening of his hand at the War Office. It was at this point that Jones and most of those who had been members of the Provincial Surgeons' Club ceased to be informal advisers to Keogh and, instead, assumed official power. Meetings of the Army Medical Advisory Board were convened, as well as meetings of the Council of Consultants. To both bodies Moynihan was appointed Chairman.[56]

Yet crucial as this moment was for establishing the power of the modernists in medicine, and for consolidating Jones's position in particular, it cannot in itself be held responsible for the wartime making of orthopaedics. More fundamental, though not separable, was the state's positive economic and political valuation of rehabilitating the disabled. Behind this, the principal agent was not Jones, but the Joint War Committee of the Red Cross and Order of St John.

The Joint Committee was organized by the War Office in October 1914 as a body to delegate the wartime management of all auxiliary hospital accommodation. Although there were drawbacks to such a delegation of power, it was expedient in order to establish some control over the competitive enthusiasms of the various voluntary agencies.[57] From the perspective of taxpayers, medical men, and the disabled there were further distinct advantages. The Committee was able to support medical and welfare innovations in a way the state could or would not, and – as in the case of orthopaedics – its initial funding often primed the pump for undertakings by the state which might not otherwise have occurred.

At the head of the Joint Committee was the politician, philanthropist and 'benevolent autocrat' Arthur Stanley, the Governor of St Thomas's Hospital, and the chairman of the executive committee of the Red Cross. During the war he was to become one of Jones's influential friends and patrons – a relationship which was probably helped by the fact that Stanley was lame, and by the fact that he was also the chairman of the Royal Automobile Club, to which both Jones and Moynihan belonged.[58] (At the war's end Stanley appointed Jones the Director of Orthopaedic Surgery at St Thomas's, and he was to be responsible for the generous Red Cross funding of orthopaedic hospitals for crippled children.)[59] Like Keogh, Stanley toured the country with Jones inspecting the military orthopaedic centres that were financed wholly or in part by the Joint Committee, and purchasing others.[60] But this was subsequent to Jones's appointment as Military Director of Orthopaedics; prior to that there was another figure on the Joint Committee who, then and after, was to be far more important than Stanley in publicizing, enabling and sustaining Jones's orthopaedics: King Manoel of Portugal.

Manoel became preoccupied with the plight of the disabled after establishing a private officers' hospital at Brighton in 1914.[61] In November 1915, under the auspices of the Joint Committee, he set off on a tour of France and Belgium to determine how the allies were coping with their disabled. Unaware of Jones's experiment at Alder Hey, Manoel concluded that the work of the Allies was far in advance of anything in Britain. The Joint Committee thereupon organized a mission of three medical men to investigate Manoel's claims,[62] the confirmatory findings of which served to underwrite his authority in the rehabilitation of the disabled. Although it was not until June 1916 that he was officially appointed a representative of the Joint Committee (and not until January 1917 that he became the head of the Committee's newly constituted Orthopaedic Hospitals' Department), from the beginning of 1916 Manoel effectively controlled the Committee's considerable spending on the treatment of the physically disabled. He also undertook regional fund-raising which vastly increased that spending.[63]

In ways other than strictly financial, Manoel was to be as important for the wartime development of orthopaedics as Jones. Indeed, as far as the military orthopaedic centre at Shepherd's Bush is concerned,

Manoel's role was probably greater. Although Jones had written to Keogh in February 1916 suggesting the creation of such an institution, it was Manoel, not Jones, who literally moved into Shepherd's Bush to direct its daily affairs when it was taken over by the War Office on 1 March 1916, some time *before* Jones's appointment as the Director of Orthopaedics.[64] At this point Shepherd's Bush was not exclusively for orthopaedics; it only became so in May 1916 partly as a result of the introduction (again by Manoel not Jones) of the 'curative workshops'. Manoel borrowed the idea for these from the Anglo-Belgian Hospital in Rouen and the Canadian Hospital in Ramsgate, and when the War Office refused to permit their installation at Shepherd's Bush, it was Manoel who persuaded them otherwise after obtaining £1000 for the purpose from the Joint Committee.

It was upon the success of the curative workshops that the military came to believe, and ultimately to invest, in Jones's orthopaedics. Of the first 1300 men who entered Shepherd's Bush and passed through the workshops, no less than 1000 were sufficiently rehabilitated to return to military action.[65] Thereafter, as at Alder Hey, approximately 75 per cent of the intake was rehabilitated, the remaining 25 per cent, according to Jones, being sufficiently restored to 'take their place as useful workers and efficient members of organised society'.[66] In the context of the then urgent demand for troops, it was no less significant that the workshops also allowed many of those who had been rejected for military service on the grounds of physical defects – Category E recruits – to be made fit for service. Many of the first cases admitted to Shepherd's Bush were in fact sufferers from club foot, claw foot and flat foot who required only a minor operation and a few weeks rehabilitation.[67] But perhaps the most important feature of the workshops to win the favour of the War Office was their 'very considerable economy to the State'.[68] In the name of occupational therapy, not only did they manufacture all the splints, surgical boots, and other appliances required for the hospitals to which they were attached (thus rendering further savings on transport costs), but they were also able to turn a profit by supplying the Ministry of Pensions with orthopaedic appliances.[69] At Shepherd's Bush, those assigned to the workshops also performed all the maintenance work for the hospital, and carried out the wiring and plumbing for the installation of the electrotherapeutic and hydrotherapeutic departments. They even manufactured surgical instrument tables, including an entire Abbott's operating table with all its pulleys and movable parts. Here was productive therapy indeed; and it was all the more commendable for being conducted without the commanding officers having to resort to military discipline. Where 'persuasion and example' could not lead patients to take up the work ethic for their own physical and psychological good, a system of rewards and privileges sufficed.[70] Ironically, in this former Poor Law institution was achieved precisely those economic and ideological goals that in the long history of the Poor Law had rarely ever been attained.

The military-cum-economic and ideological valuation of the curative workshops was not the only leaven to the orthopaedic conscience of the state. Working to the same end were political fears over the consequences of neglecting those who had become disabled. Although the extent of disaffection among those who found themselves on the social and economic scrapheap as a result of their military service remains unclear (at least for Britain),[71] there was a widespread belief that each of these wounded men 'represented a centre of unrest, and that unless something could be done to improve their condition, or at least to have them feel that the government had done its best for them ... [they] would have become centres of revolution'.[72] Goldthwait claimed to have received this message in 1917 'at least a half a dozen times, by persons high in authority'.[73] To what extent this fear was attributable to the rhetoric of Jones and his colleagues is difficult to tell, but it is easy to see how a national showpiece for the rehabilitation of the disabled could have suited the interests of government. Insofar as the disabled might not be just 'a focus for seething discontent', but also, 'a menace to successful recruiting',[74] such an institution could seem doubly advantageous. Besides serving to check the haemorrhage of manpower at the front, it could enhance recruiting by providing evidence of the government's commitment to the humanitarian gospel of rehabilitation.[75]

A more particular, possibly related, reason for the government to encourage such a 'demonstration institution' may have been the petition sent to the Under-Secretary for War, H. J. Tennant, early in 1916 on behalf of the bonesetter Herbert Barker.[76] Signed by numerous aristocrats, senior army officers and others of social note (including H. G. Wells), the petition sought to grant Barker the liberty to treat cases of disablement 'which [have] not yielded to orthodox methods'. Thereby Barker could gain official recognition for his 'manipulative surgery'. Although he was proposing only to treat some of the Category E would-be recruits, not the war-wounded, his initiative served to draw attention to the larger problem. Questions were asked in the Commons, and although Tennant managed politely to decline

Barker's proffered 'services for the state', 58 MPs supported Barker (including the Minister of Labour, John Hodge) and organized the Injured Soldiers' (Parliamentary) Committee. Consequently, it became all the more essential for officialdom and orthodox medicine to be seen to be acting in concert, positively and constructively.[77] Whether or not this partly accounts for the large amount of work initially done on Category E recruits at Shepherd's Bush, it is clear that what was intended as a means to advance the interests of Barker ultimately did as much or more for Jones.

MANAGING SCIENTIFICALLY

Thus far our focus has been on the changing political and economic circumstances that, by enabling the rise of an orthopaedic conscience, encouraged the practice of military orthopaedics. In so doing, we have sought to undermine notions of the inevitability of the specialism's making as premised simply on the medical demands of war, and/or on the individual 'greatness' of Jones. However, to stress that the process was not inevitable in these ways is not to suggest that it was therefore historically arbitrary. Since Jones's expertise in the handling of trauma had been shaped and structured by the economic rationalizations of the late Victorian and Edwardian period, there is nothing odd in the fact that his clinical methods should have become favoured in a context in which an increasing proportion of the army's labour force was becoming disabled and when, therefore, concerns with efficiency and economy in the military were becoming analogous to those of modern industry. In effect, by 1916, the war had become the Manchester Ship Canal project writ large: a socioeconomic context in which, as the value of labour rose in proportion to the fall in its supply, the state (as employer) had come to perceive medical expertise as crucial to its manpower problem. It is no coincidence that the other areas of medicine that were to contribute most to this 'industrial strategy' were cardiology and neurology – areas also occupied by hitherto marginal 'new men' given to physiologically mediated 'economic' understandings of the body.

All three of these areas of medicine became privileged in the military when their physiological basis for conceptualizing and dividing up patient populations came to be seen as economically relevant. Within the 'new cardiology', for instance, the reconceptualization of the coronary condition known as 'soldier's heart' – the third largest medical problem of the war, after wounds and shellshock – resulted in enormous savings in military manpower and state pensions. Formerly, 'soldier's heart' had been understood in terms of permanent structural or anatomical impairment, but this was redefined during the war in terms of physiological dysfunction, and relabelled 'effort syndrome'. With this new definition, patients could be segregated according to *degrees* of functional inability (literally by their capacity for work) and all but the worst cases could be understood as capable of recovery through appropriate medical/physiotherapeutic means. The cost benefits of this 'successful' redefinition of the condition served, in turn, further to validate the functionalist physiological approach to it.[78] Similar economic validation occurred through the functionalist reconceptualization of asphyxia from gas poisoning, and from the functionalist interpretation of shellshock – the treatment of which, significantly perhaps, gained the name 'medical orthopaedics'.[79]

Revealingly, the American orthopaedist, Robert Osgood, defined orthopaedic surgery during the war as 'the specialty of a principle and not of a portion of anatomy'.[80] Similarly, the anatomist and Hunterian Professor at the Royal College of Surgeons, Arthur Keith, entitled his lectures on orthopaedics delivered at the College in 1917–18, 'The Anatomical *and Physiological Principles* Underlying the Treatment of Injuries to Muscles, Nerves, Bones and Joints' (my emphasis).[81] But the physiological principles behind modern orthopaedics were always manifest more in practice – including the organization of practice – than in any body of theoretical knowledge. In effect, the functionalist understanding of the locomotor system and its rehabilitation provided a metaphor for the administration of integrated services for the restoration of locomotor function. Thus was medical work to be organized. As during the Manchester Ship Canal project, Jones required clear lines of communication down a hierarchical chain of command; efficient division of labour within and between Casualty Clearing Stations and base hospitals; the efficient segregation and transportation of cases; standardization of supplies and clinical procedures; uniformity in the surgical control over patients; and continuity in patient-care and after-care. Like scientific management in factories, this integrated regulatory system was designed to process its goods as efficiently and economically as possible. Elaborate systems of uniform record cards were devised to establish, control and monitor quality. Like a scientifically managed factory, this system – both at the level of administration and of the handling of individual patients – was seen to depend on organized team work, or on the harmonious cooperation between general surgeons, neurologists, nerve injury experts, physiotherapists, limb-fitters, occupational therapists and so on, all of whom within military orthopaedics, now came under the command of Jones.

It was this wartime organization of orthopaedics, symbolized by the concept of 'team work', that informed Bertrand Dawson's subsequently much vaunted 'plan' for the organization of medicine (first elaborated in 1918).[82] As taken up by him, and as celebrated by others in the new consultant elite, team work signified the integrated organization of the supposed interdependent parts of health care. Through team work, it was proclaimed, the efficient operation of large-scale medical services could be effected. Implicitly, therefore, the concept was a critique not just of uncoordinated haphazard civilian medical services, but of the power relations existing in the civilian voluntary hospitals where (in contrast to military medicine) lay managers still exercised control whilst competing against each other for charity funding. At the same time, team work signified a hierarchical division of medical labour with the new consultant elite at the top.[83] It was corporatist, not collectivist; indeed, when the idea of 'teams' to carry out surgical operations in France was first proposed (by Moynihan in January 1915), it was 'with a view to preventing junior and inexperienced MOs [in CCSs] undertaking serious operations on their own responsibility'.[84]

As previously suggested, the physiologically aligned team work organization of modern orthopaedics did not derive from the principles of scientific management as set forth by the American engineer Frederick Taylor. Rather, Taylor's ideas and the new 'physiological' orthopaedics emerged together in a socioeconomic context in which economic metaphors were drawn into physiology as much as physiological metaphors were drawn into economics (thus rendering the perceived insides and outsides of humans mutually constitutive). Nevertheless, during the war there were ways in which Taylorism came directly to bear upon the management of orthopaedic cases. For example, when Goldthwait and the American orthopaedists set up operations in France in 1917, it was arranged that they should visit Paris to see Dr Pedro Chutro, a Brazilian exile specializing in fractures who, after carefully studying Taylor's work, developed a surgical technique which eliminated 'a great deal of waste, not only of time in connection with the surgery, but with reference to dressings, and the care of the wounded ... following ... operation[s]'.[85] Similarly influenced by Taylor, and influential in applying his principles to medicine, was the French fracture expert, Jean Camus. 'Convinced of the necessity of employing combined treatments for the sequelae of wounds', he reported, 'I have endeavoured to group them harmoniously establishing an entente between the different departments and to co-ordinate them by the use of uniform cards and central control.'[86] Camus's work, along with that of his countryman Jules Amar on *The Physiology of Industrial Organization and the Reemployment of the Disabled*, was translated into English at the end of the war.[87] Meanwhile, as physiological principles were structuring the work and the workplaces of doctors and patients alike, so engineering principles were being applied to the reconstruction of the physiological efficiency of the human body. In Germany especially, 'medicine's most dramatic achievement during the war' was thought to be the accomplishment of orthopaedic engineers in simulating locomotor functions within dozens of different types of artificial arms and legs.[88]

Although there was little formal adoption of Taylorite principles in British medicine during the war (or for that matter in industry), the application of and insistence upon principles of efficient management was stepped up from the time that the 'physiologs' came into military power in 1916. Command orders on the proper sorting, treatment, labelling and transport of fractures, for instance, flowed from AMS headquarters from the moment of Jones's appointment. To enforce such orders, some of Jones's closest surgical colleagues were promoted in rank and sent off with Moynihan on tours of inspection of the base hospitals in France.[89] Obtaining such military medical authority was obviously not regarded simply as a mandate to specialty-building; rather, it was seen as a means to assert and affirm a reorganization of medicine in which fields of expertise would constitute the essential parts of an efficient integrated whole. The reorganization of any branch of medical work was perceived as reflecting (and, therefore, necessarily contributing to) medicine's overall greater functional unity and efficiency. The specialization in medicine that became apparent during the latter stages of the war[90] was a consequence of striving towards these administrative ideals, rather than any 'natural' outcome of patient numbers or conventionally understood scientific and technological breakthroughs. Appreciating this, Fielding Garrison and others were later to reflect on the 'truly remarkable' achievement in medical administration during the war, regarding it as far outshining the merely 'clever' and 'respectable' medical innovations and inventions of those years.[91] Of course there were those (such as Sir George Makins in France) who resented some of the implications of this striving, such as the 'criticism which reaches us from Authorities and others in the U.K.' on the less-than-perfect administration of cases of fractures of the femur.[92] But by 1917 few in medicine dared challenge the view that such 'scientific' managerial means were the 'proper' medical solution to the problem of military manpower.

ORTHOPAEDIC MANPOWER

Within the organization of military orthopaedics itself, however, there was a formidable obstacle to meeting these ideals of efficiency: the lack of suitable personnel. Because of this, Jones could not have asserted the interests of an orthopaedic occupational group even had he wanted to. He had little choice but to speak in terms of raising an 'orthopaedic conscience'. But once that consciousness was raised, and the stage was set for the national network of orthopaedic hospitals, the problem of staffing became acute. In all, there were scarcely more than a dozen men in Britain upon whom Jones could draw, and not all of these could be counted as wholly sympathetic to his objects and leadership.[93] The kind of men he sought – 'well-trained operative surgeons of mechanical mind who will work in team and . . . be loyally prepared to develop the work they are most fitted for'[94] – were few and far between. Moreover, with the increasing demand for medical men at the front, and the introduction of conscription, it was difficult for him to retain the few recruits he had.[95]

Jones was thus compelled to turn elsewhere for the estimated 50 men he needed. For the largest of the military orthopaedic hospitals, that purpose-built at Leeds with 1800 beds, Harry Littlewood, a consulting surgeon to the Leeds Infirmary whose main area of interest was abdominal surgery, was seconded. Littlewood's appointment was likely made at the recommendation of Moynihan, as probably was that to the orthopaedic centre at Aberdeen of John Marnock, the Regius Professor of Surgery there, who was best-known for his work on gastric ulcers. Arthur Mitchell, who was appointed the assistant director of orthopaedics for Ulster, was another whose main interest was abdominal surgery. Although Jones made a wise decision in appointing as his personal assistant-inspector the diligent Major A. M. Paterson (d. 1919), it is symptomatic of the restrictions upon recruitment that this former professor of anatomy at the University of Liverpool was then 54 years old and in poor health.[96]

Properly qualified electrotherapists and physiotherapists were equally hard to come by. To head the physiotherapy department at Shepherd's Bush, Jones recruited G. Murray Levick, a young man who had recently served as surgeon and zoologist on Scott's Antarctic expedition. Levick surpassed expectations and went on to become an active member of the BOA, as well as one of the first persons in British medicine to venture into the private market for physiotherapy.[97] Similarly surpassing expectations was the young W. Rowley Bristow, who was recruited from the Physical Exercise Department at St Thomas's Hospital to head the electrotherapy department. Essentially a medically qualified technician, Bristow is said to have begun his distinguished career as an orthopaedic surgeon after he observed a surgical operation at Shepherd's Bush and asked if he might try his hand at it.[98]

Had it not been for the timely arrival of American orthopaedic surgeons, however, the military orthopaedic enterprise could barely have been sustained. The first batch of 20 arrived in Liverpool with Goldthwait at the end of May 1917, less than a month after America's declaration of war on Germany. By helping to meet the shortage of trained personnel at the front, the American recruits made it possible for the military orthopaedic centres to flourish and expand, ultimately enabling Jones's principles of patient-care to be put into practice.[99]

The arrival of the Americans was not simply fortuitous. According to Jones, it was the British liaison officer in America for the AMS, T. H. John Goodwin, who in 1917 'placed before the American authorities a statement of our difficulties'.[100] But the idea, if not the actual plans, for recruiting American orthopaedists for the British centres seems to have emerged some time earlier, and to have owed much to the good relations that Jones had fostered with the Americans since the 1880s. Franklin Martin (one of the seven members of the Advisory Commission of President Wilson's Council of National Defense) had written to Jones in July 1916 to ask if he required surgeons, 'because I am in touch with the right sort of men – the best material in the United States'.[101] Plans appear to have gone forward, for in March 1917 Jones expressed his fear that America's entry into the war might jeopardize the arrangements: 'From letters I posted to you', he wrote to the War Office, 'it is clear that the imminence of War may stop an American supply.'[102]

Ultimately, more than 400 Americans passed through the ortho-paedic centres in Britain, most of whom were trained for three or four months before being transferred to the American hospitals in France.[103] By arrangement with Jones, the original 20 places were kept filled with replacements, and a few of the most competent of the recruits were retained in Britain as instructors for the subsequent arrivals. Thus at any one time after October 1917 (when Goldthwait returned with a further 42 orthopaedists) there were never fewer than 60 or 70 Americans in the centres, all of whom were paid by the American government.

The first batch of 20 that Goldthwait accompanied to Liverpool were hand-picked (no less than 12 of them being Bostonians or graduates of Boston) and apparently had 'a considerable amount of experience in civil orthopaedic surgery and some experience in industrial surgery'.[104] It was principally from this first batch that the core of the American permanent staff at the centres was drawn. Murray Danforth (1879–1943) was not untypical: a graduate from Johns Hopkins in 1905, he spent the next five years in general practice

Plate 6.1   Sir Robert and Lady Jones with American orthopaedic surgeons at Jones's Liverpool home, 1917

Source: Manchester Medical Collection, Platt Papers, John Rylands University Library, Manchester. (By courtesy of the Director and University Librarian.)

in Providence, Rhode Island, before deciding to specialize in orthopaedics. At the same time as commencing postgraduate studies at the Massachusetts General Hospital, he began work in the orthopaedic department of the Rhode Island Hospital, and in 1913, immediately upon his appointment as orthopaedic surgeon, he set off for Liverpool to study under Jones and to visit the orthopaedic clinics of Europe.[105]

Of the subsequent recruits, however, most were raw medical graduates with little initial interest in orthopaedics. These were clearly not 'the best material' that Martin had referred to, though for Jones they were, indeed, the 'right sort' insofar as they were 'young, malleable and eager'.[106] The War Office, too, may have thought them the 'right sort', for in August 1917 it 'abruptly notified' part-time consultants working in military hospitals (who were drawing salaries whilst continuing with private practice) that their services were no longer required. According to Burdett, their places were taken by American doctors, many of whom were said to have orthopaedic training.[107] Possibly this compounded the frictions and frustrations that some of the American orthopaedic recruits were already experiencing,[108] and possibly it helps explain why Osgood, the American deputy to Jones between August 1917 and February 1918, felt the need to imbue these young surgeons with the feeling that they were privileged to be serving 'under a great master, dealing with new forms of reconstructive surgery and gaining a new perspective'.[109] Some, indeed, were given a free hand to experiment: at Alder Hey one of them carried out a study of the clinical end-results of operations on nerve lesions, while another spent most of his time in laboratories at Liverpool University isolating and publishing on unusual nerve tissue.[110]. In the 200-bed hospital in Ulster, another of the recruits did special work on stiff hands and puzzling foot conditions, while at Cardiff, Winnett Orr pioneered a new system of orthopaedic record-keeping, developed tendon transplant techniques, and established a printing department in the curative workshops (where he edited the *Weekly Welsh Orthopaedic News*).[111] Like Orr, who later became editor of the *Journal of Orthopaedic Surgery* and an AOA president, most of these recruits eventually acquired what Osgood was pleased to call 'the orthopaedic bee in their bonnets' and were 'eager, when conditions are explained to them, to climb on to the band wagon and play a part in the tune'.[112] Presumably, what was explained to them was not just the military medical situation, but, also, the postwar potential of orthopaedics to capture a significant share of acute general surgery.[113]

PHANTOM LIMBS

Jones said later of the American orthopaedists: 'They came to us in our extremity; they filled a gap which seriously threatened to sterilize our reconstructive efforts, and they filled it with distinction and success.'[114] And it was the Americans, Osgood in particular, who, before the war's end, mustered the necessary confidence and unity among the British orthopaedists to enable the organization of the BOA.

Plate 6.2   British Orthopaedic Association Inaugural Meeting, Roehampton, 2 February 1918. Thirteen of the twenty founder members.

Sitting (left to right): Thomas H. Openshaw (1856–1929), Ernest Muirhead Little (1854–1935), William Edward Bennett (1865–1927)

First row: Robert B. Osgood (1873–1956), Thomas Porter McMurray (1888–1949), A. S. Blundell Bankart (1879–1951), David McCrea Aitken (1876–1954), Harry Platt (1886–1986), Reginald Cheyne Elmslie (1878–1940), E. Laming Evans (1871–1945), Naughton Dunn (1884–1939)

Back row: William Henry Trethowan (1882–1934), Walter Rowley Bristow (1883–1947)

Source: Manchester Medical Collection, Platt Papers, John Rylands University Library, Manchester. (By courtesy of the Director and University Librarian.)

But it was the Americans themselves who gained most from the war. Better organized than the British before the war, they came away from it with professional profits more than doubled. Merely by pointing to the British experience, they were able to justify an enormous role for themselves in the surgery of war. According to Goldthwait, Jones's cable for American orthopaedic surgeons indicated to American military medical command the importance of trained men, and 'from that time on the orthopaedic surgeon . . . was treated as something apart from the regular Medical Staff organization'.[115] Thus, although the AOA carefully prepared its ground before formally offering its services to the American Surgeon-General in May 1917,[116] there was little question in America about the right of this hitherto minor specialism to exercise a large hand in military medicine and eventually to have total responsibility for all bone, joint and muscle cases.[117] The fact that the AOA was already organized and that its senior members were active in the College of Surgeons allowed it to more or less dictate policy – the Army's Orthopedic Advisory Board being composed 'mainly of ex-presidents of the AOA, and of those representing the Orthopedic section of the AMA'.[118] Less than six months after America's entry into the war, E. G. Brackett, who was on the General Medical Board of the Council of National Defense and Chief of the Division of Orthopedic Surgery, AEF (constituted in August 1917), submitted a report to the Board which was far in advance of anything that Jones had dared to propose to comparable authorities in Britain:

> Arrangements have been made for orthopedic care of the soldiers to begin at the time of injury and to be carried on continuously until the soldier is returned to active duty, or, disabled, is returned to industrial life. The arrangement has been made overseas to attach orthopedic surgeons to the medical force near the firing line, and at the different hospitals back to the base orthopedic hospital . . . Thirty-five thousand orthopedic beds are planned, . . . [and] soldiers permanently disabled, or at least unable to return to duty, will be returned to the reconstruction hospitals established in the United States . . .[119]

The Board made no objection to these extensive plans.[120] Indeed, according to Franklin Martin, they had been anticipated:

> We had appreciated the importance of the matter, for on June 22 there had been held in Washington a conference, as a result of which our Committee on Rehabilitation of Maimed and Crippled had been

appointed . . . Our rehabilitation committee's work led later to a recommendation to the Secretary of War that a comprehensive reconstruction board be appointed; Mr Baker instructed the Surgeon General to call a conference on January 14, 1918 and a bill was drafted providing for vocational rehabilitation.[121]

As we shall see in the next chapter, the success of this legislation in America and its absence in Britain was to have far-reaching implications for the development of orthopaedics in the respective countries.

Of even greater consequence was the ability of the American orthopaedists to exploit the wartime situation to attract and train medical recruits. 'In view of the large number who would be called upon for orthopaedic service', the official history reads, 'it was the opinion that instruction should be instituted in the universities and hospitals to give additional training to those who should take up the work.'[122] By September 1917 arrangements had been made with the postgraduate departments of Harvard and the New York medical schools and hospitals to establish courses of instruction in orthopaedics. These were soon standardized and extended to Philadelphia, Chicago, Boston, Oklahoma and Los Angeles; at the Walter Reed General Hospital, Washington, the course became a part of the Army Medical School curriculum. Before the war was over, some 700 officers had passed through these different schools,[123] recruits enough to plant the specialism in virtually every city in America after the war.

As noted above, the majority of these trainees spent some months in the British orthopaedic centres gaining practical experience, and they were joined by large numbers of surgeons from Canada, Australia and New Zealand, also taking advantage of the training facilities. Britain thus became an extensive postgraduate school for orthopaedics, but one in which British students were conspicuous by their absence. To Jones, this seemed among the greatest tragedies of the war; in December 1918 he lamented to John Goodwin (who had succeeded Keogh as Director-General):

We have less than forty British operative surgeons, about half of whom are first-class . . . Many of our Colonial friends have already availed themselves of [the centres] . . . the New Zealanders have sent their best surgeons in sufficient numbers to staff fully three large orthopaedic centres. These men have undergone a very thorough course of training, lasting six months, and have expressed in very appreciative terms the advantages they have derived from this experience. During the whole time these classes have been in progress hardly any young Englishman has had the time or opportunity to gain similar knowledge, so that the classes have consisted mainly of Americans, Colonials, and the junior staffs attached to the centres . . .[124]

Whereas in America the wartime situation not only encouraged the recruitment and specialist training of GPs, but led to talk of raising the standards of admission to orthopaedic surgery, in Britain nothing of the kind occurred.[125] Specialization remained largely confined to hospital consultants, and the principle of patient referral from GPs was kept intact, even though the practice temporarily suffered after the war partly as a result of the breakdown of established referral networks, and partly as a result of the glut of demobilized medical men.[126] The cut-throat competition stemming from the latter affected orthopaedics in particular, for the specialism could be seen as encroaching on the traditional patch of GPs. For consultant trainees in hospitals, too, orthopaedics remained unattractive for, in addition to low remuneration in its practice, it still lacked 'the dramatic appeal of the brilliant operation'. Furthermore, it had as yet no clearly '"established relation" with general surgery'.[127] Tellingly, the BOA, by its second annual meeting in 1919, had added only two new members to its founding twenty-four.

In other words, the war did little to 'Americanize' British medical education. Although Jones was still hopeful in the early 1920s that orthopaedic departments would come to be attached to university teaching hospitals, that undergraduate teaching programmes in orthopaedics would be established in the outpatient departments of the large voluntary hospitals in London, and that the RNOH and the Shepherd's Bush Hospital would be equipped for the training of postgraduates, such hopes were all but extinguished by 1924 when the Ministry of Pensions (who had taken over the control of the orthopaedic centres) relinquished its lease on Shepherd's Bush.[128] In April 1925 Shepherd's Bush was restored to its original function as a workhouse and Poor Law infirmary.[129] Most of the other centres, similarly housed in workhouses, schools, asylums and mansions-on-loan, met the same fate.[130] In 1919 Harold Stiles hoped 'that other large schools would follow Edinburgh's lead and institute lectureships in orthopaedic surgery', but few of them were to do so before the Second World War (even though in 1926 the General Medical Council recommended that orthopaedics be included in the undergraduate five-year curriculum as one of the branches of surgery).[131] Nor, outside Liverpool, where Jones and McMurray arranged with the University in 1921 for a twelve-month course leading to a Master's degree in orthopaedics, was there any examinable postgraduate instruction in the specialism in Britain.[132]

In America the expansion of teaching in orthopaedics during the war was linked and became fastened to the universities and their connected teaching hospitals (with salaried posts). In Britain, however, the war-time institutional structure was wholly vulnerable to postwar marginalization and eclipse. By its very nature, it could only be as lasting as the war itself, and with only one exception – the Wingfield Military Orthopaedic Hospital in Oxford under the direction of Girdlestone – the temporary edifice fell to the axe of postwar austerity. Only in Liverpool and Newcastle upon Tyne were wartime foundations built upon, the Ministry of Pensions funding orthopaedic clinics in the principal teaching hospitals.[133]

Thus in the immediate aftermath of the war, although the status of British orthopaedics had risen substantially, the professional scope for the specialism was scarcely larger than before the war. In fact, there is evidence of shrinkage as a result of the now heightened fears of old-guard generalists in surgery. At St Bartholomew's Hospital, for example, the number of beds for orthopaedics was cut by half (from eight to four), despite a waiting-list of over 200. The situation only improved with the threatened resignation of Reginald Elmslie – the appointee to the department in 1913 who had worked with Jones at Shepherd's Bush during the war.[134] At Guy's Hospital attempts were made by a general surgeon to prohibit W. H. Trethowan (the surgeon appointed to the orthopaedic department in 1913) from carrying out meniscectomies, an operation for the separation of knee-joint cartilage. Trethowan refused to budge and eventually won out, but the dispute is typical of the enduring tension between traditional general surgeons and the new consultant surgeons pursuing modern orthopaedics.[135] As Bristow tried to explain to the readers of *JAMA* in 1927, 'In the teaching hospitals of England, from tradition, the encroachment of specialization is viewed with alarm.'[136] But as Bristow, Elmslie, Trethowan and other founder members of the BOA knew only too well, it was above all the new orthopaedics that affronted traditionalists. Openly expansionist (in part *because* of its generalist orientation), its scope seemed to some to be almost limitless. As one complainant wrote to the *BMJ* in 1920, 'I cannot see anything that is excluded by the modern orthopaedic surgeon.'[137] Indeed, shortly before the end of the war – in July 1918, only two months after Goodwin had appointed Jones (along with Moynihan, Stiles, and Dawson) to the revived Army Medical Advisory Board with the rank of Major-General – a committee of the Council of the Royal College of Surgeons was formed with the specific intention of circumscribing orthopaedics. Presided over by Makins, who was also the President of the RCS, the committee stated that it regarded with

> mistrust and disapprobation the movement in progress to remove the treatment of conditions always properly regarded as the main portion of the general surgeon's work from his hands, and place it in those of 'Orthopaedic specialists'; and thus to educate the layman to the belief that the British surgeon is incapable of dealing with the majority of the most serious injuries the body may sustain.[138]

Jones defended the orthopaedic centres and had the Army Command on his side, but he saw that the tide was turning and how, without strategic concessions, orthopaedics might again be 'reduced to a side show'.[139] Thus, in accord with the recommendations of the committee of the RCS, the name of the centres was changed from 'Orthopaedic' to 'Special Military Surgical Hospitals', to avoid what the RCS claimed was the 'implication that only specialists are capable of carrying on the surgery practised in [the centres]' and to avoid 'the recognition of a class of practitioners who may, or may not, be competent general surgeons'.[140] Jones tried to reassure the profession that there need be no sharp demarcation between generalists and orthopaedic specialists, and by 1920 he was insisting that the general surgeon must have 'an absolute right to treat any case and as many of any type of cases as he desires'.[141] Meanwhile, suggesting to his orthopaedic associates that one of the benefits of the war was 'the better understanding which has developed between the general surgeon and ourselves', he admonished them

> to foster this good feeling in order to attract the best men to the surgery of the extremities. So long as we remain an exclusive cult, claiming a special perfection, we deserve some of the neglect and scorn with which we have been treated in the past.[142]

Arguably, Jones had done his war work too well: rhetorically, by early exploiting the political fears over disabled soldiers; and practically, by conducting the surgery and rehabilitation of the disabled so effectively as to undermine future means of rousing the public against dissolution of the orthopaedic centres. Although *The Times* rejoiced in March 1922 when Shepherd's Bush was temporarily saved from closure, the two-year reprieve was far from proving 'how near to the people is the care of the sick and suffering'. On the contrary, it was the elected representatives of the local ratepayers who had threatened the closure by demanding more rent for the hospital from the Ministry of Pensions. Significantly too, this dispute never became more than back-page news.[143] Not until the mid-1930s – over the issue of fracture clinics – was the press to take up the cause of orthopaedists under the banner of 'War Gains Lost'. In the early 1920s the public wanted to forget the recent past, as the closing down of Shepherd's Bush in part reflects. With over three million unemployed after the war and a massive national debt, there was distraction enough.[144]

Among other reasons why the movement suffered after the war was, ironically, the success with which some of the orthopaedic experience gained during the war was incorporated into general surgical practice. There was little doubt about this in the minds of those who were in positions to compare the prewar and postwar treatment of fractures. William Brander, for instance, Principal MO for the Hospitals' Division of the London County Council, remarked in 1936, 'that the experience of War surgery effected great improvements and that the standard of treatment [of fractures] has continued to improve consistently since the War'.[145] Hey Groves, who was then out to prove otherwise, came to the same conclusion after examining fracture services in seven LCC hospitals.[146] Such improvements might be explained in part by the fact that some of those involved in Jones's war work were subsequently appointed consultants to general hospitals. But the main reason was the acceptance of the lessons of wartime orthopaedics by general surgeons – especially young men, who were concerned to carry out the latest and best treatment. While the cardinal principles of segregation, unity of control, continuity of treatment and after-care were only fully implemented in a few hospitals, general surgeons all over the country attempted piecemeal to introduce methods along these lines and, on the whole, were responsive to the new surgical techniques and knowledge. But few of these young surgeons had any interest in specializing in orthopaedics.[147]

Likewise, with regard to the widespread wartime use of physiotherapy to speed up convalescence, it was difficult after the war to erase the benefits that had been witnessed,[148] but at the same time difficult to realize any specific gains for orthopaedics. Experts in physical medicine, such as R. Tait Mckenzie and R. Fortescue Fox, who were involved in the wartime orthopaedic rehabilitation Command Depôts,[149] began after the war to seek their own separate professional space within medicine. Greater professional confidence was also acquired by the over 2000 masseurs who had been employed in the Military Massage Service.[150] Although it was not until the Second World War that physical medicine achieved autonomy, increasingly physiotherapists and allied professionals were in competition with orthopaedists for control over patients and over the ill-defined territory of rehabilitation. The British Association for the Advancement of Radiology and Physiotherapy, for example, was established in 1917 specifically to clarify, extend and protect such territory.[151]

Finally, at the same time that orthopaedists were suffering the backlash of certain general surgeons and failing in their bid to prevent the disabled from being treated merely 'as local sick under local doctors',[152] unorthodox practitioners began to seek their revenge. In 1918, as a result of further representations by the Injured Soldiers' (Parliamentary) Committee, an Army Council Instruction was issued directing that 'no obstacle will be placed in the way of an officer or soldier who desires to avail himself of the services of a practitioner in manipulative surgery who is not possessed of a medical qualification'.[153]

Overall, then, orthopaedists gained remarkable little, materially, from the war. The argument used rhetorically by Keogh and others to legitimize the expansion of orthopaedics during the war – that military conditions were necessarily 'entirely different' from those of civil life – proved only too true when peace was restored.[154] Even in America there was a sharing of some of these postwar realities. While senior general surgeons there claimed that they were as competent as orthopaedists in caring for fractures and deformities, physiotherapists challenged the authority of orthopaedic surgeons to control their training.[155] As a senior AOA member stated many years later,

> There need be no attempt to deny that the orthopedic surgeons of the United States had to have some of their early enthusiasm dampened by learning that the role which they were to play [after the war] was not that which they had, at first, conceived it to be, . . . they had . . . the opportunity to learn that their place in the whole scheme of rehabilitation was not permitted to become what it might and should have been.[156]

In light of these constraints, it is difficult to endorse the view that the Great War constituted the making of modern orthopaedics, if by 'making' is meant the consolidation and safeguarding of a sought-after occupational niche and the self-interested achievement of monopoly in the medical market. Had it not been for initiatives taken in the 1920s and 1930s to recoup some of the lost professional gains of the war years, orthopaedists might have come to regard the war very differently. In the short term, at a practical level, the immediate prewar and postwar circumstances were more alike than they were different. Only if by 'making' is meant the process by which professional identity is attained can the war be said to have performed this function for orthopaedists on both sides of the Atlantic. Amply testifying to this is the formation of the BOA in 1918, and the postwar speeches of the presidents of the AOA.

On the British side, however, the experience of the war signified more than merely the professional cohesion of a few orthopaedists, and it would be wrong narrowly to construe the wartime history of orthopaedics in terms which are essentially those of the historiography of medical specialization in America. Above all, in Britain, the wartime experience was important for advancing the socioeconomic values and professional interests of the emergent consultant elite as tendered through notions of managerial efficiency in medicine. Specialization was at best secondary, at worst contrary, to these aims of restructuring through rationalization. From the latter perspective, rather than that of specialization, the wartime organization of orthopaedics could serve as a resource in constructing medicine's future, not only for Jones and his orthopaedic colleagues, but for a broad spectrum of medical modernizers.

7

INDUSTRY AND LABOUR, PART I
BRITAIN AND AMERICA, 1920s

It was widely believed when the war was over that the experience of orthopaedic surgeons with the war-wounded would be transferred to the wounded 'soldiers of industry'. Contemporaries agreed that 'the problem of the workman disabled by industrial accident was similar in many respects to the problem of the disabled soldier'.[1] It was common knowledge that 'disabled soldiers ... as a result of systematized orthopaedic treatment [had] been able to return to their work'. '[E]vidences of the efficiency of such treatment are broadcast, and ... the Local War Pensions Committees and the Ministry [of Munitions] have unassailable information of the wonders that have been achieved.'[2] According to another source,

> the workmen of this country are sharply alive to the fact that during the war nothing, from the moment they were wounded, was spared in the endeavour to re-establish their fitness for service. They know what hospitals mean, and their appreciation of what orthopaedic surgery has done for them and their comrades is ever present.[3]

With an estimated 5 per cent of the country's workforce disabled through war service,[4] the issue of 'rehabilitation', or 'reeducation' as it tended to be called before the mid-1920s,[5] concerned a wide spectrum of society – politicians, employers and labour leaders not least. Most agreed with Sir Alfred Keogh in 1918, that

> the relation of disabilities to earning capacity is one of the most important problems of the day. A large part of this subject comes within the domain of Orthopaedic Surgery, and indeed, the relation of physical infirmity to industrial work has become of no little moment.[6]

In industries with large numbers of accidents, notably coal mining and ship building, hopes were entertained that special orthopaedic clinics might be established.[7] Indeed, in 1919 Dr Frank Shufflebotham, a medical adviser to the Ministry of Munitions and a medical referee under the Workmen's Compensation Act for the North Staffordshire District, called for 'specific legislation' to establish such clinics for colliers.[8] Quoting him, the journal *Engineering* thought that his recommendations did not go far enough; Shufflebotham's scope, they claimed, was 'dangerously narrow'. Such arrangements

> must be extended to the whole body of our industrial workers if we would have the man-power of the nation a[t] maximum. . . . There may be something to say for organising the trade of coal-mining on a quasi-army footing, and, therefore, the creation of a mining-medical service which would carry out recommendations for after-treatment of injuries, but there is much more to say for the stirring-up of an effective desire to apply the wonderful methods and equipments of orthopaedic hospitals to cases of disablement from the wide range of accidents in any of the industries on which our general and individual prosperity depends.[9]

In view of the obstacles thrown in the path of orthopaedic expansion in the voluntary hospitals at the end of the war, one might have expected orthopaedists to seize these opportunities in industry. After all, this was an open domain where new claims for expertise could be staked without fear of vested medical interests. Jones's experience in the war and his earlier involvement with the medical services for the construction of the Manchester Ship Canal ideally positioned him to guide orthopaedics into this area. Such a move would have been all the more appropriate in that, as the *Lancet* once tellingly remarked, it was *from* industry that the principles of Jones's orthopaedic organization had originally been drawn.[10]

But it was not to be. Despite the encouraging rhetoric and the apparent opportunities, Jones and his colleagues were unable to transfer military orthopaedics to industry. In part, this was because orthopaedic surgeons were few; though, as we shall see in the next chapter, their small numbers did not prevent them from building an empire around the treatment of crippled children. Yet it would be wrong to suppose that they decided simply to ignore the opportunities presented by industry. Rather, as with all their involvements after the war (including, ultimately, industrial injury), social, political and, above all, economic circumstances determined the limits of the possible.

This chapter explores the circumstances that constrained orthopaedic inroads into industry in the 1920s. The study of these can help explain the alternatives pursued, as well as shed light on the structural obstacles to the reorganization of medical services as a whole in interwar Britain. Both issues can be illuminated, moreover, through comparison with the different circumstances surrounding orthopaedics in post-war America.

According to an editorial in the *Lancet* in 1919, socially aware medical men in America realized that 'at the back of industrial unrest lies the health . . . of the people'. Furthermore, they were alert to the economic fact that medicine could profit from attending to the needs of industrial workers.[11] Although the editorial did not specify which American medical men were alive to these facts, orthopaedists could be numbered among them. While Jones and his colleagues at a BMA meeting in 1919 discussed reconstructive surgery in a context of sinking hopes for the future of the specialism in hospital medicine without reference to industry,[12] their American counterparts at AMA meetings focused intently on the relations of orthopaedics to industrial surgery. 'Except as a compound comminuted fracture of the femur has been produced on the battlefield by bullet, shrapnel or shell,' Albert Freiberg stated in his address to the AMA Section on Orthopedics in 1918, 'it does not materially differ from the one produced in railroad yard or steel mill, as concerns the prevention of deformity and the preservation of function, . . . the principles are the same.'[13] According to another, in a keynote address to the same body in 1920, the war had been a brilliant preparation for a predestined opportunity, for 'the orthopedic surgeon has arrived at the threshold of the greatest opportunity in his career – *the invasion of the field of industrial medicine*' (my emphasis).[14] The new *Journal of Orthopaedic Surgery* encouraged this view in editorials proclaiming the 'definite place' of orthopaedics 'in the care of those injured in the industries';[15] Robert Osgood perceived in it an 'alluring . . . vision of opportunity, . . . wide and varied';[16] Leo Mayer, in an article on 'The Orthopedic Surgeon and Industrial Accidents', presented the readers of the *Journal of the American Medical Association* with the argument for a less haphazard treatment of these cases by 'experienced experts';[17] while Fred Albee staked the boundaries of the field through his massive textbook *Orthopedic Reconstructive Surgery: industrial and civilian* (1919).

In part, these aggressive claims to the territory of industrial injury by American orthopaedists were possible because attention had already been directed there before the war. The wartime experience had made plain 'the writing on the wall', as the President of the AMA Section on Orthopedics put it.[18] As early as 1909, the Section on Orthopedic Surgery of the New York Academy of Medicine had held a symposium on the question 'What Shall We Do With Our Cripples?', which cast light on the appalling production of crippling through industrial accidents. Speakers included a former Commissioner of Labor and an expert on accident insurance in Germany.[19]

Such interests reflect broader social and ideological concerns in pre-war America, in particular with the organization, control and safety of the work-place,[20] and with the supposed relationship between industrial injury and national prosperity. Although, in fact, less than a quarter of physical handicap was attributable to industrial accidents (as examination of army recruits and other subsequent studies were to reveal),[21] the common view of the early decades of the century was that industry was the most prevalent source of adult crippling. In financial and other ways this was seen as a problem of the first magnitude: according to the above-mentioned Commissioner of Labor, some 19 000 Americans were involved in accidents in factories in 1907, of whom 'between 4,000 and 5,000 were either permanently injured, or reported to be so seriously injured as to be practically permanently crippled'.[22] Statistics from the Department of Labor for 1917 indicated that of 875 000 non-fatal accidents reported for that year, some 74 530 workers were permanently disabled.[23] The contemporary preoccupations which framed this problem are clearly indicated in the name alone of the Economic and Efficiency Commission of New Jersey, the body that acted to secure the first state Workmen's Compensation Bill in 1911. Essentially these same interests lay behind the formation in 1912 of the National Safety Congress, and the Division of Industrial Hygiene within the United States Public Health Service. In New York in 1913, the Institute for Crippled and Disabled Men was established,[24] followed in 1914 by both the Conference Board of Physicians in Industry[25] and the American Public Health Association's Section of Industrial Hygiene. By this date the physical examination of employees had become common in major industries[26] and in 1915 the AMA recognized industrial medicine by holding a symposium on the subject within its Section on Preventive Medicine. Among other agencies involved with the problem before the war were the Bureau of Mines, the Department of Labor, and the American Railway Surgeons'

Association,[27] many of which were brought together in October 1917 by Franklin Martin and organized into the Advisory Committee on Industrial Medicine and Surgery. This further enhanced the credibility of industrial medicine which was accepted that same year as a specialty by the AMA's House of Delegates.[28]

The immediate prewar years also witnessed the organization of the first comprehensive scheme for the treatment of industrial accidents within a single company – at the US Steel Corporation in Pittsburgh (formerly the Carnegie works). Begun in 1909 under the direction of the trauma surgeon and fracture expert William O'Neill Sherman, this company accident hospital was to become nearly as well-known internationally as Jones's clinic in Liverpool.[29]

In Britain, however, there were few such prewar developments. As we have seen, medical consciousness of and organization around accidents was minimal, and concern with such issues as compensation costs and the earning capacity of injured workers scarcely featured in medical debates. In the BMA's report of 1912 on the treatment of simple fractures, for instance, therapeutic matters excluded all else. Few statistics on industrial injury were ever gathered and, excepting the Birmingham survey of 1911, few of the local inquiries into the causes of crippling ever included statistics on those crippled by industry.[30] Safety First campaigns and the like were all interwar phenomena.[31]

Although the absolute numbers of those disabled through industry were doubtless higher in America than in Britain, there is no reason to suppose that the differences in the response to industrial injury hinged on this. Despite rhetoric on 'speed-mad' America, accident rates within industries were probably much the same in the two countries.[32] Differences in the structure of medicine and in the nature of medical insurance and workmen's compensation had more bearing on the relative responses to the problem, as will be discussed below. More fundamental, though, was the state of the economy in general and the different demands upon labour in particular. Overall, there was far less concern in Britain than in America with labour efficiency and economy. Even before the economic slump of the early 1920s, British labour was plentiful and relatively cheap, while formal conceptions of scientific management in industry were weak (partly because of the smaller scale of industry). Even the Industrial Fatigue Research Board, which had been established during the war to conduct time–motion studies and so increase output in the munitions factories, had modified its concerns by 1921. Still preoccupied with 'increasing the efficiency of the human factor in industry', it was now suggesting that 'the word "efficiency" ... is not to be interpreted as equivalent merely to productive efficiency, but as the physiological quality which results from favourable conditions of work. The word is in fact almost equivalent to "fitness".[33] Insofar as industrial medicine was perceived as the complement to scientific management in maximizing profits in the long term, and as offering a possible 'cure for Bolshevism' in the short term, the same 'necessity ... demanded by the existing social order' seemed less acute in Britain than in America.[34]

The differences in opportunities for orthopaedists owed much to the different legacies of the war. American orthopaedic surgeons, too, had suffered some of the backlash of general surgeons, but on the whole, they did far better out of the war than their British counterparts. Their separation from general surgery allowed them to set up courses in civilian universities to train surgeons in the modern methods of orthopaedics. By 1923 there were over 300 full-time orthopaedic specialists in America.[35] Of the 400 or so others trained up during the war, many were to find employment in the purpose-built Orthopedic Service Reconstruction hospitals. That at Colonia, New Jersey, which was modelled on the work of Jones at Alder Hey and Shepherd's Bush, was directed by Fred Albee, subsequently the co-founder and editor of *Rehabilitation Review*.[36] Although the wartime creation of the Division of Special Hospitals and Physical Reconstruction took some of the control of physiotherapy out of the hands of orthopaedists, and the Federal Board for Vocational Education (established in 1917) led to the separation of medical from vocational rehabilitation, the 'functional restoration' of soldiers and others *in hospitals* remained with the orthopaedic surgeons.[37] The Sears-Smith Vocational Rehabilitation Act of 1918 and the Federal Vocational Rehabilitation Act of 1920 largely fulfilled the ambition expressed by Goldthwait and others during the war to extend military rehabilitation services to civilians – especially to 'that great body often spoken of as the industrial army'.[38] To a considerable extent, this enlarged the professional sphere of orthopaedists.[39]

With this expansion came a recasting of the specialty's image. Whereas before the war 'the term "orthopedist" was applied quite as freely to the man who made and sold braces as to the medical man who ordered them', and orthopaedic surgeons were thought of as 'basically mechanicians' or 'surgeons only in name',[40] after the war this was seldom the case. Albee, in the Preface to his 1919 textbook on industrial orthopaedics, identified the 'metamorphosis', as from an almost exclusively conservative therapy [to one which] while in no wise *detracting* from the *importance of conservative methods*, aims to assemble and bring to the attention of the profession in a practical manner those surgical procedures which have contributed so largely to the reclamation of the cripple and to the rehabilitation of the physically incompetent, and which offered to the great armies engaged in the titanic world struggle an alleviation of their sufferings and reconstruction of their physical deformities.

But the emergent 'corrective surgical' practice – encoded in Albee's dictum 'NEVER TRAIN AROUND A DISABILITY THAT CAN BE REMOVED'[41] – was more than a technical carry-over from wartime experience. It was a part of a response to recent changes in the medical market-place consequent upon the growth of third-party (highly cost-conscious) insurance companies, employers and workmen's compensation bodies. In this new medical market, faced with competition from paramedics in physiotherapy, on the one hand, and GPs, on the other, American orthopaedists were virtually compelled to act less conservatively. Robert W. Johnson Jr, who had worked under Jones and Girdlestone during the war, observed in 1920 how 'industrial insurance has done much to encourage reconstruction', but realized, too, that

> More careful consideration must be given to the Time Element in Industrial Reconstruction than in Military, for its economic soundness must be clearly established before the procedure is undertaken. The surgeon cannot indulge in 'good surgery' at the expense of either employee or employer.[42]

A few years later, a specialist in industrial surgery writing against the more conservative techniques used by GPs was to state more emphatically:

> In industrial work where compensation laws are in operation, corrective surgery becomes imperative. First, human justice requires it; second, the injured man must regain his poise and his feeling of self-reliance by returning to work; third, pain and discomfort must be eliminated; fourth, the insurance company should properly discharge its obligation to employer and to society in general by returning injured men fit for their former work as quickly as possible.[43]

Eschewing a conservative surgical outlook was thus an appropriate strategy in the economic context. It enabled American orthopaedic surgeons more readily to enter into the industrial injuries market – a fact born out by Albee's own appointment in 1919 as the first Chairman of the New Jersey Rehabilitation Commission.

In Britain, however, the situation was radically different. The sharper reaction of the old-guard general surgeons led to different perceptions of orthopaedics and its potential. British orthopaedists numbered less than 5 per cent of the American total, so different strategies were in order. Moreover, the field of industrial medicine remained commercially undeveloped and professionally low-status. As the *Lancet* pointed out in its editorial in 1919,

> In this country, the [medical] profession neglects social economics and is far from awake to its responsibilities for paying to the industrially employed that close attention necessary to increase our national productivity. . . . [T]he profession is content to tinker at occupational injuries and disabilities without any appreciation of the loss to the workers in time and wages, and to the community in their productivity.[44]

In 1939 there were still only 50 full-time and 250 part-time doctors employed in British industry, few of whom were surgeons, and many of whom were women.[45] Like public health medicine, industrial medicine suffered from the centralizing tendencies of high-status hospital medicine.[46] Furthermore, unlike work with crippled children, the care and cure of disabled workers was unattractive to most industrialists, philanthropists and local authorities. Typically, Jones's suggestion to the West Derby Board of Guardians in 1923 that the Poor Law Infirmary at Alder Hey be turned into a central hospital for traumatic surgery to take advantage of its wartime rehabilitation facilities was rejected, while the suggestion that it be turned into an orthopaedic hospital for children was greeted enthusiastically.[47] Finally, in Britain, in contrast to America, there was the habitual drag of medical traditionalism. As one dissident medical officer remarked in 1922, the British 'have not written so much regarding the treatment of industrial accidents and the organization of factory medical service as has been done in America'; in that country 'medical arrangements were started . . . *de novo*, not grafted on to pre-existing and probably inadequate arrangements as is the case in an older country'.[48]

Crucially, Britain enacted no rehabilitation legislation comparable to that in America, despite having nearly a million more soldiers who had been disabled.[49] Between 1917 and 1924, state intervention was out of fashion,[50] and in the worsening economic climate and falling employment after 1921, government had little reason to accept calls for legislation which would enable the disabled to regain their former earning capacity.[51] In Britain, unlike America, there was *not* 'work enough to make provision for the [industrial] cripple possible without having to meet the charge of displacing the able-bodied'.[52] Clearly, the thesis that Richard Titmuss elaborated with the Second World War fresh in mind – that war stimulates public welfare by revealing social inequalities and injustices – holds little water with reference to the Great War, at least in this connection.[53] The Ministry of Pensions merely supplied artificial limbs, offered minimal physiotherapy and waited for the number of eligible claimants to decline,[54] while the Ministry of Labour issued a schedule of occupations suitable for the disabled and drafted the wholly ineffectual 'disabled employment appeal' for presentation by the Prime Minister to all employers.[55] A government commission, appointed in September 1920, had suggested that local authorities set up sheltered workshops, similar to those established in Leatherhead by Lord Roberts where medical and occupational rehabilitation was practised.[56] But the Chancellor of the Exchequer (the Conservative, Austen Chamberlain) opposed the idea and it was dropped. Ironically, it was Lord Roberts's Workshops that inspired the first American rehabilitation legislation in New Jersey, 'whereunder treatment, training, and placement in industry were available to all State citizens'.[57]

The incorporation of medical 'treatment' into the legislation on workmen's compensation in America was of course the other difference in Britain, and the one that ultimately bore heaviest on professional developments. Such politically sanctioned medical intervention was America's distinct contribution to solving the problem of industrial disablement, and it was widely endorsed by the medical profession who found in it an easy income. By the mid-1920s almost every state had introduced such legislation.[58] But in Britain the various Workmen's Compensation Acts, dating from 1897, left medical treatment up to the worker, or to the benefit club to which a worker might belong.[59] For minor illness and injuries the legislation provided weekly payments (to a maximum of 25*s* in 1920) at the employer's expense, but encouraged workers to seek lump-sum cash settlements for serious disablement resulting from occupational diseases or accidents. While employers and government lived in fear of workers' economic 'abuse' of the system, workers put all of their effort into 'getting their due' in compensation cases. Efforts were made to alter the legislation, so as to include medical and vocational rehabilitation, but they tended to be frustrated, not least because of the existence after 1911 of National Health Insurance.

Like workmen's compensation itself, NHI was initially intended to prevent those earning less than £160 a year from becoming pauperized though ill-health. 'Adequate medical attendance' from 'panel' doctors, along with weekly cash sickness benefits, was provided.[60] In these respects NHI provision for insured working people went further than some countries' workmen's compensation schemes, though it did not provide for specialist services (except for tuberculosis). Furthermore, recipients of workmen's compensation benefits were prohibited from claiming NHI benefits, so that many injured workers (an estimated 3500 a week by 1939) also claimed out-relief in money or kind from Poor Law authorities.[61] Thus financial compensation for industrial illness and injury remained separate from medical benefits. For these reasons, commentators on the medical insufficiencies of workmen's compensation tended to argue not for its wholesale reform, but rather for an extension of NHI. William Beveridge, in his famous Report of 1942, would be merely the last in a long line of spokesmen who objected to the 'splitting [of] social insurance into separate sections', and called for a more comprehensive service to bring the hospital treatment of accidents and post-medical rehabilitation under national health insurance.[62]

A further obstacle to the 'Americanization' of workmen's compensation in Britain, not unrelated to the split in services, was the reluctance of workers to undergo full medical treatment while receiving compensation. In the opinion of the BMA spokesmen who gave evidence to the Holman Gregory Committee on Workmen's Compensation in 1919, 'compensation should either be reduced or stopped altogether if the injured workman unreasonably refuses to undergo [medical] treatment'. Workers often failed to seek proper care for injuries, it was claimed, because they still believed that hospitals were 'places where doctors practised cheaply on human material', and because they perceived that the injury, once received, was a means of getting an income out of an employer without working for it. But by these same spokesmen's own admission, most of the provision for accidents was poor in quality, 'particularly with reference to orthopaedic [i.e., reconstructive] work'.[63] As the TUC and Labour Party observed in 1922 in a pamphlet on hospitals,

The chances of permanent recovery are greatly impaired by entire lack of after-care and supervision. The invalid partly recovers and force of circumstances then drives him or her back to work too soon, or to work, the nature of which . . . will again inevitably lead to a fresh breakdown. A complete health service for the community ought to make some provision for following up cases discharged from hospitals, and for securing change of occupation where the individuals would otherwise be certain to suffer relapse after relapse leading to permanent incapacity.[64]

Not surprisingly, workers were not keen to receive second- or third-rate medical care which might compromise their chances for good lump-sum settlements and leave them only suited for 'light work'. In most cases to undergo medical treatment was to have one's compensation payments 'reduced or stopped' altogether.

Finally, there was the less-than-progressive attitude of the employers themselves, linked as this was to the attitude of the private insurance companies. For most employers, lump-sum settlements for cases of disability were more convenient and cost-effective than indefinite and incalculable weekly payments, and/or involvement in rehabilitation schemes. Their compensation expenses were covered by insurance, and the insurance companies appear to have had little interest in persuading industrialists to change their ways. Again this was in contrast to America where insurance companies had introduced 'merit rates' enabling those employers who took steps to safeguard their employees from accidents to pay less for accident insurance. A commentator in the *Lancet* in 1926 remarked on such schemes, 'it is interesting to note that very much the same [insurance company] influences are now in America endeavouring to speed up the complete recovery of the injured workman and to lessen the time during·which he draws compensation allowance'.[65]

These American practices did not originate entirely with the insurance companies. Doctors specializing in industrial injury – orthopaedic surgeons among them[66] – were keen to persuade the insurance companies that it was cost-effective to utilize specialists, rather than GPs, in compensation work.[67] But in Britain this argument was little heard before the mid-1930s. Only then was the apparent unwillingness of insurance companies to act in 'medically responsible ways' (i.e., become third parties in specialist medical care) labelled as 'criminal'. Of course, by attacking the insurance companies, the specialists who made these claims simultaneously attacked the structure of British medicine which inhibited the growth of their own professional interests. R. D. Gillespie, speaking at the annual meeting of the British Association for Dermatology in 1937, was typical of such specialists in expressing his surprise 'that insurance companies do not conduct . . . extensive inquiry into the question of compensation for injury and sickness resulting from it'. Characteristic, too, was the innuendo in his rhetorical questions: 'is it that they do not realise the existence of the problem? . . . Or – let us whisper it darkly – is it that they should have no real desire that the position should be improved?'[68] The orthopaedic advocates of fracture clinics in the 1930s were similarly to argue that the private insurance companies ought to be induced or compelled by 'legal enactment' to support fracture and rehabilitation services. Ernest Hey Groves, for example, wrote in 1935:

Consider what happens when a motor accident occurs in which the car is smashed and the driver's bones are broken. The car is sent to the works and the man to the hospital; the car is paid for, but the man's bones have to be mended as a matter of charity! It would be only just and fair for insurance companies to be asked to pay for the repair of the broken bones. But they certainly could only do this if compelled by legal enactment, and the innovation would be an addition to their present liabilities. On the other hand, if insurance authorities could be convinced that organization of fracture treatment would effect so much saving in compensation as more than to cover the cost of the treatment, then the matter would bear quite another aspect.[69]

Neither orthopaedists nor other specialists seem to have considered the possible ill-effects (professional or otherwise) of allowing powerful private insurance companies to enter into medical care as independent third parties. On the contrary, along with lobby groups such as the British (Voluntary) Hospitals' Association, who insisted that the voluntary hospitals should receive payment for treating injured workers receiving workmen's compensation, they saw only potential benefits from a new source of funds.[70] Dazzled by American hospital experience in this area, they accepted as more or less self-evident that 'the almost universal application of insurance and compensation to [industrial injury] . . . cases [leads] . . . to a demand for more accurate diagnosis, prognosis and treatment'.[71]

But there was little reason to worry about the potential ill-effects of third-party funding. As pointed out at a meeting in the Ministry of Health in 1930, British insurance companies were not interested in arguments grounded in efficiency and economy:

> [they] simply say that so far as they are concerned, it is a question of finance, and the premiums are so regulated that they cover even the most expensive case. Any attempt on the part of the doctors or the hospitals to reduce the period of incapacity would merely mean that employers would press for a lower premium and the Insurance Company would be no better off than before. We cannot hope, therefore, for much help in the way of securing improved treatment from the Insurance Companies.[72]

Only in a few isolated cases did insurance companies provide funds for fracture and rehabilitation services in the 1930s. For the most part they stuck to the argument that by the terms of the legislation on workmen's compensation they were unable 'to do anything more than pay the compensation so fixed', as the Federated Employers' Insurance Association told Hey Groves in 1936.[73] An agreement had been reached with the Home Office in 1923 that bound the private insurance companies to pay out a fixed percentage of their aggregate premiums in compensation claims, or else grant a refund on the next premium due.[74] (The situation was thus far different from in those countries, such as Austria, where a state monopoly existed with regard to compensation insurance.) A representative of one of the companies pointed out in 1935 that 'while certain insurance corporations might very well be inclined to put up considerable sums for rehabilitation work, they were held back by the consideration that their action would be benefiting other companies which were not contributing at all'.[75] The insurance companies, like employers and trade unions, had little economic incentive to medicalize workmen's compensation. Industry feared greater costs, labour feared monetary loss.

Economic motives were thus as central to the *absence* of industrial rehabilitation schemes in Britain in the 1920s as to the provision *for them* in America. Despite claims to the contrary, and Carlylian niceties about 'cash payment not [being] the sole relation of human beings',[76] it was not in humanitarianism that the two countries differed. It was the prevailing forces of capitalism and the associated legislation which determined the value of human labour and the benefit or otherwise of restoring labour power. As American industrial physicians and surgeons were willing to admit: 'Big business is not in the habit of adopting any type of welfare service that cannot pay its own way.'[77] Indeed, more than one American orthopaedic surgeon proudly contended that it was 'largely due to the influence of the industrial surgeons that attention has been brought to the economic side of the question'. He looked hopefully to the time when industrial surgery would be established on a 'purely commercial basis'.[78]

Further research into the history of orthopaedics in America is needed to establish exactly how, when, and to what extent the factors identified here served to advance the interests of orthopaedic surgeons.[79] It is clear, however, that in Britain the absence of these predisposing factors inhibited the expansion of orthopaedics in the area of industrial injury. In the absence of rehabilitation centres, accident hospitals, and fracture clinics under orthopaedic control, most industrial injuries continued to find their way to the 'surgeries' of GPs and to general surgeons in the voluntary hospitals – albeit, the latter sometimes in orthopaedic clinics.[80] In rural mining districts and in some small-town settings where there were few hospitals but often many industrial injuries, opportunities for the expansion of orthopaedics were greater, but the incentives less, for these were places where there was even less scope for private consulting practice. To have developed orthopaedic clinics within industry (had this been possible) may have meant forfeiting professional autonomy and becoming financially dependent on industrialists. As we saw, it was for this reason that Jones turned down the offer to act as company surgeon to the Manchester Ship Canal project.

For all these reasons, few initiatives on behalf of the industrially injured came from aspiring members of the BOA in the 1920s. Such 'orthopaedic' initiatives as there were came, rather, from outsiders to the specialism – persons even more marginal to the medical establishment than the orthopaedists themselves. One such person was the above-mentioned Frank Shufflebotham, who had argued that Ministry of Pensions' orthopaedic clinics should be made permanently available to working people.[81] Another was James Rutherford Kerr, who had been surgeon to the Pilkington Glass Company since 1907, but who in 1916, after acting as the surgeon-in-charge of the Hôpital de l'Alliance at Yvetot in France, was commissioned by Pilkington Brothers to organize the Pilkington Orthopaedic Hospital for the rehabilitation of disabled sailors and soldiers.[82] This was subsequently transformed into a full-scale company hospital. Relatively isolated individuals like these, along with institutions such as Manor House Hospital, London, or the Glasgow Orthopaedic Clinic constitute intriguing parts of the neglected history of workers' health care in the early twentieth century, as well as the history of the use of the word 'orthopaedics'. Manor House Hospital, whose origins were also in the wartime hospital at Yvetot and which maintained links with the Pilkington Hospital until the latter's closure in 1925, was a uniquely worker-funded hospital, democratically managed by the Industrial Orthopaedic Society.[83] The Glasgow Orthopaedic Clinic, on the other hand, was established in 1925 by the Scottish Local Board of the Chartered Society of Massage and Medical Gymnastics to provide (at a minimum charge of 2*s* 6*d*) for those 'suffering from crippling diseases who are unable to pay the ordinary fees for the special and, in most cases, prolonged treatment at the infirmaries'.[84] These institutions and those involved with them helped to keep 'orthopaedics' associated with the treatment of industrial injury, but they played little part in the professional organization of the specialism in Britain, nor in the expansion of specialist opportunities. More important for this during the first half of the interwar period at least were the traditional subjects of orthopaedic attention, crippled children.

8

COLONIZATION AMONG CRIPPLES

To some extent, all professional agendas are set by circumstances outside the immediate control of the professionals involved. But some agendas are thus determined more than others, and this was especially so for orthopaedists in the increasingly cold economic climate of the postwar years. Despite gains in status during the war, the specialism lacked autonomy, and its practitioners were too few to constitute an effective lobby in medical politics. Whatever orthopaedists might have wanted to do, they were more or less directly constrained by their immediate socioeconomic and political context. In particular, they were bound by the social and ideological concerns of the agents and agencies through whom they were compelled to negotiate their interests. In more ways than one the postwar agenda was to be set for them, not by them.

Among those who set this agenda were social policy professionals such as George Newman, then Chief Medical Officer to both the Board of Education and the new Ministry of Health. Keen to extend prewar developments in the care of crippled children, Newman was soon arguing that wartime advancements in orthopaedics now made it possible to deal 'with crippled children in a satisfactory and scientific way'. 'The surgery of the War', he asserted in 1919,

> has, in developing the treatment of the crippled soldier, enormously advanced the possibilities of mending the crippled child, and has opened out opportunities which were not dreamt of when the [Elementary Education (Defective and Epileptic Children)] Act of 1899 was passed.[1]

By the mid-1920s this view had become widespread in official circles. The Duchess of Atholl, for example, while Parliamentary Secretary to the Board of Education in 1925, found it

> a comforting thought that as a result of all that our men suffered in the War we have made a great advance in orthopaedics, which may

be of incalculable benefit to the children. Wonderful results are seen even in the case of men who had received terrible injuries, and I can therefore well believe what wonderful results can follow such treatment when applied to young children, and I do rejoice to think that now we have a definite skilled and organised attempt to give the advantage of all the knowledge and experience that has recently been gained on this subject.[2]

Since by this date more than half of the senior members of the BOA had resident or consultancy positions at children's hospitals,[3] and the leaders of the specialism were themselves making claims about the applicability to crippled children of the lessons of the war, the postwar orthopaedic re-engagement with the crippled child has been regarded as a wholly logical step, positively taken. There was of course continuity with wartime orthopaedics, inasmuch as the work at the military centres had consisted of rectifying deformities that might have been cured had they received treatment during childhood. For the most part, however, orthopaedists had little option but to return to their prewar work with the chronic deformities of children. Once returned to that work, though, they were soon enthusiastically pursuing it; under circumstances which once again placed child health high on the social and political agenda, it was professionally advantageous to do so. Indeed, no sooner had the postwar agenda been established, than Jones and his colleagues became involved in an ambitious scheme to recreate the power and glory of their military empire with a similarly organized 'national scheme for the cure of cripples'.

THE PROPOSED NATIONAL SCHEME

The National Scheme was elaborated by Jones and G. R. Girdlestone in the *British Medical Journal* in October 1919,[4] and it drew heavily upon the prewar organization for the care of crippled children as established around the hospital at Oswestry. The latter system was now writ large in the call for a national network of central orthopaedic hospitals (COHs) and affiliated after-care centres.

Such a scheme was suited to a specialism whose aspirations were out of proportion with the number of its practitioners. Through it, an orthopaedist attached to a COH would be assured of sufficient numbers of patients to sustain specialist interests by single-handedly meeting the needs of patient populations often dispersed over large areas. The scheme was also intended to facilitate the integration of orthopaedists into traditional centres of medical authority and status. Although the COHs were to be separate from existing voluntary hospitals and located in rural areas wherever possible, there was to be an affiliation between them which was not meant to 'stop short at administrative affairs'. Clinical cooperation was seen as 'even more important' and, in general, a spirit of mutual help – 'team work' – rather than rivalry was to run through all the relations between the COHs and other hospitals and the medical community as a whole. A great step would be taken towards this 'ideal co-operation', it was suggested,

> if the general hospitals would institute an orthopaedic section of their out-patient department and appoint one of the surgeons of the C.O.H. to take charge of it. Out-patients coming up could then be given provisional grouping, and later, after examination, be referred readily, for opinion or transfer, from general to the orthopaedic side or vice versa. (p. 458)

For London and other cities where orthopaedic hospitals or departments already existed, Jones and Girdlestone argued that it would be necessary 'merely . . . to bring the existing institutions into the general organization so that co-ordination might be obtained'.

In this, as in other respects, no effort was made to conceal the fact that what was being sought was essentially the 'civilianization' of military orthopaedics. Girdlestone in particular was concerned not to forfeit the military base he had established in Oxford.[5] There was now a 'unique opportunity', he and Jones argued,

> [for] before many months there will be a diminution in the needs of the soldier cripples. Then will come a time when [the pensions hospitals] will have to be cut down, and their staffs gradually dispersed, unless the Ministry of Health has made its plans in advance and obtained the necessary financial provision for taking them over. (p. 460)

Naively, as we shall see, Jones and Girdlestone imagined that the Ministry of Health would respond as the War Office eventually had during the last two years of the war. They expected the same kind of reasoning to apply. In fact, their argument for special hospitals for crippled children was no different from that used by Jones for the segregation of orthopaedic cases at Alder Hey in 1915 – namely, on the basis of the purported extent of the problem, and the inability of nonspecialists in general hospitals to deal properly with it. Because 'hitherto there have been very few orthopaedic hospitals in England', it was argued, crippled children had to be sent to general hospitals where, 'owing to the pressure on beds', on the one hand, and to the lack of 'specially trained surgical and nursing staff', on the other, they had not been properly cured and the problem of crippledom had scarcely been touched. Accordingly, 'the demand [of these patients] for a very great deal of personal attention on the part of the surgeon [was] . . . one of the most convincing arguments for specialization'.

But the National Scheme appropriated more than just wartime rhetoric for state support for orthopaedic specialization. It also sought to transfer military structures and practices to the civilian sphere. Orthopaedists were to be employed part-time at the COHs on a salaried basis, just as for much of the war consultants had been employed in the territorial hospitals. Furthermore, the proposed COHs and clinics were to mirror the regional organization of the Ministry of Pensions' clinics for dealing with disabled soldiers. Like the proposed pay structure, which was radical for peacetime,[6] the 'regionalism' was an early expression of such thinking in medical planning in interwar Britain. Indeed, in 1948, under the National Health Service, Girdlestone would proudly recall it as 'the first of the regional specialist services'.[7] In actual fact, the NHS's Regional Health Authority structure was to develop largely around the long-standing voluntary (teaching) hospitals in regional capitals, whereas Jones and Girdlestone's scheme, like contemporary ones for tuberculosis, involved new institutions planned to meet the projected needs of populations within specified geographical areas. But, in general, the National Scheme exemplified the tendency, noticeable in both British and American medical planning after the war, towards what Daniel Fox has styled 'hierarchical regionalism' – that is, towards efficient, equitable, accessible and comprehensive medical services, regionally coordinated through a division of medical labour descending from hospital-based consultants and specialists to outlying general practitioners and allied health workers, much as had been the case during the war.[8]

THE CENTRAL COUNCIL FOR THE CARE OF CRIPPLES

In December 1919, only two months after the publication of the National Scheme, Jones and his colleagues became involved in establishing what became the Central Council for the Care of Cripples

(CCCC). In part, this developed out of the Education Sub-committee of the Joint Parliamentary Advisory Committee that had been chaired by Mrs Humphry Ward – a body composed solely of women and MPs, whose object had been to consider all the legislation coming before Parliament concerning women and children. The orthopaedists Elmslie, Bristow, Jones and Girdlestone became involved with this committee in 1918, by the end of which year Girdlestone had succeeded in obtaining funding and permanent office space for it from the Red Cross.[9]

No records appear to have survived of the negotiations by which the Advisory Committee became the CCCC, and not until the mid-1920s is there evidence of soul-searching over the latter's precise function.[10] The orthopaedists seem to have had in mind an organization similar to the Central Association for the Care of the Mentally Deficient (established in 1915), or that for crippled children established in Germany before the war by Konrad Biesalski.[11] Such a body would press for medical and educational facilities for crippled children, as well as coordinating the various charity groups already involved. Among the founding members of the CCCC, besides the orthopaedists and Mrs Ward, were Mrs Monro and Mrs Ursula Townsend of the Invalid Children's Aid Association, and Mrs Wilton Phipps and Lady Lawrence of the London County Council Education Committee. Early recruits included Margaret Beavan, Lady Beatrix Wilkinson, the President of the Church of England's Children's Union of Waifs and Strays Society, Dame Florence Barrie Lambert, Honorary Director of the Central Council for Infant and Child Welfare, Arthur Black, the Secretary of the Shaftesbury Ragged School Union Society, and Sir Arthur Stanley and Sir Napier Burnett of the Red Cross, together with the orthopaedists Harry Platt, H. A. T. Fairbank and W. H. Trethowan.[12]

Jones was the chairman of the CCCC's executive, and one of the first tasks undertaken by the organization was 'a house to house inquiry' to 'procure accurate figures as to the number requiring [orthopaedic] treatment'. The findings would serve to justify the provision of 'sufficient and suitable facilities for [the] treatment of orthopaedic cases'.[13] At the same time the CCCC sought to pressure local education authorities to implement immediately (rather than wait until 1927 when it would become compulsory) the permissive legislation for special provisions for cripples put forward in the Fisher Education Act of 1918 – what Elmslie referred to as that 'powerful weapon in the hands of orthopaedists'.[14] Specifically, as Jones informed H. A. L. Fisher, the President of the Board of Education, the CCCC wished 'to provide a national scheme for the provision of orthopaedic treatment of cripples and to co-ordinate with this a system of education for them'.[15]

PRESSURE POLITICS

Initially, Jones and Girdlestone hoped that the very existence of the CCCC would pressure the Ministry of Health into taking up the National Scheme. Had it done so, this might have solved the complex administrative and financial situation surrounding institutions for the care of cripples. Before the Local Government Act of 1929 (when block grants for health services became available to local authorities), funds were channelled through a bewildering array of governmental agencies. The Board of Education, through the special provision for physically defective children, was one source of funding, and the Home Office was another if industrial training were offered or compulsory child custody was involved. In cases where mental handicap was certified, funds could be procured from the Board of Control (which took over the duties of the Lunacy Commissioners in 1913); the Board of Trade gave assistance to some of the voluntary associations concerned with the after-care of cripples; and the Ministry of Pensions could be involved when orthopaedic treatment for children was carried out in their clinics.[16] At the local level, services could involve Poor Law Boards of Guardians, Education Authorities, School Medical Officers, MOsH, workers in Maternity and Infant Welfare centres, and Public Health Authorities – the latter especially through the Public Health (Tuberculosis) Act of 1921. Additionally, there were the various local and national voluntary organizations, some of them operating in quasi-official capacities. Such a variety of agencies with different remits, most of them working within different administrative boundaries, makes the appeal of the rationalized structures of wartime medicine easy to understand. Indeed, it was in light of the wartime organization and the consequent perception after the war of the 'chaos in our health services', especially services for women and children, that Lord Rhondda, the President of the Local Government Board, urged forward the prewar plans for the creation of a central Ministry of Health.[17] Even with the establishment of the Ministry in 1919, however, integrative ideals continued to be frustrated by competing local and national, intra-governmental, and voluntaristic interests.[18]

George Newman was sent a pre-publication copy of Jones and Girdlestone's National Scheme in May 1919, and Lloyd George and

H. A. L. Fisher were among those to receive one of the 300 carefully placed offprints.[19] Jones and Girdlestone also twice visited Newman at the Ministry in 1920 to plead their cause. Their hopes were high, for Newman was known to be enthusiastic about the rationalization of health care. His *Annual Report to the Board of Education* for 1919, for example, made much of 'the purely haphazard and disjointed' kinds of provision available, and underlined the urgent need for a 'one-portal system' of access.[20] However, even after the first meeting with Newman, in February 1920, it was clear to Jones (as he confessed to Girdlestone), 'that the Ministry may not react to pressure'.[21] The second interview, in March 1920, largely confirmed the impression. At this, Jones and Girdlestone sought the Ministry's endorsement for the seven points summarizing the policy and aims of the CCCC – the first of which was that the organization would work under the Ministry of Health and the Board of Education.[22] They made plain that they were seeking 'a capital grant from the Ministry on the same scale as that now made to sanatoria'. But Newman could only reply that, while his 'heart and soul' were with the general aims and objectives of the National Scheme, there were 'obvious difficulties' in trying to coordinate this action under the Ministry. 'Was it intended', he asked, 'to give this Central Committee [the CCCC] executive powers, and if so, how could the Ministry delegate to it statutory powers?'[23]

If Jones and Girdlestone were unprepared for the rebuff, it was because they had little notion of the political realities behind the new Ministry of Health (which officially came into being in June 1919). Supposing it to be a rationalizing and funding body, they did not realize that the local authorities were to remain the crucial factor in postwar medical planning and policy. Here one must recall the institutional context: the Ministry of Health, though it replaced the Local Government Board in 1919, continued to oversee most aspects of local government throughout the interwar period. In medical matters, the Ministry acted mainly *through* the local authorities. Although Newman was fully cognizant of the constraints that local authorities imposed on large-scale rationalizations, he was compelled to work within those constraints as CMO to both the Ministry of Health and the Board of Education. Attentive also to the political, professional, ideological and above all the economic issues involved in a scheme such as Jones and Girdlestone's, he was thinking pragmatically, not dismissively, when he advised them to seek their support from the local authorities and the voluntary agencies and, as it were, to bend these bodies to their will from within.

Newman was as good as his word when it came to mustering support for the general organizational and therapeutic objectives of the national plan. In his *Annual Report to the Board of Education* for 1923, for instance, he spent 13 pages outlining the scheme, claiming that 'it represents a new understanding of the problem and its solution'.[24] In subsequent reports he referred to the scheme as the 'approved' one. Thus, gradually, Jones and Girdlestone's scheme did come to infiltrate and divert the existing medical and welfare structures. But it remained politically impossible for Newman to commit the Ministry of Health to the rationalizing and funding role that Jones and Girdlestone had envisioned. And after April 1920, when the depression set in, this role became financially unthinkable. As Sir Arthur Newsholme observed in 1922 in a speech on 'The Betterment of Child Life':

> At the present time all efforts at betterment of human conditions which imply expenditure of communal funds are being curtailed, by apparently undiscriminating 'rationing', on the basis of a percentage reduction of expenditure. Action on these lines is obviously irrational. It implies the bankruptcy of government policy. Ministers of Health have made themselves responsible for reductions in expenditure on active child welfare, and have thus given the cue too successfully to local authorities to economise without regard to the relative importance of their several activities.[25]

Newsholme, formerly the Principal MO to the Local Government Board, had lost out to Newman for the post of CMO to the Ministry of Health when the latter absorbed the former in 1919; he therefore had added reason to be critical. Nevertheless, the cutbacks and consequent erosion (or at least stagnation) of central government intervention in health and welfare schemes were real, and the new and untried Ministry of Health had reason to fear that more assertive action on its part would lead to a decline in subscriptions to voluntary organizations and to hospitals especially.[26] Such were the realities that lay behind Girdlestone's reaching into his own pocket in 1922 in order to purchase the Wingfield Hospital in Oxford from the Ministry of Pensions.[27]

CHARITY AND THE STATE

In the face of the recession, Jones and his orthopaedic colleagues had less and less reason to regret their affiliations with the CCCC, which now came into its own as an organization for the rationalization of charity funds. The founding in 1919 of the Save the Children Fund and a variety of other such child-centred agencies all contributed to the success of the orthopaedists' cause. They were also well-placed to exploit the expansionist ambitions and surplus revenues of the British Red Cross.[28] Some of the better-heeled welfare agencies for cripples (notably the Shaftesbury Society, the ICAA and the Church of England's Children's Union, whose ordinary incomes in 1930 were £70 000, £100 000 and £200 000 respectively) could also be tapped for the funding of new hospitals and homes. Further funds became available from the Carnegie Trustees.[29]

As in the prewar period, relying on charities for organizing, managing and financing orthopaedic schemes did not necessarily conflict with seeking assistance from the state. Jones was willing to have his name appear on the letterhead of the ultra-voluntarist People's League of Health; he was no less willing to write to the Minister of Health, Neville Chamberlain, in 1927, urging him to push on with the municipalization of the Poor Law infirmaries.[30] As he saw it, any programme directly or indirectly likely to assist orthopaedic interests was worthy of endorsement.

Indeed, orthopaedists sought the best of both worlds, even though the cost might be a doubling of their anxieties. Thus, after the mid-1920s when the depression had eased and more and more local authorities were inaugurating 'approved orthopaedic schemes',[31] they would share the charities' fear, not of state assistance, but rather of state control, bureaucracy and the dreaded stultification of voluntary work.[32] (According to one informant, the charitable bodies were becoming increasingly obstreperous at this time because 'the squire and the clergyman, or rather their womenfolk' regarded their charitable activities as a means to regaining from the state some of their lost administrative powers.)[33] Orthopaedists delighted in the state's imposition of uniformity in the organization and provision of care, and welcomed fees from the state (paid regularly and unfailingly), but they resented the impositions of officialdom. They complained about the delays that occurred in passing patients through official channels, and the restrictions placed on the types of cases they could treat at orthopaedic clinics – 'only those for whom Parliament has made local authorities responsible'. Such constraints were held to have negative therapeutic consequences, as for example upon the cherished principle of continuity, for 'when a patient reaches the limit of school age he can no longer attend [the clinic] and is lost sight of'.[34]

Girdlestone, discussing the pros and cons of 'purely official' versus 'purely voluntary' clinics in 1926, expressed a widely shared sentiment when he endorsed the need for both. 'The voluntary muscles act on and through the official bones, while the latter provide the underlying form, stability and permanence. . . . neither can do its best without the other.'[35] Since orthopaedic expansion was dependent as much on private as public funding, it would have been highly impolitic to have said otherwise – not least because the voluntary CCCC was itself largely dependent on the goodwill of the ICAA until 1935 when the car magnate, Lord Nuffield, began lavishly to endow it.[36]

In discussing such questions we should not assume that central government held very different views. The Treasury would willingly have retained the government's position as the 'junior partner' with the voluntary organizations in the provision of health and welfare for cripples. But, quite apart from what it perceived as 'a certain sharpness of utterance which makes the problem of working those societies into a many-sided health service none too easy',[37] it was increasingly obvious that to rely on voluntary funding was also to permit radical disparities in the provision of care between regions. In particular, in the deprived industrial North, not only were there fewer voluntary agencies, but the incidence of crippling from tuberculosis and rickets was much greater. It was said of Leeds in the mid-1920s, that it 'would be like the town of Hamelin after the visit of the Pied Piper if all the children suffering from rickets were taken away'.[38] Much the same could have been said of other northern cities, such as Manchester, where atmospheric pollution apparently blocked out some 40 per cent of the sunlight.[39] In such locations the scope of the problem was well beyond the resources of private agencies.

In the face of these inequalities, progressives in the welfare ministries (Health and the Board of Education) pressed local authorities to set up schemes 'on the rates'.[40] As J.J. Butterworth, the MOH for Lancashire, emphasized at a CCCC conference in London in 1928, 'the County Council only stepped in and established clinics of its own because it found the voluntary resources inadequate'.[41] In fact, the Lancashire County Council set something of an example: well before the Local Government Act of 1929 facilitated new local authority initiatives in health care, it not only established orthopaedic clinics in most of its major towns, but (in 1926) established the country's first orthopaedic hospital for children to be provided entirely by a public authority.[42] By 1935, although it was now nearly two decades after the Oswestry after-care clinics had been established, the Lancashire

Plate 8.1 Biddulph Grange Orthopaedic Hospital, Staffordshire, in the 1930s. Purchased by the Lancashire County Council in 1925, this was the first orthopaedic hospital for children to be provided entirely by a public authority.

Source: Manchester Medical Collection, Platt Papers, John Rylands University Library, Manchester. (By courtesy of the Director and University Librarian.)

network of children's orthopaedic hospitals and free clinics had become a model of comprehensive statutory medical provision.[43] Without undermining voluntary efforts, it demonstrated how local authorities could level services up and so provide a more even distribution of quantity and quality.[44]

By the 1930s, then, although no National Scheme under the Ministry of Health had emerged, there were coordinated networks of services, more or less guided by official public bodies at municipal and regional levels (see map). By the mid-1930s, some 40 orthopaedic hospitals and 400 orthopaedic clinics were in operation in Britain, run either by voluntary agencies or by municipal and county councils, or by combinations of them.[45] And whatever the funding, medical interests now predominated where educational and other social concerns had once prevailed.

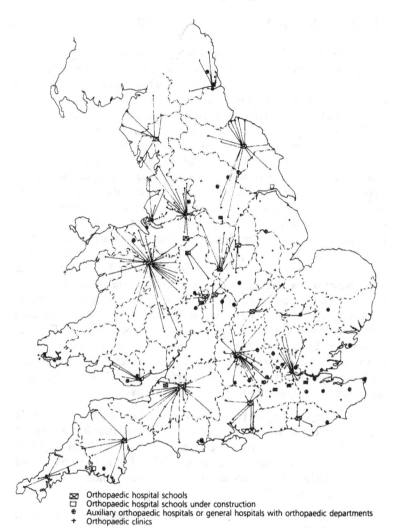

Map 8.1   Orthopaedic schemes in England and Wales, 1934

Note: The scale of the map does not allow space to show various other children's hospitals where orthopaedic cases were taken, such as Great Ormond Street, Wray Crescent Hospital Home, etc.

Source: Frederick Watson, *The Life of Sir Robert Jones* (Hodder & Stoughton, London, 1934).

THE POLITICS OF PREVENTIVE MEDICINE

The 'progress' of child orthopaedics in the interwar period was obviously not guaranteed. But, as with the advancement of orthopaedics during the Great War, neither was it wholly arbitrary. To a large extent, it was bound, not just to local medical negotiations, but also to crucial shifts in medical policy as arbitrated by the state and the profession. In many respects the history of orthopaedics provides a unique opportunity for the investigation of such shifts.

Yet, at first sight, the idea of medical policy playing a determinant role in the postwar expansion of the specialism seems strange: as we have observed, one of the most powerful arguments for the extension of orthopaedic facilities was the claim that the war – and the war alone – had been responsible for an 'immense advance . . . in the knowledge and practice of the rectification of congenital and acquired deformities'. Such claims, however, were little more than propaganda – in this particular case, as issued by the Glasgow Children's Hospital in 1930 in a campaign for 'a completely equipped orthopaedic department'.[46] In truth, not only were some prewar research programmes arrested in 1914, but some major therapeutic techniques devised under wartime conditions were abandoned by peacetime orthopaedics.[47] The real transfer of knowledge, it could be argued, was not from war to peacetime, but the other way round, though this argument was seldom no less propagandist. Jones, for example, in an address of 1920 to a conference on crippled children, proclaimed that

> the advances in surgical procedure which did so much to alleviate the sufferings of the War, were largely the result of knowledge acquired from the treatment of the crippled child, and the methods employed in orthopaedic military centres were closely allied to those which had been practised in various children's hospitals where orthopaedic supervision existed.[48]

The notion of the war as responsible for promulgating a better social and psychological comprehension of cripples which, in turn, encouraged the development of orthopaedic schemes was simply a more sophisticated version of this same support-seeking rhetoric. In 1928, in the CCCC's *Cripples' Journal*, Jones wrote that the war had provided 'a severe counterpoise to the traditional spirit of aversion or indifference . . . produced in all sections of society by the sight of their own flesh and blood stricken with deformity'. Now, he contended, while physically handicapped persons had come to feel that there was something even heroic in their condition, the rest of society had become accustomed to witnessing the complete recovery of function of the severely disabled by orthopaedic means. Thus there emerged the 'profound realization of the cripple as made, not born, and also as curable'[49] – the popularization, in effect, of the professionalizing rhetoric adopted by Jones and his colleagues in the late nineteenth century. As echoed in the publicity material issued by the Glasgow Children's Hospital in 1930: 'Nowadays the prevalence or otherwise of physical deformities in a community marks its degree of aesthetic and medical knowledge since all but the rarest congenital defects are either preventable or else amenable to successful treatment.'[50]

The perception of the cripple both as 'made not born' and as 'curable' did indeed take on new significance after the war, but this was not primarily a result of the rhetoric about achievements of wartime orthopaedics. Very largely, it was a consequence of a burgeoning discourse on Preventive Medicine – a discourse into which orthopaedic rhetoric readily fitted, and out of which emerged major policy documents. An extensive memorandum on the subject by Newman in 1919 was in fact among the first publications of the new Ministry of Health.[51] But Preventive Medicine was also enthusiastically pursued by the medical profession as a whole, as well as by the Labour Party and local authorities. Invariably capitalized, it signified more than merely the scientization and medicalization of 'public health' after its encounter with bacteriology (though it signified that as well).[52] As Newman insisted in his published memorandum, Preventive Medicine had to be understood as 'something wider than "Public Health"' – something more than merely the avoidance of disease. It was, he emphasized, 'the *removal of the occasion of disease and physical inefficiency, combined with the husbanding of the physical resources of the individual*'.[53]

The stress on the individual was fundamental, for although Preventive Medicine did not wholly forsake environmental health concerns, essentially it reconceptualized and reconstituted 'public' health in the personal terms of curative medicine – hospital-centred and science-based.[54] In part, of course, this reconceptualization was simply the result of the late nineteenth-century discovery of new cures and prophylaxis for diseases which formerly could *only* be attacked through environmental measures. Koch's tuberculin cure for tuberculosis announced in 1890 was perhaps too uncertain to bring to a halt social campaigns such as those against public spitting or the use of public drinking cups, but the widespread introduction of an effective antitoxin for diphtheria in the mid-1890s and, more dramatically, the commercial development after 1909 of Salvarsan as a 'magic bullet' for syphilis, anticipated an age in which person-orientated technological fixes could be expected for most of mankind's ills.[55]

Yet it would be wrong – not to say wildly anachronistic – to construe Preventive Medicine in these terms. 'Fixes' as such, hardly entered into the discourse; on the contrary, as suggested by Newman's references to 'physical inefficiency' and 'husbanding', the emphasis was on disease-prevention and body-restoration through the pursuit of physical culture. 'To develop and fortify the physique of the individual and thus to increase the capacity and powers of resistance of the individual and the community,' was widely acknowledged to be the first object of Preventive Medicine.[56] As such, Preventive Medicine was essentially holistic, combining preventive and curative techniques at the same time as exposing the interdependence of the health of individuals and communities. Furthermore, its vast implications for medical education and medical organization served to legitimize policy-thinking in terms of large-scale coordinations and integrations of health-care services and medical labour.

The latter interests are well reflected in the title of the Labour Party's *The Organization of the Preventive & Curative Medical Services & Hospital & Laboratory Systems under a Ministry of Health* (1919). But they are perhaps best witnessed in the famous Dawson Report of 1920, where commitment to the 'close co-ordination' of preventive and curative services fits hand in glove with the proffered model of integrated health care.[57] Dawson's proposed Primary Health Centres were to be places where the skills of GPs would be combined with those of specialists, and, equally important, where there would be 'the widest range of remedial and athletic activities, designed to keep the population in a maximum state of physical fitness'.[58] The Secondary Health Centres (which were essentially the existing general hospitals staffed by consultants and specialists), were intended to deal quickly with difficult cases referred from the Primary Centres. They stand at the centre of the regionally conceptualized medical map contained in the Dawson Report, while domiciliary as well as 'supplementary' medical services (such as those for maternity, geriatrics, tuberculosis, and mental health) fill in the web-like structure of medical services. Significantly, perhaps, teaching hospitals and associated medical schools schematically connect to the network as a whole in a manner suggestive of the thread that attaches a spider to its web. Although the 'primary' and 'secondary' nomenclature in Dawson's plan was drawn

from education,[59] most of the rest of it was modelled on the wartime
hierarchy of military medicine – primary health centres with their
limited beds and laboratory facilities closely resembling Casualty
Clearing Stations.

Interwar orthopaedics did more than merely fit neatly into the
philosophy of Preventive Medicine, it was part and parcel of it. Jones
and Girdlestone's orthopaedic scheme was as much an expression of
the philosophy as Dawson's scheme. The functional relation between
the COHs and the after-care clinics was scarcely different from that
between Dawson's secondary and primary health centres, and the
concern with comprehensive and coordinated services within and
between 'health regions' was no less emphatic. In the therapeutics of
open-air hospitals, moreover, the preventive and the remedial were
rendered indistinguishable from the curative. Indeed, modern orthopaedics perfectly represented the contemporary dissolution of the
boundaries between preventive medicine and public health, on the
one hand, and preventive medicine and curative medicine, on the other.
Furthermore, orthopaedic practitioners were among the most outspoken on the need to eliminate the social and environmental causes of
crippling,[60] while at the same time, they demonstrated the advantages
of modern scientific medicine. Such technological fixes as vitamin D to
conquer rickets, and milk pasteurization to counter tuberculosis were
celebrated by them. And, not least important, they conducted *surgery*,
though, crucially, as we have seen, this was conservative surgery, with
techniques elaborated and defended humorally, in terms of an
integrative or holistic approach to a body balanced both in itself and
in relation to its environment. Given these characteristics, it is small
wonder that many besides Newman were to proclaim modern
orthopaedics the most 'remarkable instance of the association of
preventive and curative medicine',[61] and that Dawson was to
incorporate it into his scheme as a supplementary consultant service.[62]

The progress of orthopaedics in interwar Britain can thus be
understood in terms of a growing consensus on Preventive Medicine
among policy-makers. A fuller understanding is to be reached,
however, by taking into account the deeper divisions in ideology that
the apparent consensus masks. Dawson and Newman may have been
equally interested in promoting national health and efficiency, but they
worked from different political agendas. Dawson, in common with the
Royal Colleges, the BMA and the voluntary hospitals, was principally
concerned with halting the intrusion of the state into medical decision-
making and private practice. In particular, he wanted to undermine the

role of local authorities in health care and to ensure the marginalization of state-salaried MOsH so as to promote an autonomous
medical profession (though preferably one that had the backing of the
Ministry of Health).[63] Preventive Medicine, in its alliance of remedial
and curative practices, was seen by Dawson as a way of directing policy
away from the growth of municipal services under state control by
incorporating independent GPs into a system more representative of
the profession's interests.[64] Newman, on the other hand, as a state
employee and former MOH, was less concerned with doctors' interests
than with setting up an effective medical system which, above all,
would make economic use of the existing state and charitable
resources. For him, the rhetoric of Preventive Medicine, with its
emphasis on the union of environmental health with curative
medicine, was an ideal means to rationalize health care.

The fate of orthopaedics in interwar Britain both reflects and
illuminates these tensions. Nowhere is this better indicated, perhaps,
than in Newman's response to the pre-publication copy of Jones and
Girdlestone's National Scheme. Newman's copy reached him at the
Ministry of Health via Dawson, to whom Jones had mistakenly mailed
it. 'My Dear Orthopaedist', Newman replied to Jones:

> When one thinks of it, it was very fitting that the scheme should go
> to him [Dawson], as demonstrating that he is not the only person
> who thinks imperially! I shall do all I can to press the claims of the
> child and adult needing orthopaedic treatment, but I am afraid it will
> be a long time before we are able to ignore the local authorities and
> govern England in the Teutonic way you suggest. I believe long
> contact with Royalty and an extravagant War Office has bred in you
> a new amplitude of desire! But I shall do my best.[65]

Although Jones's mistake in mailing the scheme to Dawson may reveal
his greater sympathy for Dawson than for Newman, Newman exulted
in the mistake. By commenting on Dawson's 'imperial' and Jones's
'Teutonic' aspirations, Newman warned Jones of the need to fit in with
the local authorities, and not to expect the autonomy that Dawson
sought for the profession.

It is not known if Jones ever intentionally sent Dawson a copy of the
National Scheme. Dawson obviously knew about it and doubtless
would have been aware of its publication in the *BMJ* (some six months
before the publication of his own 'plan' in May 1920). But it is not
surprising that he paid no special attention to it in his Report. While he
could fully agree with Newman that Jones's orthopaedics was a

wonderful 'model of "team" work, and [of] the complete integration [of interdependent medical functions]',[66] for the political purposes of his scheme orthopaedics had little to offer. It was just a new specialty which fitted rather poorly into the existing professional elite; linked to the School Medical Service and the Infant and Maternal Welfare Clinics, it was, at best, 'supplementary' to Dawson's consultant and GP-promoting, free-market architecture. It would have been counter-productive for Dawson to have celebrated a union of preventive and curative medicine in which most of the coupling would probably depend upon, and serve to enhance, state controls in medicine, if not in fact the power and prestige of local authorities and MOsH. Indeed, as can be seen in retrospect, the orthopaedists' struggle for survival in the 1920s and that of MOsH was to a large extent symbiotic, while independent GPs could be openly hostile to local authority orthopaedic schemes (see below). Dawson had good reason, then, to do no more than mention orthopaedics in his Report.

Conversely, to the Ministry of Health, and to Newman in particular, orthopaedics was worth exploiting, primarily because it conserved scarce resources. From the government's point of view, open-air hospitals and conservative surgery were valuable precisely because they were cheap. It was largely on the basis of financial estimates, in fact, that Jones and Girdlestone made their pitch to Newman in 1920 – their estimate of the cost of a complete open-air orthopaedic hospital of 250 beds being between £10 000 and £20 000, or the very low figure of £108 per bed. Newman agreed with them that 'in the past there had been a tendency to make hospital buildings unnecessarily elaborate and costly', and in his annual reports to the Board of Education he made much of the orthopaedists' economic good sense. For hard-pressed local authorities, there were clearly advantages to a scheme that called for the erection of inexpensive open-air hospitals and hospital schools on cheap land away from urban centres. (Again, wartime experience with tented and hutted hospitals helped to confirm the feasibility and practicality of such provision.) Newman further pointed out that the orthopaedic clinics were not only a long-term hedge against future expenditure on decrepit adults, but were also a short-term economy, since the majority of them were wholly financed and staffed by voluntary bodies.[67]

Legitimacy was thus given to the existing emphasis on conservative surgery in orthopaedics. In Girdlestone's publications, as in those of Gauvain, Tubby, A. Rollier and other specialists in children's orthopaedics in the interwar period, there were strong echoes of the nineteenth-century writings of Fergusson and other conservative surgeons. Tuberculous lesions, for example, were regarded as the local manifestation of a general disease which was amenable to treatment by fresh air, sunshine and improved diet.[68] 'It is delightful to me', Gauvain wrote to Girdlestone in 1924, 'to see an orthopaedic surgeon taking the wide view you do and regarding the patient first and the lesion as an incident. I am awfully pleased you have arrived at that as I do regard it as fundamental and I really don't think the ordinary surgeon, though he would deny my statement, ever gets to that stage.'[69] Among 'Jones's men' it was commonplace to remark, as Bristow always reminded his pupils at St Thomas's Hospital, that 'we treat patients, not disease'.[70] This whole-person rhetoric became sufficiently well-known for it to be publicly exploited in the cause of orthopaedic facilities. To quote again from the 1930 publicity material of the Glasgow Children's Hospital: 'In the scientific world the desire to make the body into an efficient and harmonious whole finds its most important expression in orthopaedics.'

Such was the context in which some orthopaedists actually came to style themselves 'orthopaedic physicians'.[71] Harry Platt was not among them, but in discussing the symptomatology of poliomyelitis in the *BMJ* in 1924 he expressed an ideal connected with this holism which was widely shared among his colleagues:

It may be suggested that, as a surgeon, I am traversing ground outside my immediate province, but there is an ever-increasing body of surgeons who are best described in the striking phrase of Sir Berkeley Moynihan as 'physicians doomed to practise surgery,' or who fulfill the spirit of Harvey Cushing's conception that 'the physician has become his own surgeon.'[72]

Cushing's friend Sir William Osler may have been thinking of this when he remarked during the war that 'the new orthopaedics . . . which has grown in a remarkable way within the past twenty-five years . . . is more than surgery'.[73]

It is telling that the strength of the postwar commitment to conservative surgery among British orthopaedists and its association with Preventive Medicine was not to be found among American orthopaedists. Partly for the contextual economic reasons touched on in the previous chapter, American orthopaedists expressed the view that 'if we essay to be trusted with the knife, if we call ourselves surgeons, we must be so in very fact'.[74] Some, indeed, came to style themselves 'bone and joint surgeons' precisely so as to distinguish themselves from the lingering association of orthopaedics with non-invasive surgical techniques.[75] Presumably this also stood behind the titling of the American *Journal of Bone and Joint Surgery*, which superseded the *Journal of Orthopaedic Surgery* in 1922. Overall, as an AOA president recalled in 1935, 'in the fifteen year period following the War, the [invasive] surgical side of orthopaedics has been explicitly stressed'.[76] Jones had detected this drift as early as 1924, but hoped that it was only a temporary swing of the pendulum away from the earlier – equally bad – emphasis on indefinite treatment by apparatus.[77]

The invasive surgical emphasis never did diminish in American orthopaedics, despite the reservations of certain senior practitioners,[78] and ultimately the British followed. By the late 1940s, even Girdlestone was granting the legitimacy of 'well-timed operations', though basically he remained, as Platt confessed of himself at this time, 'under the influence . . . of the Liverpool school which was so powerful an antidote to impulsive surgery and emotional therapy'.[79] But this was long after the heyday of children's orthopaedics in Britain in the mid-1920s, which was characterized not by new operating theatres, but by dozens of new open-air hospitals. At the zenith of the medicalization of the crippled child, British orthopaedists denied that they were principally 'operating surgeons'.[80] Although by the criteria of the 1950s to 1970s, it seems odd that an aspiring group of surgical specialists should have advocated a non-operative, low-technology therapy, in the economic context of the 1920s and 1930s this professionalizing strategy made perfect sense.

Thus the history of child orthopaedics in the interwar period provides not only a barometer of the development of Preventive Medicine, but also a working model of the practice of preventive medicine. As the practice can be seen as bound to the assertion of particular professional interests within a particular economic context, so the conceptualization and policy deployment of Preventive Medicine can be seen as textured by the differing social and ideological interests of the state and the medical profession generally. For Newman especially, Preventive Medicine provided a vehicle for pushing through more specific reforms, and out of that particular concern, much of the peculiar constitution of interwar orthopaedics is to be accounted for. Positioned between the interests of the state, charity organizations and the profession, orthopaedics serves to remind us of the balance of political forces within British medicine at this time and the inherently political nature of 'progress' on any front. It only remains to be shown why, in relation to this balance of forces, the position of the specialism was compelled to change.

A DECLINING SPHERE

The problem of the cripple was not of course solved in the interwar period. New institutions, such as the Queen Elizabeth's Training College for the Disabled, at Leatherhead, opened by Dame Georgina Buller in 1935, and the St Loyes College for the Disabled, opened at Exeter in 1937, together with new organizations such as the Surrey Voluntary Association for the Care of Cripples, established in 1935, all testify to the problem's persistence. In 1948 Beveridge reported that

> the physically handicapped people of Britain present a problem large in scale, distressing in the degree of suffering of individuals, and still more distressing in the deficiency of provision to prevent or alleviate suffering.[81]

That it was now 'handicapped people', not just children, who constituted 'the problem', however, and that it was vocational and occupational rehabilitation, not just hospital care and schooling, that was called for, demonstrates how the issue had changed. The corresponding transformation of orthopaedics from community-orientated concerns with the chronic diseases of children to hospital-orientated concerns with the acute conditions of adults will concern us in subsequent chapters. Here it only needs saying that the increasing age of the physically handicapped population before the Second World War was taken to show that the right kind of orthopaedic training and treatment had not previously been available among the very young. By 1933, adults and adolescents appeared to constitute an estimated 80 per cent of the 200 000 cripples in England and Wales reckoned by the CCCC.[82]

But by then the rhetoric for the extension of the programme was being tempered by diminishing professional expectations. The fact that the major sources of crippling in children – rickets and tuberculosis – were in steep decline had become widely apparent. Even by the early 1920s this was becoming clear. Elmslie, examining 2294 children in attendance at London schools for the 'physically defective' in 1921, compared his findings with the results of his similar examinations in 1907 and 1912, and observed a 'noticeable . . . diminution in the

number of children suffering from deformities due to tuberculous
disease and [an] increase in the number suffering from infantile
paralysis'. The apparent increase among the latter he saw as
'probably a relative one due to diminution of other conditions', while
the real decline of cases of surgical tuberculosis he attributed to two
causes:

> (1) that there is now much more accommodation for in-patient
> treatment of this disease, so that a larger proportion of the children
> actually under treatment are in country hospitals; and (2) that the
> general level of the treatment has improved so considerably that
> more of the children become fit for transfer to ordinary elementary
> schools.[83]

Although measures to eliminate the causes of tuberculosis of the
bones and joints in children were not undertaken until after the Second
World War, a decrease in the incidence together with greater public
attention to hygiene and diet, on the one hand, and earlier notification
and treatment on the other, effectively reduced the number of children
either dying or requiring hospitalization (Table 8.1). By the mid-1920s
it seemed to Jones that 'the days of exclusively tuberculous hospitals
for children are numbered', and he noted with approval that 'such
important institutions as Alton and Leasowe – hitherto tubercular –
have appointed orthopaedic surgeons on their staffs, and take into
their wards every type of cripple'.[84] In America too, the decline of

Table 8.1   Tuberculosis mortality in children under 15, 1898–1927
(per 100 000 living in each age group: England and Wales)

Year	Under 1 year	0–5	1–5	5–10	10–15
1898	765	364	230	62.2	51.6
1900	678	328	224	63	56.6
1905	515	280	199	65.7	55.1
1910	391	217	171	57.9	54.5
1915	289	187.9	162	61.3	63.9
1920	146	121.3	106.5	43.2	48.7
1925	125	100	92.4	35.4	39.1
1927	109	91.6	76.6	33.2	33.8

Source: GLRO:PH/GEN/3/11. Compiled by the Research Department of the
Brompton Chest Hospital, London.

tuberculosis of the bones and joints was marked; Osgood pointed out in 1927

> When I finished my medical course, less than thirty years ago, I am sure that half the beds in the orthopaedic wards of the Boston Children's Hospital were occupied by cases of bone and joint tuberculosis. At two periods lately we remarked on ward rounds that for the moment not a single bed was occupied by a case of bone or joint tuberculosis.[85]

To some extent cases of poliomyelitis could take up the slack (see Table 8.2). But the capricious and sporadic nature of that disease in Britain before the major outbreak of 1947, and the fact that it did not respond as well to therapy as rickets or tuberculosis, rendered it a less satisfactory area for professional investment and rational planning.[86] Only in London was the patient catchment sufficient before 1947 to sustain facilities. Thus Gauvain could argue in 1939 that the 'treatment [for poliomyelitis] is primarily and in fact almost exclusively orthopaedic', and then criticize the existing arrangements for the hospital organization and transfer of such patients.[87]

Table 8.2   Poliomyelitis cases in London and in England and Wales, 1926–37

Year	England & Wales		London*	
	Cases	Deaths	Cases	Deaths
1926	1158	176	85	13
1927	801	112	79	10
1928	452	112	45	8
1929	531	· 85	62	4
1930	506	108	29	4
1931	339	63	51	4
1932	656	103	79	11
1933	711	135	60	9
1934	589	82	66	6
1935	630	97	81	7
1936	528	66	36	7
1937	758	94	95	7

* 'London does not appear to have suffered any epidemics of this disease, even though districts adjoining London may have been visited by such epidemics, as for instance Broadstairs in 1927 and adjoining districts of Essex during the autumn and winter of 1938' (Enquiry, p. 10).
Source: GLRO:PH/HOSP/2/33 (Sir Henry Gauvain's Enquiry into Polio Cases in London).

In the case of rickets, the dissemination after 1925 of knowledge about the role of vitamin D in its prevention both diminished the problem and improved its treatment.[88] Unlike tuberculosis and polio-myelitis, the disease was not notifiable, so statistics are hard to come by, but the available evidence suggests a steady falling off (Tables 8.3, 8.4) even during the worst years of the Depression. This is perhaps not surprising since the government laid particular stress on nutrition and the research into it during this period precisely in order to divert attention from the realities of the impact of unemployment on health.[89] Such was the sharp end of the ideology of Preventive Medicine in which concerns with socioeconomic and physical environments came to be marginalized.[90] Except during the Second World War, ever fewer cases of rickets required hospitalization, while less severe cases were increasingly treated in light-therapy clinics run by the School Medical Service.[91]

Table 8.3   Percentage of 'defects' among schoolchildren in Manchester at routine inspection periods, 1919, 1931

	Aged 5		Aged 13		Aged 15	
	1919	*1931*	*1919*	*1931*	*1919*	*1931*
Malnutrition	1.47	0.8	1.5	0.9	0.8	0.3
Anaemia	4.69	1.2	5.4	1.3	4.2	0.7
Chronic Bronchitis	8.57	3.7	4.1	1.3	2.2	0.6
Rickets	10.10	0.9	2.7	0.3	1.7	0.001
Squint	3.52	1.8	2.9	1.2	1.8	1.0

Source: J. Meakin, 'The Health and Welfare of Manchester School Children, 1870–1945', MEd thesis, University of Manchester, 1977, p. 153.

Table 8.4   Malnutrition and rickets among Manchester schoolchildren, 1929–33

	% malnutrition	*% rickets*
1929	1.38	0.62
1930	1.26	0.56
1931	0.78	0.54
1932	0.75	0.51
1933	0.63	0.33

Source: Dr H. Herd, School Medical Officer, cited in Board of Education, *Annual Report of the Chief Medical Officer for 1934*, pp. 13–25.

At the same time that the causes of crippling in children were coming under control, orthopaedists were claiming that 'the orthopaedic scheme for the discovery, treatment, training and employment of the cripple [was] becoming a very complete one'.[92] They were right. Already by 1926 some 23 of the 40 English counties had central orthopaedic hospitals and 24 had established orthopaedic clinics. Nationally, 144 out of 318 local education authorities had adopted the 'approved scheme' for the treatment of crippled children; 45 hospitals were involved (the majority of them specifically 'orthopaedic'), 16 of which were also recognized as special schools; and 70 orthopaedic clinics had been provided by education authorities and another 80 by voluntary bodies.[93] A decade later 254 local education authorities in England and Wales were providing orthopaedic services, and arrangements existed for dealing annually with 7000 residential and 6000 day-school cripples.[94] As previously indicated, there was hardly a centre of population in Britain by the mid-1930s which did not have some medical and educational provision for crippled children.

For crippled children, this was good news indeed; less obviously so for orthopaedic surgeons. True, the orthopaedic medicalization of the crippled child had been accomplished. By 1925, the idea that 'the Hospital must come first' had triumphed over the educational side of the problem,[95] and orthopaedic surgeons had firmly established themselves as the experts in the field. Indeed, rheumatologists among others looked on enviously: orthopaedic clinics, they claimed,

> provide an object lesson of what can be done . . . , and suggest that similar methods might prove equally successful in the field of rheumatism . . . It is certainly not the least important feature of the orthopaedic scheme that it has succeeded in bringing into a harmonious working relationship the Orthopaedic Hospital, the Voluntary Hospital, the County Council, the Educational, Maternity and Child Welfare Authorities, the Medical Officers of Health; in short, every form of organisation, voluntary, municipal and state, that has for its end the treatment and conquest of suffering.[96]

But neither the overall 'logic of organization' for which Jones and Girdlestone had argued in their National Scheme, nor professional power and status for orthopaedists within the medical establishment, had been fully achieved. Partly this was a matter of incomplete organization, especially in London. But partly it arose from the professional limitations of any such scheme.

As we have seen, the Ministry of Health failed to take up the coordinating role that Jones and Girdlestone had envisaged, and although COHs with radiating clinics did indeed come to be established, their rationalization was far from an accomplished fact. Not uncharacteristic was the situation described by Ernest Hey Groves for Bristol in 1930:

Ten years ago it was urged that there existed in the district a pressing need of an open-air school hospital for chronic and curable disease, chiefly affecting children. This need is now being met by three perfectly independent bodies. The municipal authority is completing a hospital for surgical tuberculosis, a charitable organization is opening a hospital for non-tuberculous diseases, whilst an existing children's hospital is enlarging its premises and making open-air wards in order to cope with exactly the same type of case as the other two. So that before long there will be three institutions each of 100 beds competing instead of cooperating, requiring three sets of administrative officials, three sets of apparatus, and, what is worst of all, three sets of experts. This lamentable and indefensible absurdity was foreseen, and an attempt was made to prevent it, but when the providers met together it was evident that they only agreed about one thing, and that was the determination to maintain their independence, each one of the other.[97]

Well into the 1930s, there were similar complaints about competition between private and public bodies, and counter-productive jealousies between voluntary workers, hospital managers and surgeons.[98] Girdlestone's view from the spires of Oxford of 'real team work between the hospitals, general and orthopaedic', and of 'both doctors and hospitals becoming ready to give each patient the best possible service regardless of their old established and vested interests',[99] was not shared elsewhere. In Kent, where the County Council initiated an orthopaedic scheme in 1925, organized around three hospitals, opposition was mooted by general practitioners. As the MOH recalled,

In the first place, objections were raised on the ground of expense; the medical profession took the position . . . of an over-burdened taxpayer. . . . The idea got abroad that we were going to build two large main hospitals for cripples. When I pointed out that we should not need a new building, but could use an existing one, that little trouble seemed to end. Then I was up against the difficult proposition that I was suggesting another specialist service for an important county. My only truthful reply could be that the problem had never been dealt with efficiently. As to how many of the 1,300 [crippled] children had received treatment replies varied, designating treatment from a bottle of medicine to mechanical contrivances.[100]

Even more lamentable from the point of view of Jones and his colleagues was their failure to make headway in London. Here, too, independent action was the norm, and the place of orthopaedics in the major teaching hospitals little advanced. The fact that Gauvain in his discussion of poliomyelitis in 1939 laid the blame on the teaching hospitals for delays in referring cases to the orthopaedic hospitals[101] suggests that the teaching hospitals were still unwilling to acknowledge some of the orthopaedists' claims to expertise or to assist in their professional expansion. Organizationally, the ICAA continued to be the effective orchestrating body for crippled children in London, but its primary interest continued to lie with the needs of individual cases, rather than with regional schemes or the rationalizing of medical services. It was partly because of this that London had no residential school within its boundaries, so that some £35 000 had to be expended annually on transporting physically handicapped children to day schools.[102] Partly for this reason, also, nothing had come of the suggestion made by Jones in 1920 to appropriate for children's orthopaedics the ex-Ministry of Pensions' hospital for facial injuries at Sidcup.[103] The London County Council's 900-bed St Mary's Hospital for Children at Carshalton in Surrey, established in an ex-fever hospital in 1909, continued to be London's major centre for cases of crippling. Although its Medical Superintendent, W. T. Gordon Pugh, became a member of the BOA, and in 1926 was elected President of the Orthopaedic Section of the Royal Society of Medicine, Carshalton was never remotely like Oswestry. In 1926, only 40 per cent of its cases were described as 'orthopaedic', and it had no network of affiliated after-care clinics.[104]

In London, as in other urban centres, it was usual for hospital outpatient departments to serve as orthopaedic clinics, but this did not lead to an increase in the number of inpatient orthopaedic beds or to organized provision under orthopaedic control.[105] Efforts to change this situation and enlarge the professional space of orthopaedists were not helped, moreover, by opposition to the ambitions of the Liverpool school. In 1927 matters came to a head when the Secretary of RNOH alleged that at Chailey Jones was poaching crippled children from the catchment area believed to belong to RNOH surgeons, and that he was denigrating the latter's professional abilities. Embarrassed, Jones was compelled to forfeit his honorary appointment at the hospital and thus lose his toe-hold in the only major centre for orthopaedics in Britain outside Liverpool.[106]

More generally, there is reason to suspect that in Britain, as in America, consultancy work within the children's orthopaedic hospitals engendered a feeling among orthopaedists 'of isolation' and 'loss of contact'. The AOA president who voiced these feelings in 1924 urged his colleagues to remind themselves that they were 'a part of general surgery; a part of general medicine'.[107] British orthopaedists, more conscious of their ambiguous place *between* surgery and medicine, may have felt that 'social orthopaedics', while providing them with job opportunities, situated them closer to MOsH and GPs than to hospital consultants and teachers.

Certainly, the experience with crippled children had not secured for them a permanent niche in the general – especially teaching – hospitals. In peacetime, as in the Great War, British orthopaedists had proved adept at mobilizing lay support, voluntary and statutory, and so creating novel medical structures. Their scheme for children, like that for soldiers, would provide a model for would-be reformers of medicine's mainline structures. But however exemplary the orthopaedists had proved, they remained, as yet, marginal to the established body of British hospital medicine.

9

THE FRACTURE MOVEMENT*

Fractures, like crippled children, were a major interwar issue. By the 1930s, provision for their treatment – or, rather, the lack of it – was nearly as topical as tuberculosis and maternal health. The subject was extensively discussed in the press, became a subject of government inquiry, and in various ways forged new links between medicine, industry, trade unions, and local and national government. The BMA, the British Hospitals' Association, the Federation of British Industry, the Federated Employers' Insurance Association, the Trades Union Congress (TUC), and the London County Council (LCC), along with the Ministries of Health, Labour and Pensions were only some of the more visible parties to become actively involved in the issue.

This chapter concentrates on the movement among orthopaedic surgeons to gain hospital control over fracture treatment. Although this cannot be fully understood without the wider social and economic context, the present focus on professional interests is essential to understanding how and why the issue came to attract the attention it did in interwar Britain. Among orthopaedists, the campaign for the control of fractures became paramount in large part because of declining opportunities in the area of crippled children. Some orthopaedists had come to fear that 'when there will be no cripples, there will be no orthopaedic surgery'.[1] Others appreciated that if 'the gospel of orthopaedics [failed to] . . . animate more than the surgery of crippledom' it would 'remain apart [and] tend to a dead and static perfection, having the semblance but not the reality of advance, like waves on the sand of the desert!'[2] Thus the movement for the control of fractures had important implications not just for the status of orthopaedics as a specialism, but for its survival. As Harry Platt, one of the central figures in the movement, reflected: 'In the years between the wars the field of orthopaedics was to those of us then young an expanding universe, and we fought the battle for the control of

---

* Earlier versions of this chapter appeared in *Medical History* (volume 31, 1987) and in J. V. Pickstone (ed.), *Medical Innovations in Historical Perspective* (Macmillan, 1992).

fractures with gusto. . . . Our opponents often accused us of adopting the attitude of the German philosopher Nietzche [sic] – "that a good fight sanctifies a cause".'[3]

For medicine as a whole, the fracture issue signified and illuminated crucial changes in the division of labour and in the diffusion of ideas on economy, efficiency and expertise. The effort to organize fracture treatment was part of a more general move to place accidents and trauma on a par with disease entities. But because the treatment of accidents and trauma – like orthopaedics itself – cut across the organ geography of other medical and surgical specialties, these changes were to raise serious problems of conceptualization as well as of hospital organization. Thus the campaign for the efficient treatment of fractures in the interwar period was to become central to the politics of British hospital and health services reform.

PLATT AND THE PILOT FRACTURE CLINIC

The roots of the postwar orthopaedic involvement in the fracture movement can be traced back to the work of Harry Platt at the Ancoats Hospital in Manchester. Along with Meurice Sinclair (St James's Poor Law Hospital, Balham, London), Platt was an exception to the general picture of the loss to orthopaedics of fracture treatment after the First World War. But whereas Sinclair was marginal to mainstream orthopaedics before the 1930s, when he would have considerable influence on LCC thinking around fracture services,[4] Platt was at the very heart of the specialism. A founder member of the BOA, he saw his work at Ancoats as a demonstration of orthopaedic specialization from the start. In this, as in much of his other work, he was greatly influenced both by Jones and by American models.

After graduating in medicine at Manchester in 1911, Platt had gone to London to 'round off' his education by gaining experience at the specialist hospitals of St Peter's (genito-urinary), St Mark's (colo-rectal), and the Royal National Orthopaedic.[5] In seeking special skills at these non-teaching hospitals in London, Platt was like other provincial would-be consultant physicians and surgeons at the time. He was also typical in not seeking to become a specialist. Nevertheless, it was while he was at the RNOH in 1913 that he was invited by Elliot Brackett, then the head of the Orthopaedic Service of the Massachusetts General Hospital, to further his training in Boston. This was a more unusual experience and it was to prove decisive in Platt's turning to orthopaedic specialization and to his willingness to face lean years as a prelude to a successful consulting practice.[6] His 'Orthopaedic Surgery in Boston', written just before he returned to Manchester in April 1914 to secure the appointment of honorary consultant surgeon to the Ancoats Hospital, well reflects the enthusiasm he acquired not only for Boston and orthopaedics, but for the reformist thrust of American medicine and surgery in general.[7] Though he lived to be 100 Platt never lost those enthusiasms.

The Ancoats Hospital was a small voluntary hospital – a typical place for a junior appointment. But in many ways it was ideally suited to a surgeon with orthopaedic interests. Both of the other surgeons, John Morley (later Professor of Surgery at Manchester University) and W. R. Douglas (later a renowned cancer surgeon at Manchester's Christie Hospital) were also young, ambitious and willing to try out new ideas and techniques. Here, too, Dr A. E. Barclay set up Manchester's first radiology department.[8] The location of the hospital, in one of Manchester's most populated industrial districts, rendered it essentially an accident hospital with a large turnover of fracture cases.[9] And, not least important, the Hospital had a relatively liberal board of managers.

Ancoats was thus a propitious setting in which to try out American-style reforms, and within months of his appointment at the hospital Platt was conspiring with Morley and Douglas to do just that. Instead of following the traditional routine, in which the surgeon on duty took responsibility for all incoming cases, the three colleagues divided up the cases (and the hospital's 75 beds) according to surgical specialties.[10] Thus Platt was able to establish therapeutic control over all incoming fracture and other orthopaedic cases, and establish uniformity in their treatment.

Interestingly, none of this brought rebuke from the lay managers of Ancoats, although when he was first appointed Platt had had to promise not to indulge *only* in specialist work. Since the specialization he was proposing could be seen as meeting the needs of the public – especially the needs of the locality's industrial workers whose contribution schemes were an important source of the hospital's revenue[11] – the lay governors had little reason to fear that the specialization would be only self-serving. It may have been important, too, that the governors could see the specialization as emerging from the cooperative efforts of their own consultants, rather than as something imposed from the outside challenging their autonomy. The Secretary of the hospital, far from expressing worry over the new division of labour, was apparently more concerned about the effect of orthopaedic specialization on Platt's own financial future, the limited scope for private practice being appreciated. This negative economic aspect of orthopaedics may also help explain the willingness of Morley and Douglas to comply with Platt's reformism, as may the specialism's reputation as an unchallenging area of medicine and a professionally risky career option. Evidence of incompetence in orthopaedics, unlike other areas of surgery, was visible and enduring – a single case of malunion having the potential to ruin a surgeon's reputation.

The strategic potential of small hospitals like Ancoats for rising specialisms was also illustrated at the nearby Salford Royal Hospital, where Platt's friend and wartime colleague, Geoffrey Jefferson, was allowed to concentrate on neurosurgery.[12] But it was much more difficult for orthopaedists to carve out specialist niches in the major teaching hospitals where their interests threatened the territory, incomes and intellectual rationales of senior general surgeons. As we have noted, the orthopaedic departments established in some of the London hospitals just before the war were mainly for chronic cases, rather than for the victims of accidents, and were under the control of those who sought to become general surgeons first and foremost. One such defender of this tradition and one of the staunchest opponents of the orthopaedic specialists in the 1920s was George E. Gask, who in the 1900s had himself been Chief Assistant to the Orthopaedic Department at St Bartholomew's Hospital.[13]

During the war Platt was involved with Manchester's military orthopaedic centre at Grangethorpe; his plans for Ancoats were delayed a little, but his experience was enlarged and his commitment to 'progressive orthopaedics' strengthened.[14] After the war, between 1919 and 1921, he was able to establish at Ancoats what is now generally regarded as the world's first segregated fracture service under orthopaedic control. Essentially, this carried over into civilian practice the main wartime lessons of effective fracture treatment: segregation, expert supervision, team work, continuity of treatment and appropriate after-care. To these features Platt added detailed record-keeping on the social and medical condition of patients, a technique he had learned in Boston.

Such were the basic principles of the 'ideal fracture service' reported by Platt in the *Lancet* in 1921;[15] they were to be reiterated in the barrage of papers and lectures issued by orthopaedists from the mid-1920s to the 1940s; and they were to receive the backing of the medical establishment in the BMA's highly influential Report on Fractures

(1935) – a report written mainly by orthopaedists during the time that Platt was President of the BOA.[16] As a result of the BMA's Report, a motion was passed in the House of Commons in April 1936 which led to setting up the government's Inter-Departmental Committee on the Rehabilitation of Persons Injured by Accidents, under the chairmanship of the distinguished civil servant Sir Malcolm Delevingne. Through the Delevingne Committee's interim and final reports of 1937 and 1939, official endorsement was given to the principles first implemented by Platt.[17]

CAMPAIGNING FOR THE CAUSE

How did orthopaedists argue their case for the hospital treatment of fractures? In general they appealed for a return to the wartime organization of fractures or, as they put it, the organized application of knowledge already possessed. This was in part a tilt at the old-guard generalists in surgery who, in the orthopaedists' view, had stolen their wartime heritage and built an army of cripples in their ignorance of good orthopaedic practice. It was also an attempt to forge an alliance with GPs on whom orthopaedists relied for referrals. These issues came together in a lecture that Jones delivered in Liverpool in May 1925 and subsequently published in the *BMJ*.[18] The lecture was prompted by Platt and by Platt's orthopaedic colleague, Rowley Bristow of St Thomas's Hospital (the person, according to Platt, who was largely responsible for spearheading 'the attack on the idealogical [sic] barricades of the London teaching hospitals').[19] Although in many ways Jones only reiterated the complaints against fracture treatment that had been heard intermittently since the middle of the nineteenth century, friend and foe alike came to regard the lecture as a 'slashing attack on the "Methods by which fractures are dealt with at the big teaching hospitals".'[20]

Jones opened his lecture with the observation that the existing treatment of fractures was 'a blot upon our surgical escutcheon'. 'Hopelessly wrong', he maintained, was the situation in 'the big teaching hospitals' where outpatient ambulatory fractures were treated by unsupervised junior medical officers and where inpatient fractures – of little interest to general surgeons who resented them for blocking beds – were left to the care of under-trained house-surgeons. 'Even more pathetic', he argued, was the fate of fracture cases sent to Poor Law infirmaries, most of which institutions were totally unfit for the modern treatment of fractures. It would, he concluded, be 'far better for our hospital authorities to say, "We are not prepared to treat fractures", than that they should take on responsibilities which they cannot meet. . . . We cannot disguise the fact that great numbers of adult cripples are manufactured by want of adequate provision.'

Such 'plain speaking', as Jones himself called it, struck an alarmingly familiar chord among some. Jones protested that it was 'not a mere matter of the encroachment of specialism, but a call to our sense of proportion and sense of duty', but his polemics clearly hinted at a return to the wartime division of medical labour, with organized clinics under specialist control.

The lecture was meant to raise controversy, and Platt and his friends ensured that it did. Among other things, they arranged for the subject to be discussed at a special session on Surgery and Orthopaedics (chaired by Moynihan) at a BMA meeting in Bath in August 1925.[21] An illustrious audience of orthopaedists, some from as far afield as Boston, Montreal and the Mayo Clinic, were called upon to attend.[22] It was here that George Gask made his début as the defender of generalism over surgical specialism. Gask, then Director of the Professorial Surgical Unit at St Bartholomew's Hospital, was hardly a conventional opponent of specialization. Nevertheless he rose to challenge the assumption that greater efficiency would result from greater division of labour.[23] His retort was that the extended division of labour which served the needs of the munitions factory hardly applied to the manufacture of good general practitioners.[24] But the orthopaedic ideologues were not much interested in the proper education of GPs; on the contrary, they linked bad fracture treatment with the fact, made clear at the meeting, 'that probably one half of the fractures in this country were treated at home or in cottage hospitals by general practitioners', and concluded that fractures needed to be treated under their own expert supervision.[25] Gask was made to emerge from the proceedings at Bath as the 'quixotic [defender] . . . of a losing cause'.[26]

The 'incompetence' of GPs in fracture treatment was to figure in some of the rhetoric for specialist fracture treatment in the 1930s.[27] But this line of argument never moved to centre-stage (nor, unlike in America, were statistics gathered on the extent of GP fracture treatment, or on their clinical results).[28] Since orthopaedists relied on GPs to refer crippled children to their clinics, there was reason not to antagonize them.[29] Indeed, orthopaedists produced primers on modern fracture treatment for their use and, in time, spawned a vigorous market for such texts.[30] GPs, for their part, did not generally feel threatened by the advent of fracture specialists.[31] 'Panel' GPs could refer a patient to an orthopaedic specialist without losing that patient from their patient list and, hence, without forfeiting National Insurance remuneration. Overworked GPs were also mindful of patients' desires for specialist care, of the complications involved in difficult fracture cases, of the legal and professional risks in cases of mal-union, and, not least, of the fact that complicated cases entailed extra costs which might go unremunerated from local National Insurance Committees.[32] More idealistic GPs could construe the orthopaedists' campaign for fracture services as akin to their own interest in improving community medical services. Sir Henry Bracken-bury, for example, who spoke for GPs and who served on the BMA's Fracture Committee, seems to have seen orthopaedists and GPs as allied underdogs fighting together against the reactionary general surgeons who were powerful in hospitals.[33] It is perhaps unsurprising, therefore, that over half the cases treated at the fracture clinic at the City Lodge Municipal Hospital in Cardiff in 1937, for instance, were sent by private practitioners.[34] By 1939, the Delevingne Committee was of the opinion that only a 'small number of fractures were treated by general practitioners', and the Committee was confident that the numbers 'may be expected to decrease in proportion to the spread of fracture schemes'.[35]

The orthopaedic advocates of fracture clinics had reason, then, to continue to believe that the old-guard general surgeons were their main antagonists. Confirming this belief, Dawson Williams in a strongly worded editorial in the *BMJ* in 1927 abjured as 'baseless' the 'stock argument used by the opponents of specialization . . . that by transferring any department of surgery from the general surgeons to the specialist, the student might become indifferent to the specialty and cease to receive instruction in it.' 'The plain fact is', he insisted, 'that, as things are at present in London, the student gets little instruction in the treatment of fractures . . . for the simple reason that such cases are not usually admitted to the wards of the great teaching hospitals.'[36]

But gradually this adversarial tone was muted and head-on confrontation over specialization and education avoided. Instead, attention was directed to the issue of the practical means by which to achieve the best possible results in the treatment of fractures. This was how Jones had pitched his reply to Gask in 1925, and by 1928, when the budding proponent of fracture services, Ernest Hey Groves, delivered his presidential address to the BOA 'On the Treatment of

Fractures', it was this practical matter that was construed as 'the problem'.[37] The solution, it was increasingly stressed, was 'not new knowledge', but rather, 'the organized application of the knowledge we already possess', namely, the wartime 'segregation of cases, the training of team workers, and the systematic tabulation of results'.

This call for a return to wartime practices underlines what was largely absent from the campaign for fracture clinics – the rhetoric of new technologies. Although fracture clinics made use of certain technologies, just as they relied on a variety of surgical skills, such technologies were not central to the professional or therapeutic identity of orthopaedists or to their call for the hospital treatment of fractures. Within that call, techniques for the internal fixation of fractures – such as had been central to the prewar debate over the proper treatment of simple fractures – scarcely featured.[38] Nor did X-rays, which did not become routine in hospital fracture treatment in Britain until the late 1930s.[39] Much of the orthopaedic armamentarium was in fact remarkably mundane – so much so that it was possible for plaster of Paris to symbolize modernity. At the heart of the orthopaedists' specializing strategy were not techniques and mechanical gadgets, but scientific management and the recasting of networks of professional power and control.

THE CAUSE IN CONTEXT

Jones's lecture and the ensuing debate at Bath are illustrative of how specialization was negotiated in interwar Britain. But wider issues were also involved. Close to the surface of the debate, and in the fabric of its rhetoric, lay the pressing question of the reform of the health services.

The early 1920s were, as we have previously observed, a period of retrenchment in health care, but the financial crisis that lay behind the cutbacks also provided a context for strong arguments for rationalization. In health services, as in agriculture, transport, industry and trade unions, wartime experience had cemented a faith in the application of system and uniformity to problems of complexity and 'waste'. In 1921, even the Voluntary Hospitals' Committee under Lord Cave, which recommended state subsidies to voluntary hospitals in order to shore up the existing system, urged the hospitals to take more concerted action in the future and, among other rationalizing measures, to standardize their accounting procedures.[40] In the following year, in a pamphlet on *The Labour Movement and the Hospital Crisis: a scheme for a hospital service*, the TUC and Labour Party put forward a sweeping programme for the equitable distribution of health provision under a state medical service. The pamphlet's social and political impact is open to question,[41] but one notes the overlap with the rhetoric of orthopaedic rationalizers in references to 'continuity of treatment', 'team work', and the 'unification of control' over all hospital facilities and services. It is also difficult not to note the striking difference between the priorities for rationalization in this document compared to that by Dawson in 1920. In the Dawson Report, it will be recalled, the emphasis was on a restructured service in which suitably retrained GPs would provide primary medical care and act as filters for referring patients up the health-care hierarchy to hospital consultants. Focused on GPs, the Dawson Report made little reference to accident and emergency services, and its only reference to ambulances was in connection with transporting patients from primary to secondary health centres.[42] In the TUC and Labour Party's vision, however, accident services occupied the front line of medical care.[43] Indeed, the promotion of a unified, coordinated accident service provided an effective means of criticizing the unequal, haphazard and confused hospital system, and of pointing to the advantages of a coherent state organization. Here, as in few other places in medicine, the need for regionally coordinated and rationalized facilities could be shown to be as vital and urgent as it had been in the war. And, as in the war, doctors would need salaries, for there was little scope for private practice in the treatment of acute injury.[44]

The case for the rationalization and reform of accident services also found specific, if less overtly political, expression in 1924 in a report prepared by a special committee of the King's Fund. The *Report on the Disposal of Ambulance Cases* dealt only with London and was primarily concerned with the costs, distribution and availability of hospital beds for accident cases and with the proper 'relationship to be observed between . . . rate-supported institutions and the Voluntary Hospitals'. Nevertheless, it exposed the nation's capital as lacking an effective accident service.[45] The LCC's Ambulance Service, which by 1923 was attending some 24 626 calls a year with its seven ambulances, was constantly faced with the problem of where to take patients. The majority of Poor Law infirmaries had no accommodation for accident cases, while most of the large voluntary hospitals were grossly underprovided. (At the Charing Cross Hospital, for instance, where 980 accident cases were received in 1923, there were only four accident beds.)[46] When contrasted with the coordinated accident services in some American cities, and with the elaborate system for dealing with fracture cases at the Massachusetts General Hospital since 1922,[47] the situation in London appeared antediluvian. Moreover, such poor accident facilities threatened to become an embarrassment, compounding the already existing financial problems of the voluntary hospitals. The Liberal MP and Chairman of the Lambeth Board of Guardians, Frank Briant, pointed out to the King's Fund Committee, 'Of course, quite wrongly, the average person in the street imagines the Hospital exists for accidents. To a large extent I do not think it does; but I think it would have a very bad effect upon the finances of Hospitals if the general public had a general idea that accidents had not to be taken there, or they would not deal with them.'[48] Although the committee of the King's Fund stuck to its limited brief, it was evident from their report that there was an urgent need for 'uniformity of procedure', and for the coordination of accident services, if not for the greater control of these services under a single specialist group.

However, there was little hope of implementing such changes in the financial climate of the early 1920s. Not until the 1930s did possibilities open, less as a result of economic improvements or of stepped-up political interests and commitments to rationalization, than of mounting concern over the incidence, handling, and costs of accidents. Particularly important were accidents involving motor vehicles. It is ironic, of course, that motorization, which in the form of ambulance transport was vital to the postwar plans for accident services, was now to be regarded as a major cause of the injuries that were to require speedier transport. 'Before the era of the motor car', commented the *Lancet* in 1926, 'the receiving officer of a large general hospital might be embarrassed by a run of Pott's fractures on the first snowy evening of the year; today a fine Sunday evening may overwhelm any cottage hospital with a glut of complicated injuries.'[49] Major surgical problems, it seemed, were now more likely to occur miles away from essential facilities.

For England and Wales non-fatal street accidents involving motor vehicles increased four-fold between 1913 and 1932, from 38 000 to 162 000 (Table 9.1), with the steepest rise (unsurpassed until the 1950s) occurring between 1927 and 1934. And, as the Ministry of Transport was acutely aware in 1928, 'There are few questions, as reference to the daily press will show, which excite more constant and widespread interest.'[50] Although the number of domestic and industrial accidents was greater, the sociomedical profile of road accidents was considerably higher, in part because these accidents were not confined to

Table 9.1   Fatal and non-fatal street accidents caused by vehicles 1913–32

Years	England & Wales Fatal	England & Wales Non-fatal (so far as reported)	Scotland Fatal	Scotland Non-fatal (so far as reported)	Northern Ireland* Fatal	Northern Ireland* Non-fatal (so far as reported)
1913	1 743	38 050	219	3 240	137*	1 254*
1919	2 239	43 305	249	3 957	140*	1 436*
1920	2 386	49 317	318	4 417	133*	1 176*
1921	2 328	55 153	350	4 790	167*	1 329*
1922	2 441	62 504	327	4 987	Information	
1923	2 694	74 290	285	5 832	not available	
1924	3 269	87 867	362	6 717	79	1 004
1925	3 535	102 704	436	8 798	59	1 239
1926	4 236	108 846	567	10 638	73	1 321
1927	4 581	117 239	614	11 509	100	1 338
1928	5 353	129 199	625	12 405	110	1 357
1929	5 817	132 529	688	12 767	124	1 293
1930	6 317	136 077	757	13 642	113	1 344
1931	5 855	159 257	644	15 321	111	1 471
1932	5 800	161 952	687	15 567	118	1 647

Note: In 1913 accidents caused by pedal cycles are not included.
* Inclusive, in 1921 and prior years, of particulars for territory which is now the Republic of Ireland.
Source: *Board of Trade: Statistical Abstracts*, HMSO, 1934, p. 293 (compiled from the Annual Returns to the House of Commons).

specific 'hospital areas' and therefore presented severe obstacles to medical planning.[51]

Allied to this troublesome geographical aspect of motor vehicle accidents was the financial spectre they raised: the problem of recouping from insurance companies the costs of treating the victims, especially those injured away from home. The BMA reckoned that doctors were paid in only one out of five cases, and in 1931 it was estimated that some 25 000 victims of motor accidents treated as inpatients in voluntary hospitals had cost nearly one-quarter of a million pounds.[52] The problem was taken up by Moynihan in the House of Lords in 1933 and it was partly solved by the Road Traffic Act of 1934.[53] But the general issue of recovering costs for accident victims who were covered by insurance for *other than* hospital medical services remained contentious. It was partly because of this problem that widespread interest arose in the Vienna Accident Hospital.

Erected in 1925, at a cost of £10 000, the Vienna Accident Hospital was a fully-equipped 125-bed establishment which was entirely maintained by the Austrian National Insurance Company at a cost of around £16 000 per annum. The idea for the hospital belonged to Lorenz Böhler, a general surgeon who had developed a special interest in the treatment of fractures during the Great War.[54] Like Platt and his colleagues, Böhler had become a propagandist of segregated fracture treatment, and by the late 1920s, a visit to his 'mid-European fracture synod'[55] had become a badge of progressivism among orthopaedists – a badge often signified by the adoption of Böhler's methods of plaster of Paris fixation. (Platt and Bristow began this practice after their visit to the hospital in 1929 on behalf of the BOA.)[56] Like an earlier generation of visitors to Jones's clinic in Liverpool, those who went to 'Böhler's Clinic' were impressed by his various surgical and manipulative techniques, many of which were praised as 'revolutionary', but they were even more astounded and delighted by his meticulous organization.[57] Böhler himself regarded the latter as the secret of his success, and specifically likened it to the progress of rationalization that in recent years had rendered industry, transport and agriculture more efficient. It was by means of the subdivision of the labour process, he insisted, that the parts of medical treatment were simplified – thereby suited to mass application – and greater efficiency achieved. In Böhler's clinic 'team work' reigned supreme, with all the steps before and after the surgeon's work carried out by nurses, junior doctors and orderlies.[58]

Above all, it was the economic advantages deriving from this organization that gained Böhler his fame – his impressive accumulation of statistics to prove this point being unique for the time. As the *Lancet* hastened to comment in an editorial in 1926, not only had Böhler shown the clinical and social advantages of specialized fracture treatment in restoring patients to their full earning capacity, but he had also demonstrated to the insurance companies that surgical specialization could render enormous savings. That the Vienna Accident Hospital had 'saved the insurance societies something in the neighbourhood of £18,000' was just the sort of 'astonishing' fact to stimulate interest in the whole question of accident services and, at the same time, draw attention to the 'grave scandal' that in Britain the insurance companies 'contribute nothing towards the treatment of their injured clients in hospital'. 'Both equity and self-interest alike', the *Lancet* maintained, 'should lead the [insurance] companies to support an accident department in every hospital'.[59]

Not surprisingly, it was precisely this conclusion that was taken up by the orthopaedic advocates of segregated fracture services. Ernest Hey Groves, who was more insistent than most on the need to convince British insurance companies 'that organisation of fracture treatment would effect so much saving in compensation as more than to cover the cost of the treatment',[60] translated Böhler's work on fractures into English. In his preface he noted that Böhler had 'demonstrated that the proper treatment of fractures is not only a scientific problem or a philanthropic duty, but also a business proposition. In other words, it pays to treat fractures well!'[61]

The Austrian National Insurance Company which financed the Vienna Accident Hospital in fact maintained a state monopoly on workmen's compensation insurance; Austria had no private insurers in this field as Britain did. To a degree, therefore, the use of Böhler's Clinic as a model for the economic management of accidents in Britain was inappropriate, and it was necessary to take some care in its rhetorical deployment. Additionally, there were two obvious reasons why the British proponents of segregated fracture services needed to be highly selective in their use of Böhler's model. First, the Vienna Accident Hospital was intended not just for fractures, but for accident cases of all kinds. References to the hospital did not therefore automatically serve the interests of those seeking to expand their professional space on the basis of fracture treatment alone (which may be why orthopaedists always referred to the hospital as 'Böhler's Clinic'). Although fractures predominated over other injuries in certain industries and industrial regions in Britain, in general only about 10 per cent of injuries involved fractures.[62] Thus arguments for the segregation of accident cases did not necessarily promote a commanding role for orthopaedists, nor even necessitate their employment in accident services.

The other reason why orthopaedists had to be selective in their use of the Böhler model was that the Vienna Accident Hospital was an institution separate from other hospitals and medical schools. As such it hardly furthered the orthopaedists' ambition of securing a permanent niche *within* the major teaching hospitals. The last thing that British orthopaedists wanted (as one of them put it) was to end up like Böhler – 'cut off entirely from general hospitals and cut off entirely from the general surgical and medical staffs [and students]'.[63]

Yet it was not by rejecting outright the idea of accident hospitals that those interested in segregated fracture treatment established the priority of their claims. Rather, it was by putting accident hospitals forward as one option among others that they exploited the wider social and political issue of accident services, and also served their own interest in hospital fracture clinics. While explaining the difficulties involved in establishing separate accident hospitals they revealed the economic viability of 'accident departments', meaning primarily hospital fracture clinics under orthopaedic control.[64] This was how Hey Groves presented the case for fracture clinics to the LCC in the mid-1930s.[65] Around the same time, in an address on 'Broken Bones and Money Wasted' to the National Safety Council, he proceeded further along the road to making an economic virtue of necessity: 'To avoid all the muddle and inefficiency it was not necessary first to provide a large special hospital devoted to accidents. The principles of organisation were comparatively simple and cheap; they required no new buildings nor any capital expenditure.'[66] Possibly it was because of a perceived threat to this attractive solution to the problem that the LCC in 1936 regarded the author of an alternative plan for 'Units [of 100 to 150 beds] for Traumatic Surgery' *instead of* 'fracture clinics', as an idealist, and branded him a bad character interested only in furthering his own career. The 'idealist' was the young medical officer at St James's Hospital, Balham, the Australian, William Gissane; having offered to take a £200 cut in his salary in order to be able to direct such a unit, Gissane explained that his 'present enthusiasm [was] not a mushroom growth following a hard working four week visit to Dr Lorenz Böhler in Vienna, but that holiday showed me the Hospital of my ideals as a reality'.[67] In 1941 Gissane's dream came true with his appointment to the new Birmingham Accident Hospital – the only accident hospital ever created in Britain, and itself something of an accident.[68] But this was after the establishment of the Emergency Medical Service during the Second World War – that is, after government backing had already been secured for hospital fracture services under orthopaedic control. In the late 1930s a few orthopaedists began to argue for central hospitals for traumatic surgery and rehabilitation where, crucially, orthopaedists would be in charge and where teaching could be conducted.[69] But it was not until after the Second World War that members of the BOA began collectively to push for the national provision of what they called 'Orthopaedic and Accident Services'.[70] Only thereafter could be said, as by Platt in 1950, that 'the Vienna experiment . . . has proved that *within the framework of comprehensive orthopaedic schemes* there is a place for the accident hospital'.[71]

STATE AID AND VOLUNTARY PRINCIPLES

From the very start, the orthopaedic advocates of fracture clinics had sought to obtain financial and moral support from a variety of sources. Platt, in his 1921 article in the *Lancet*, had concluded that 'It is unnecessary to elaborate the obvious economic importance to industry in general of the efficient treatment of fractures. Those who are engaged actively in this work are aware of the fact that employers, insurance companies, and trade-unions are alive to the necessity for reform.' The BMA's 1935 Report on Fractures made a similar pitch, emphasizing in particular the economic importance to industry of improved fracture treatment.[72] How industry and trade unions responded to these appeals will concern us in the next chapter; here we wish to consider only some of the implications of the involvement of the state, especially for the voluntary hospitals.

Through their work with crippled children, orthopaedists were of course already accustomed to state support. In the case of some of the municipal hospitals (that is, those taken over from the Poor Law through the Local Government Act of 1929), local authority-funded fracture services were already in operation by the early 1930s. Orthopaedists encouraged and collaborated in this development, but it failed to meet their ultimate objective of securing a definite place for modern orthopaedics in the prestigious voluntary hospitals. One of their proposals, therefore, was that the local authorities 'subsidize the voluntary hospitals for this service from public funds'.[73] At the very least, they argued, the voluntary hospitals should enter into collaborative fracture schemes with the municipal hospitals.

Just as in the prewar period, however, the voluntary hospitals were reluctant to budge. Hard-pressed for funds, and without a legal basis for means-testing (unlike the municipal hospitals), most of them preferred to develop services for patients who could be expected to contribute to costs. Hospitals like that at Ancoats, which derived the bulk of their income from industrial workers, were responsive to workers' needs, but industrial injuries remained a low priority in the elite voluntary hospitals. 'One almost fears that nothing short of some social cataclysm, such as Communism, bankruptcy, or war, will be strong enough to break old prejudices', complained Hey Groves in 1933 after a frustrating and futile attempt to secure the cooperation of the voluntary hospitals in Bristol.[74] Meanwhile, in London, cooption rather than cooperation was occurring, with some of the larger voluntary hospitals arranging to send their fracture cases to LCC hospitals where segregated services had already been established, as an alternative to developing their own services.[75]

Faced with the vested interests of the voluntary hospitals, orthopaedists looked instead to the state. But the approach was pragmatic rather than ideological: continuing to occupy a middle ground between voluntarism and statism, orthopaedists advocated 'State aid, which does not involve State control'.[76] Like most hospital consultants, they dreaded bureaucracy, but they also wanted to be free of the lay managers of hospitals who could still use 'any excuse or justification for behaving as though they controlled a proprietary institution'.[77] State aid was seen as a means to bringing the voluntary hospitals further under the direction of 'medical experts', as well as a means to gaining new equipment and buildings.

The occupation of this middle ground was not peculiar to orthopaedists; other specialists adopted similar demands, but it was the orthopaedists' interests, above all, that were served by the argument for state aid, since they had so little scope for private practice. As was pointed out during discussions for a proposed fracture service in Dundee in 1944, the financial situation of orthopaedists was almost unique among specialists: 'the Orthopaedic Specialist has a far higher percentage of charity work than any other clinician – i.e., any clinical lecturers may pick up a fair amount of private work in general surgery but your orthopaedic colleague gets very very little as operations on long stay cases simply cannot be paid in private and consultations are scant'.[78] Other consultants could derive some income from the voluntary hospitals when they utilized pay-beds, and they sometimes received remuneration from the voluntaries through contributory hospitals' pre-payment schemes. But there was little such money for orthopaedists, and beyond it – in private practice – even less. Thus to establish and maintain the niche they sought in the voluntary hospitals, they needed to secure the principle of salaried specialists, at least on a part-time basis. But the idea of salaried consultants was virtually unthinkable within the voluntary hospital sector. Typically, at the meeting in London in 1938 to discuss fracture clinics, the voluntary hospitals' lobby cautioned against the appointment of full-time officers on the grounds that 'there would not be adequate competition between them and that this would ultimately be detrimental to their efficiency'.[79] Even in the LCC's municipal hospitals it was not until the mid-1930s that the employment of salaried specialists was begun on a small scale.[80] It is hardly surprising, therefore, that Hey Groves and other orthopaedic spokesmen in the 1930s did not explicitly refer to the state remuneration of fracture 'experts' when they advocated fracture clinics. To have done so would have weakened their case, and highlighted the fact that 'the voluntary hospitals are being called upon to play a prominent part in what is virtually a new type of service and that payment of whole-time directors or registrars will inevitably involve additional expense'.[81]

Only in the late 1930s, when the economic climate had significantly improved – except within the voluntary hospitals – were orthopaedists willing to go much further in their campaign for fracture clinics in the voluntary hospitals. Then, partly as a result of the alliance they formed with the TUC and the Labour Party, they were encouraged to become, as Frank Honigsbaum has noted, 'the "radicals" of the medical profession', pressing for the reorganization of the health services as a whole. Indeed, in the 1940s, as a result of their experience in seeking the hospital control of fractures, they were to emerge 'almost alone among consultants [in welcoming] . . . proposals for the state control of the voluntary sector'.[82]

EFFECTS AND MEANINGS

If measured by the number of hospital fracture clinics established before the Second World War, the orthopaedists' campaign for fracture services would have to be reckoned, at best, only a partial success. In 1937 the Delevingne Committee announced that there were 'many indications that a widespread movement for the establishment of fracture clinics has begun and is likely to make rapid headway'. In its final report of 1939, however, the Committee was forced to confess that

> progress in the general application of [the principles of fracture segregation] has not been as rapid as we hoped. The matter had been taken up in a number of places . . . [but] the hope that a *general* movement had been started and would be carried through by the hospitals themselves, both voluntary and municipal individually or in co-operation, has not been realised.[83]

By this date only four out of twelve London teaching hospitals had fully-developed fracture clinics, and there were only 74 fracture clinics in the country as a whole, 17 of which were in industrial Lancashire.[84] Among the latter was the clinic established by Platt at the Manchester Royal Infirmary in 1936, four years after his appointment as the infirmary's first honorary orthopaedic consultant. Within the LCC's hospital system the fracture unit at St James's Hospital, Balham, remained the only 'outstanding example'; there, however, it was decided that the work of Meurice Sinclair and William Gissane could be dispensed with and a three-session-a-week service instituted under a single specialist.[85] Elsewhere, 'partly organized' clinics were the norm, full implementation being held back by financial constraints and continuing disparities between hospitals, as well as by foot-dragging on the part of the Ministry of Health[86] and some local authorities, and by the shortage of qualified experts willing to undertake the work. To some extent, also, the idea of fracture clinics was overtaken by the ascending ideal of specialized industrial rehabilitation centres, where greater continuity of treatment could be expected than in most of the voluntary hospitals.[87] Thus the 'semblance of advance' in the area of fracture treatment under orthopaedic control may have been greater than the reality. In fact, it was not until the late 1940s and 1950s that orthopaedists in Britain fully succeeded in taking over from general surgeons the treatment of fractures.[88]

Yet in the final analysis, the findings of the Delevingne Committee may be less significant than the fact that, despite its brief – to report on provision for 'the rehabilitation of persons injured by accidents' – the Committee chose to concentrate exclusively on the issue of the efficient organization of fracture services.[89] Like the BMA's Report on Fractures, the Delevingne Reports, by reiterating the orthopaedists' social, therapeutic and economic rationales for efficient fracture treatment, legitimated the professional self-interests behind them. Not only did the Delevingne Committee justify the control of fractures in the hands of 'fracture experts',[90] but, by complying with the notion that the voluntary hospitals were the most appropriate place for fracture clinics, the committee also conferred the basis for the status and authority that the orthopaedists had been seeking. Furthermore, it provided a means of reproducing that authority by recommending proper undergraduate training in fracture treatment. Finally, by recommending 'departure from ordinary practice in respect of . . . remuneration' in the form of honoraria of between £300 and £500 per annum to the surgeon-in-charge of a fracture clinic, the Committee took a step towards accepting a salaried service for orthopaedists within the voluntary sector.[91]

But the Delevingne Reports have an importance beyond the professional interests of orthopaedists. They stand, with measures such as the Cancer Act of 1939 (which compelled local authorities to develop regional schemes for cancer treatment), as evidence of increasing government commitment to an organized, statutory health service which included medical specialists and their work in voluntary hospitals. Previously, where government had been involved in health-care activities (such as tuberculosis schemes and those for maternity and child welfare), the concentration had been on 'public health' conceived largely in terms of preventive, personal health services. By the 1930s, however, the focus of development for central government and for many MOsH lay with curative services (though these might still be construed as 'preventive'). Included within the negotiations for such services was the development of municipal hospitals, their staffing with consultants, and their relations with the voluntary hospitals. In these discussions, which lasted throughout the Second World War, 'medical rationalizers' played a key role – some were MOsH, some were medical academics, and several of the most active were specialists who sought hospital rationalization in order to develop more widely available services. Among the specialists, orthopaedists were conspicuous – an important example being Harry Platt, who was active on Manchester's Joint Hospitals' Advisory Board before becoming involved with the Nuffield Provincial Hospitals' Trust. That one of the first acts of the Joint Board in Manchester was the implementation of a city-wide fracture scheme, and that the first report of the Nuffield Trust outlined a plan for a unified accident service[92] is not, however, simply an indication of the influence and interests of Platt. His career was typical of several contemporary would-be consultant specialists who similarly moved through the small voluntary hospitals to regional hospital centres to involvement with the organization of the National Health Service.[93] In considering these parallel careers what emerges is the *general* importance of the orthopaedic case for fracture clinics in arguments for the reorganization of hospital medicine and the entry of outsiders. Accidents, like cancer, were on a new frontier of 'public health' – a frontier that was part and parcel of the argument for rationalization and specialization. From this wider perspective, the reports by the BMA and the Delevingne Committees appear less as orthopaedic reports in official dress, than as instances of the reformist stratagems of consultant specialists. In this sense, at least, the fracture movement did indeed 'sanctify a cause'.

10

INDUSTRY AND LABOUR, PART II REHABILITATION AND THE ASSAULT ON TRAUMA, 1930s

Rehabilitation: the restoration of privileges; of reputation; or of proper condition . . . for earning a living or playing a part in the world. (*Concise Oxford Dictionary*)[1]

The fracture issue in interwar Britain was always about more than the most appropriate place for treating broken bones. While the medical profession saw it chiefly in terms of the reorganization of services and intra-professional relations, outside the profession it was increasingly implicated in wider sets of socioeconomic and political concerns. As the economy began to recover and unemployment to lessen, so the idea of rehabilitating injured workers to 'fitness' in order to avoid 'wastage' gained a purchase in social and economic thought which far transcended that briefly obtained during the Great War and momentarily thereafter. By the early 1940s, rehabilitation was being spoken of as 'the one fashion which dominates medical thought almost to the exclusion of any rivals'.[2] By then the manpower demands of the Second World War had quickened interest,[3] but from the mid-1930s rehabilitation was already emerging as an important territory for the elaboration of professional and ideological interests in medicine, much as child health had previously served. Indeed, rehabilitation and child health were historically significant in similar ways, in that both offered a model for the development of medicine which contrasted with the standard model of individual contracts between doctors and patients. Like child health – the organization of orthopaedic care for crippled children in particular – the success of rehabilitation was seen to depend upon the integration of services and different types of health care workers within geographical areas. Accompanying the emergence of this model – as with that for crippled children – was a revival of the physiological idealization of the body with its holistic emphasis on dynamic interactions and the interdependence of subordinate parts. Although there is reason to believe that this integrative physiological body was rendered otiose at some point during or after the Second World War, its literal and social metaphorical significance before then was never greater.

This chapter, by exploring the extent to which the philosophy and practice of rehabilitation was cultivated and shaped by orthopaedists in the 1930s, illuminates some of the conditions under which rehabilitation services could be achieved and ordered. Central to this process was the rising political power of labour in this period. Modern orthopaedics had been reliant upon changes in the structure of capitalism and the valuation of labour; now orthopaedic interests came to be interwoven with the politics of labour.

In Britain as in America 'rehabilitation' was, as Eliot Freidson has remarked,

> a vague and poorly delineated concept, [whose] . . . concrete aims [were] subject to a fair degree of variation. It . . . included physical training as well as vocational education, concrete surgical repair and correction as well as psychotherapy.[4]

In both countries, too, modern orthopaedics embraced the whole range of these practices. In Britain, however, unlike in America, there were few serious contenders for the field as a specialty before the Second World War. Physical medicine, though it became a section of the Royal Society of Medicine in 1932, lacked sufficient autonomy to develop the field; many of its leading exponents, moreover, regarded themselves as aligned to orthopaedists, rather than in competition with them. Thus, although the concept and the medical practice of rehabilitation took on a meaning of its own in this period, it was (almost by default) hardly ever separable from orthopaedics. The word 'orthopaedic' was sometimes applied even to those rehabilitation schemes which were specifically intended for workers suffering from other than locomotor problems, just as during the Great War 'medical orthopaedics' was sometimes applied to the treatment of shellshock.

To a certain extent, the 'discovery' of rehabilitation in the 1930s signalled the wide acceptance of the medical messages which orthopaedists had preached since the war. But it was not quite so simple. In the first place, the practice also had roots and elaborations elsewhere. In the guise of 'after-care', rehabilitation had been cultivated for some time in the sanatoria treatment of phthisis. Indeed, Robert Jones and his colleagues had been much encouraged by its practice at the sanatorium at Papworth in Cambridgeshire. Henry Gauvain was the honorary consultant surgeon at Papworth, and Jones and Agnes Hunt were familiar with what went on there through their personal contacts with its founder and propagandist, P. C. Varrier-Jones. In 1928 Jones declared that he 'should like to associate [himself] . . . entirely with the recent statement of the Minister of Health, that Papworth is "the most perfect after-care scheme ever instituted" '.[5] By the mid-1930s, 'after-care', perceived as 'the logical outcome of the sanatorium movement',[6] was conceptually central to rehabilitation in Britain.

Second, and specifically in relation to industrial injury, after-care had been advocated in British medicine well before orthopaedic surgeons publicly took up of the cause. For example, a *Lancet* article of 1926 on the treatment of workmen's compensation cases stated:

> Our ordinary hospitals appear to have done what is usually regarded as their duty. They are chiefly concerned with the humane endeavour to save lives and to restore function to damaged limbs. Whether the man makes any economic use of his restored function does not appear to be any concern of theirs. Industry would seem to require another kind of hospital, or what might be called an 'after-care centre'. . . . There is something wrong when we find men, three or four years after a comparatively trivial injury to the muscles of the back or a straightforward fracture of the leg, still unable or unwilling to take their proper place in life.[7]

Such concern was reminiscent of that which surfaced briefly in the 1890s and 1900s over the wage-earning capacity of those who sustained fractures in industry.[8]

Finally, the orthopaedic interest in rehabilitation has to be seen to a large extent as emerging out of the work of lay groups concerned with crippled children, such as the Central Council for the Care of Cripples. It was in the *Cripples' Journal*, the main organ of the CCCC, that concern began to be expressed in the late 1920s about the neglected plight of the adult cripple. 'What chance, what thought, does the State afford [for the adult cripple]?', ran an impassioned editorial in 1926; 'it has denied him adequate medical treatment as a child, it denies him the right to work as a man. He cannot compete in the open market, and there are no jobs ear-marked for the physically defective.'[9] In 1928, in the same issue of the journal that carried Robert Jones's 'The Cripple: A Retrospect and Forecast', an editorial appeared on 'The Journal in

Transition', which took up the widening scope of crippledom in reference to the victims of accidents at work. Anticipated by the appearance in the journal of J. Whitley's lengthy and fact-filled account of 'What America is Doing for her Civil and Industrial Cripples', the editorial also noted how, in this area, 'America is ahead of Britain, and indeed of all Europe, both in legislation and in thought'.[10] Thereafter, until the *Cripples' Journal* ceased publication in 1930, the 'salvage' of the industrial cripple in all its legal, social, economic, political, and medical aspects, figured prominently.[11]

Chief among the new lobbyists for the industrial cripple was Sir Geoffrey Kelsall Peto (1878–1956), Conservative MP for Frome and subsequently for the Bilston division of Wolverhampton. He became a member of the CCCC in 1930 (rising to Chairman of the Executive Committee in 1937) and participated in the International Conference on Cripples held in the Hague in 1931. In May 1932 Peto wrote to the Board of Education and the Ministry of Labour indicating his interest in stimulating public interest in the subject of adult cripples by raising the matter in the Commons. 'Reasonable progress has been made with the treatment of crippled children and a little is being done for their training,' he wrote, 'but practically no provision is made here for the training of adult cripples.'[12] In a letter to *The Times* in March 1933, Peto lamented that not one penny of the £12 000 000 spent annually on workmen's compensation was spent on the training and rehabilitation of the maimed and crippled workmen.[13] Peto's comparison was with America:

> In the USA in 1929, 97,000 were killed in accidents. At least ten times as many are estimated to be permanently injured. This equals 8 per 1000 of population. On this basis about 350,000 are crippled through accidents in the U.K. each year. The Federal boards in the USA estimate that 80,000 cases annually require vocational training to earn their livelihood. The annual figure for U.K. should be 30,000.

The memorandum attached to Peto's letter to the Board of Education and Ministry of Labour also pointed out that those who were administratively concerned with accidents in Britain were interested only in getting 'compensations settled cheaply', while 'the trade union or legal adviser is anxious to get the maximum compensation for the injured and therefore tends to oppose any [re-]training for fear that it may reduce the damage and compensation'.

Neither the Board of Education nor the Ministry of Labour were certain how to respond. According to Dr Muriel Bywaters, one of the medical officers to the Board of Education, the Ministries of Health and Labour and the Home Office were reluctant to deal with the problem although they all realized that 'the question will have to be dealt with sooner or later, especially in relation to other countries'.[14] The Factory Department of the Home Office claimed it already had under consideration the question of providing training for persons crippled by factory accidents; the Ministry of Labour confessed to having no power in this area. 'Looking at the matter from the point of view of the unemployment problem generally', the Ministry also doubted 'if expenditure of public money on training cripples for employment could be justified'. Herwald Ramsbotham, speaking for the Board of Education, did not think that the matter came within the scope of his department either. However, Bywaters had told him that she was 'continually being told by orthopaedic surgeons that the compensation payments for an accident stand in the way of surgical treatment for industrial cases, the victim being unwilling to prejudice his compensation claim by getting cured'. Bywaters' own feeling was that it was 'necessary for a cripples' training college to be linked to an orthopaedic hospital to share the facilities for expert supervision'. For this reason, in her report to Ramsbotham, she rejected Peto's proposal that training facilities be established at Letchworth, 'some 35 or 40 miles out of London'. More generally, she thought that it was an inappropriate time to raise the subject, and that there was uncertainty over the actual extent of serious industrial injury in Britain. It was not clear to her, in the light of possible improvements in surgery, medicine and physiotherapy, how much vocational as opposed to medical rehabilitation was actually needed. Nor did she think it desirable that 'a so-called central body such as the Central Council for the Cure of Cripples should attempt [such provision]'.

Peto was thus dissuaded from raising the matter in the Commons. Nevertheless, he started wheels turning, and it was partly his efforts on behalf of accident victims that were rewarded a few years later with the opening of the training college for the disabled at Leatherhead (1935) and the setting up of the Delevingne Committee on Rehabilitation.[15] But behind these achievements lay another: the successful presentation to the Ministry of Health and the medical mandarinate of the LCC in December 1930 of the case for specialized accident and rehabilitation services. This was accomplished by the Liverpool-based Chief MO to the Cunard Steamship Company, Thomas Gwynne

Maitland (1875–1948). Educated both in medicine and philosophy, at Edinburgh, Manchester and Paris, this one-time editor of the Manchester *Medical School Gazette* was the founder in 1935 of the Association of Industrial Officers, and was subsequently regarded as a world authority on workmen's compensation. Maitland was to be one of the principal figures behind the establishment in July 1933 of the BMA Committee on Fractures (of which he was to be a member), and in 1935 he was responsible for organizing a dinner at the Reform Club attended by leading industrialists and by the King to publicize the need for specialized accident surgery and rehabilitation. As modest as he was urbane, Maitland firmly believed that his powers of influence would be diminished if they became too well known.[16] Consequently, little is known of the origins of his ideas and interests in this field, though acquaintance with Jones and knowledge of developments in America seems likely.[17]

Maitland's meeting at the Ministry of Health in December 1930 was presided over by Drs Thomas Carnwath and Montague Travers Morgan of the Ministry and was attended by Dr William Brander of the Hospitals' Division of the LCC. The success of the meeting did not depend on Maitland himself, however, but rather on his accomplices at the meeting: Hugh Ernest Griffiths (1891–1961), consulting surgeon to the Albert Dock Hospital from 1920 to 1956, and Harold Ettrick Moore (1878–1952) of the London, Midland and Scottish (LMS) Railways' Hospital at Crewe.[18] Both were pioneers in the treatment of industrial injury, for which in different ways they were already indebted to Maitland. Griffiths had benefited from the conference that Maitland held on the *Berengaria* in 1929 to investigate the best possible methods of securing rehabilitation for seamen. Little is known about the conference, but one of its outcomes was the launch of an appeal to provide a full hospital rehabilitation service at the Albert Dock.[19] Griffiths then undertook a tour of rehabilitation centres in the United States and Europe – notably, William Sherman's accident clinic at the US Steel Corporation, and Böhler's Clinic in Vienna – before completing plans for what was to be Britain's only hospital before the Second World War to deal with the treatment of injured civilians from the moment of an accident through to reemployment.[20] Although Griffiths was never a member of the BOA, he was among those who gave evidence to the BMA Fracture Committee in 1933–4, the Delevingne Committee in 1937, and the Royal Commission on Workmen's Compensation in 1939–40.[21]

Maitland was apparently also the stimulus behind the 'exceptional' work of H. E. Moore at Crewe, which he once compared favourably to that conducted at Manor House Hospital.[22] Moore's work dated from around 1926, four years after he had been appointed medical officer to the 19-bed hospital that was owned and operated (since 1899) by the LMS Railways Company.[23] Like Cunard, the LMS Railways Company was a large self-insured company with a direct financial interest in the speedy recovery of its injured workers. Although Moore never made explicit why, in 1926, he decided regularly to take into his hospital 'a small number of . . . chronic cases . . . old fractures, . . .old joint injuries, . . . old dislocations',[24] it seems that he too had been greatly inspired by the example of Böhler's Clinic in Vienna (to which Maitland may well have alerted him).[25] Other evidence, not incommensurable, suggests that Moore was himself facing unemployment at this time and was urgently in need of some convincing proofs of his and the hospital's worth. For some while, apparently, the LMS Railway Company had been contemplating closing down the hospital, and at one point it decided to grant the Town (Cottage) Hospital £1000 for enlargments for the company's use – a reflection of a general shift among large industries away from direct medical paternalism to the patronage of voluntary hospitals.[26]

Whatever the full explanation for Moore's initiatives in 1926, by the early 1930s he had sufficiently convinced his employers of the idea of 'after-treatment' to justify the hospital's being kept open (though it was eventually forced to close for other reasons). Evidence from the annual reports shows that of his first 165 cases of long-term industrial injury, no less than 115 or 70 per cent were 'considered on discharge to have made sufficient recovery to return to their pre-accident employment' after an average of only 17 days of treatment. A follow-up study of these cases conducted by the Company revealed that 65 per cent of the 115 had actually 'returned to duty and were carrying out their former work'. Presuming that the same degree of manpower 'wastage' was taking place in industry throughout the country, Moore concluded 'that employers, insurance companies, and through them the public, are incurring the expenditure of enormous unnecessary sums . . . as the result of ineffective measures for dealing with traumatic cases after the initial stage has passed'.[27]

In his lecture on 'Avoidable Wastage in Connection with Industrial Injuries' which he delivered at the Liverpool Medical Institution in December 1932 (also arranged by Maitland), Moore repeated the evidence that had 'much impressed' Drs Brander and Carnwath at the meeting at the Ministry two years earlier.[28] Again he argued that the principle culprits responsible for the 'wastage' were the general surgeons in hospitals and the National Insurance 'panel' doctors – the non-experts. Seventy-eight (47 per cent) of the 165 cases Moore treated 'had had some portion of their treatment at one or other of the General Hospitals', while 82 (49 per cent) 'had been treated under the National Health Insurance throughout'.[29] In both cases there was an evident lack of continuity, unity of control, team work, resources, and expertise. Thus Moore concluded that traumatic work should be separated from the ordinary work of general hospitals, and that there should be 'specially trained resident and nursing staff, and an entirely separate remedial department; the whole under the control of an orthopaedic surgeon'.

Although orthopaedists were ultimately to reject Moore's advocacy for separate trauma/accident centres, they were clearly delighted with his evidence overall. Up to this time few of them had been involved with accident surgery, which in Britain as in America was still very much the Cinderella of general surgery. The timely appearance of Moore's work in effect gave the orthopaedic Cinderella the hope that through alliance with independently-funded traumatology and rehabilitation centres she might improve her financial future and professional position. Among those most appreciative of this prospect was the aspiring Liverpool orthopaedist, Reginald Watson-Jones (1902–72). In 1927 he had been appointed honorary assistant surgeon to the orthopaedic department at the Liverpool Royal Infirmary, a prestigious position which Robert Jones had helped him obtain, but one which, like so many orthopaedic hospital appointments at the time, involved few inpatient beds. Watson-Jones had studied the organization of fractures under Platt at Ancoats and he was keen to develop a similar clinic at the LRI – partly as a way of increasing his bed allocation. However, in order to establish such a service he needed extensive outside funding. Aided by Moore's evidence, he was able to convince the Royal Insurance Company of Liverpool to provide the capital,[30] and thereby develop one of the few fully-organized fracture services in the country.

Others in orthopaedics appreciated the way in which Moore's facts and figures could be used to open up opportunities in the municipal hospitals. As a result of the transfer of these hospitals from the Poor Law after 1929, new space became available for specialty services, and the Ministry of Health was in favour of some of it being converted for the orthopaedic treatment of industrial injury.[31] However, in most cases progress was long delayed, partly because of the shortage of surgical consultants, and partly because it was tied to larger and more complex administrative issues – not least, the reorganization of orthopaedic services generally, and the debate over hospital fracture clinics versus separate accident hospitals.[32] Setting up committees of investigation was always easier and more financially practicable. Typically, Carnwath 'was pleased to learn' from Brander in March 1933 that the Chief MO of the LCC, Sir Frederick Menzies, had begun to reconsider the memoranda of Maitland, Moore and Griffiths, and was proposing 'to take up the question of the after-treatment of industrial accidents as soon as the hospital consultants are appointed [to a departmental committee to advise on the question of Industrial Surgery]'. But by the time the committee had been assembled, the BMA Fracture Committee was about to be set up and practical decisions were held in abeyance. Following the BMA's Report on Fractures, the recommendations of which added to the LCC's administrative problems, the Delevingne Committee was appointed, further suspending action until the appearance of its final (unbinding) recommendations of 1939.[33] Restricted in scope to fracture clinics in any case, the Delevingne Committee's recommendations were largely overtaken by the outbreak of the Second World War. Thus, as in so many other areas of interwar medicine, much more was planned and debated, and planned to be debated, than was ever practically achieved.

The few real accomplishments in accident and rehabilitation services before the war were mainly in the coalfields and owed far more to industry and organized labour than to government. The most acclaimed of these achievements were the rehabilitation centres at Uddingston, outside Glasgow, and that at Berry Hill, near Mansfield, Nottinghamshire. Both were financed by the coal industry and depended for much of their success on the missionary zeal of their surgeon mentors. The Uddingston clinic, like the Lanarkshire Orthopaedic Association under whose auspices it operated, was the achievement of Alexander Miller (1904–59), an associate member of the BOA since 1933 and the first consultant surgeon in the west of Scotland to confine himself wholly to orthopaedics. He was a pupil of the general surgeon James Russell (1880–1960), who had worked with Jones and Naughton Dunn during the war before returning to Glasgow determined to develop orthopaedics as an independent specialism. Miller became Russell's senior assistant surgeon at the Victoria Infirmary, Glasgow, where he was put in charge of the fracture department. During the Second World War he was to become a regional consultant orthopaedic surgeon in the Emergency Medical Service.[34]

The Berry Hill Rehabilitation Centre was directed by Ernest Alexander Nicoll (b. 1902), a general surgeon who, after obtaining his FRCSE and a Cambridge MD, was appointed honorary consultant surgeon to the Mansfield General Hospital with particular responsibility for trauma cases.[35] Like Griffiths and Moore, Nicoll had no training in orthopaedics as such. Indeed, while at Mansfield he had been excluded from its practice by Alan Malkin, the disciple of Jones who became Nottingham's first orthopaedic specialist in 1923, but who confined himself wholly to 'cold' or 'bloodless' orthopaedics.[36] In the manner of Watson-Jones, Nicoll was eventually able to secure 30 beds for trauma cases in a general surgical ward after he had commenced an accident-cum-fracture service at the Mansfield General on funds provided by the Midland Colliery Owners' Mutual Indemnity Company (who also provided him with a salary of £500 per annum).

The Indemnity Company, which supported three other such fracture services in the region (at Derby, Chesterfield and Nottingham) was crucial to the success of Berry Hill. Under the general management of Guy de Grouchy Warren (later to be appointed to the Delevingne Committee), it was established specifically 'to deal with injuries and compensation problems'. The company realized that a gap had been left 'between the reconstructive surgery in the hospitals and the restoration of function',[37] as a consequence of which most of the approximately 700 serious accident cases a year in the Nottinghamshire coalfield – with incapacity periods of over six months – ended up choking the pits as 'light work cases'.[38] The Berry Hill Centre was conceived as a means of closing this gap. Begun on an experimental basis with 15 beds in a former Miners' Welfare Commission convalescent home, its utility 'was soon recognised by the coal owners who agreed not only to make it permanent but to extend it to accommodate 40 patients'.[39] Rapidly it became one of the most celebrated rehabilitation centres in Britain:

> word got around, [Nicoll later reported], and we became a centre of interest not only to the medical profession, but also to industrialists, sociologists, economists, journalists and, inevitably, the politicians. . . . The Miners' Welfare Commission sent a delegation to see what it was all about. . . . We appeared in an article in the *Picture Post* magazine and later three films were made . . . [one] had a première at

the Curzon Theatre in London. It had good reviews and was translated into about 15 languages.

[Subsequently] Ernest Bevin . . . visited Berry Hill to see for himself what was going on. I was later invited to give a talk in the House of Commons to a group of MPs, with Bevin in the chair. It was here I met George Tomlinson . . . chief architect of the new Industrial Injuries Act and the creation of industrial rehabilitation units and training centres.[40]

Nicoll's individual efforts thus began to look impressive as they intersected the wider economic and political context of the 1930s and the manpower demands of the Second World War. In 1942 the Ministry of Fuel and Power requested the Miners' Welfare Commission to supply 50 per cent grants to provide similar facilities throughout Britain. A Medical Advisory Committee was appointed under Watson-Jones, and Nicoll (along with Platt and Alexander Miller) accepted an invitation to act as a consulting surgeon.

Like Moore's work at Crewe, Nicoll's at Berry Hill provided an economic rationale for specialist medical interests. Just as it was difficult not be appalled by the £12.5 million that workmen's compensation was supposedly costing industry in 1936, so, conversely, it was easy to be impressed by the supposed savings resulting from a service that greatly reduced the compensatory period between injury and return to full employment.[41] By 1945 this impression was sufficiently widespread to make the medicalized rationale capable of shifting policy priorities. Writing to *The Times* in April of that year, Minister of Labour, Ernest Bevin, observed:

> in spite of preventive and educative measures, industry is still faced with something like half a million lost time accidents each year at a cost to the community which has been computed at seventy million pounds annually. Since the speed of recovery is largely determined by the efficiency of first treatment and after care, it follows that an efficient accident and rehabilitation service should form part of every industrial medical department.[42]

Before turning to some of the politics that lay behind this impression, it is worth returning to the victims of industrial injury. While there is no evidence that workers regarded the surgeons in charge of the rehabilitation centres as making professional capital from the human tragedies of industry, there is reason to believe that they viewed company rehabilitation schemes in much the same light as company welfare schemes. They saw them in terms of management manipulation and cooption, rather than, as the surgeons would have wished, *alternatives* to the permanent loss of livelihood or the stigma of 'light work'. This is implied in Nicoll's recollection: 'Although the [coal] industry was footing the bill, I insisted that I must be seen to be quite independent – in other words, neither a bosses' man or a union man. The management committee of Berry Hill was drawn from people on both sides of industry.'[43] The same caution might be detected in the comment of Watson-Jones, in a report on a proposed rehabilitation centre for industrial injuries with which the CCCC was cooperating, that 'there was some difficulty in securing the full co-operation of the local branch of the Trades Union Council'.[44] The clearest evidence, however, is contained in the recollections of Alexander Miller, who endured 'prolonged and heated arguments' with the miners of Uddingston over the setting up of the rehabilitation centre which they regarded 'as an attempt to coerce [them] to return to . . . labour'.[45] Because the centre was financed by the coal owners until it was taken over by the Miners' Welfare Commission in 1945, Miller was 'apt to appear', as he put it, 'an ally of the compensation authorities'. Although the parent Lanarkshire Orthopaedic Association had three members of the Mineworkers' Union on its executive committee, along with three members of the Coalmasters' Association and six members of the Lanarkshire Medical Practitioners' Union, Miller was only able to obtain the cooperation of the miners by entering into agreements with them designed to protect their rights to compensation and NHI. Workers usually preferred to cling to their own medical advisers because they could more easily procure from them the certificates necessary to claim weekly benefits.[46] Miller had to promise his injured miners that the records on them drawn up in the rehabilitation clinic would be kept private, that only progress notes would be dispatched to GPs, and that no medical reports would be placed at the disposal of employers. On the issue of 'light work' – which was an especially sensitive issue because of the control it gave managers in severing disablement payments – an agreement was reached whereby, for purposes of occupational therapy, this would be designated and protected as 'alternative work', which would not jeopardize that performed by the aged or by young boys, and would be remunerated at 25s per week *in addition to* the patient's compensation allowance. Finally, it was made a condition that this alternative work, like other kinds of rehabilitation, 'should not prejudice the patient if any later disagreement occurred regarding his physical fitness and where an appeal was made to the medical referee'.

The extent of such bargaining further explains why employers of unionized labour were reluctant to become involved in rehabilitation schemes. The surgeons involved found such negotiations equally bothersome, but they also found them instructive, for they virtually compelled them to rediscover the wisdom of Thomas and Jones on the relationship between the recovery of function and financial, domestic and physical anxieties. As Watson-Jones was to put it: 'Real rehabilitation is getting into the mind of a man, finding out what his anxiety is and his worry and fear, and removing them'; or, to put it another way, 'quite simply the rehabilitation of the doctor to his patient'.[47] What Moore identified as 'the mental factor' was rediscovered by Watson-Jones in the writings of Plato and deployed against the '2000 year old error' of separating the 'psychological' from the 'physical'. BOA member William Ogilvie in 1945 proclaimed rehabilitation an 'expression of the new ideal in medicine', and welcomed it as 'a return to vitalism'.[48] Hugh Griffiths, at the Fracture and Rehabilitation Unit at the Albert Dock Hospital, put the relearnt knowledge into practice by asking the employers of his patients to promise reemployment once the rehabilitation was complete.[49] Miller, in order to cope with what he came to regard as the unappreciated 'social aspect to the problem', also found it expedient to have a well-trained social worker always to hand.

Whether these socioeconomic and psychosocial understandings made any difference to workers' attitudes to orthopaedic surgeons is difficult to tell. Most workers, supposedly, were more interested in cash than in the techniques of caring,[50] and doubtless were as aware as Ernest Bevin that the employers who provided health and welfare services tended to be among those who paid least.[51]

To the leaders of organized labour, however, 'the difference between a socialized and non-socialized accident service', as Nicoll called it,[52] did not go unappreciated. They welcomed the fact that the surgeons involved in these services were not only among the few to take a serious interest in occupational health, but were also among the few in the profession to refute allegations of malingering – allegations which had long been used to justify cuts in workmen's compensation and NHI benefits.[53] The rapport that came to be established between these surgeons and labour leaders was manifest in Manchester in October 1936 when the General Federation of Trade Unions held a conference on the institutional treatment of fractures to which Harry Platt was invited as the main speaker.[54] Representatives of most of the country's trade unions heard delegates speak of the 400 per cent increase in accidents since the introduction of NHI in 1911, and of the need to get back to the wartime national system of fracture clinics. Ben Tillett of the Dockers' Union recommended that doctors should make demands upon the government and municipal authorities in order 'to save the life and limbs of all who suffered fractures whether of industry, road, or domestic character'. This was in keeping with the tenor of the meeting and with its *raison d'être*: '[to] help forward the efforts of those medical men who have advocated special attention to the study and treatment of fractures, and who, in many cases, have had special opportunities for testing the theories that are being put forward'.[55]

It could also be argued that this was only in keeping with the outlook of the General Federation of Trade Unions – a conciliatory body that was far more interested in promoting the health and welfare of its members than in organizing strikes or drafting labour policy.[56] But the TUC, too, responded favourably to the interests of the orthopaedists. As previously noted, the TUC, in alliance with the Labour Party, had been interested in the treatment of accidents since the early 1920s. It had then spoken of the need for rehabilitation services in language strikingly similar to that adopted by orthopaedists in the 1930s.[57] The absence of such services led the TUC in 1923 to grant official recognition to the Industrial Orthopaedic Society and its hospital, Manor House, at Golders Green. 'Labour's Own Hospital', as it was promoted in the 1930s, came to be supported by some 200 000 workers, and could number among its patrons, patients and propagandists such labour leaders as Ernest Bevin, then of the Transport and General Workers' Union and Chairman of the TUC, Ben Tillett of the Dockers' Union, Ben Smith of the London and Provincial Vehicle Workers' Union, and Walter Citrine, the General Secretary of the TUC. Tom Mann, the apostle of revolutionary syndicalism, John Scurr, the Catholic socialist MP for Mile End, and George Lansbury, the leader of the Labour Party between 1931 and 1935, were also among the Society's and the Hospital's ambassadors. Dr H. B. W. Morgan, the Chief Medical Adviser to the Union of Postal Workers from 1920, a leading figure in the Socialist Medical Association and in the Labour Party Advisory Committee on Public Health, and MP for North West Camberwell (1929–31), was one of the two visiting surgeons to the hospital's busy outpatient department until his death in 1956.[58]

By the mid-1930s the TUC had reasons for interesting itself in orthopaedics beyond those relating to workers' wages and job retention, and beyond those of the early 1920s in the organization of a unified and 'classless' hospital service. Partly as a consequence of having helped to settle the dispute between GPs and the working men's medical society at Llanelli in 1936, the TUC had joined with the BMA in a Joint Committee on Medical Questions in November of that year. The Joint Committee had been a triumph, in particular for H. B. W. Morgan, then the Chief Medical Adviser to the TUC, and for Charles Hill, the deputy secretary of the BMA, and it opened a new chapter in their personal relations.[59] Within the TUC the triumph was also shared by J. L. Smyth, the secretary of the TUC's main social policy committee, the Social Insurance Committee, and by Citrine and Bevin, who had long favoured the expansion of the TUC into health matters in order to extend TUC influence.[60] But what had primarily been a marriage of convenience between the BMA and TUC soon came to grief over the issue of maternity services: the TUC favoured a clinic-centred state service conducted along local authority lines, but it was forced to acquiesce to the BMA's insistence that the service remain a part of general medicine based on private practitioners. By 1939 the unions and the Labour movement had come to regard the acceptance of the latter as the 'complete capitulation by the TUC to the BMA'.[61]

But the issue of rehabilitation was different. Both parties had an interest in eliminating the supposedly profiteering insurance companies involved with workmen's compensation. There was also general agreement over the kind of service being sought – clinic-centred, subject to third-party funding, and specialist rather than GP-orientated. There was little difficulty, therefore, at the first Joint Committee meeting in February 1937, in resolving 'that the subject [of rehabilitation] be considered by the Committee . . . and that Mr Watson-Jones be invited to appear before the Committee'.[62] The outcome was the Joint Memorandum on Rehabilitation and Industrial Injury, which was submitted to the Delevingne Committee in December 1937 by Watson-Jones and Charles Hill for the BMA, and by Citrine and Smyth for the TUC.[63] Drafted mostly by Watson-Jones, the Memorandum forcefully argued the social and economic case for rehabilitation centres. The location of the centres was left vague, although it was stated that patients undergoing treatment were 'the responsibility of the parent fracture clinic'. By implication this rendered hospitals of paramount importance; it also gave pride of place to fracture treatment, even though at the Joint Committee meeting of 20 April 1937 the TUC had stressed to Watson-Jones that 'all industrial diseases and not just fractures' should be covered in any rehabilitation scheme.[64] By further subtle implication, control of the centres was to be in the hands of experts in orthopaedic and traumatic surgery, for the centres were to exist 'as entirely independent organisations, no more under the control of the employer or the workman than the present voluntary and orthopaedic hospitals are under the control of either'. Initial funding, estimated at £7000–£10 000, was to come from a government grant, with running costs from employers.

There was not a great deal of difference between this document and the appendix on Non-Medical Factors of Prolonged Disability contained in the BMA's Report on Fractures, which had been inspired by Moore's work at Crewe. Only the means, or rather the pitch, had changed, in that the argument for support from the private sector was diminished.[65] The TUC thus had no hesitation in parading the Memorandum before its Annual Congress at Blackpool in 1938 and in resubmitting it to the Hetherington Royal Commission on Workmen's Compensation in 1939-40.[66] In doing so it reinforced the interests of the BOA, which were clearly set forth in the evidence presented to the Hetherington Commission by Griffiths, Moore and others in the orthopaedic lobby – the latter also including the orthopaedist Reginald Elmslie, who acted as one of the commissioners.[67]

The work of the Hetherington Commission was taken over by William Beveridge during the war and many of its recommendations were incorporated into Part II of Beveridge's *Report on Social Insurance and Allied Services* (1944). The famous 'Assumption B' of the Beveridge Report not only posited the need for a comprehensive health service, but specifically for 'Comprehensive Health and Rehabilitation Services'.[68] Because of this, and in view of the passage of the Disabled Persons (Employment) Act (1944) and the National Insurance (Industrial Injuries) Act (1946, replacing the Workmen's Compensation Acts), one historian has come to the conclusion that 'the BMA–TUC alliance . . . made the welfare state possible'.[69] In reality, though, neither the single nor the combined influence of these organizations was so great or so immediate. After the summer of 1939 when the Labour Party and union rank and file brought the TUC back into line over the maternity issue, the joint BMA–TUC committee did not in fact meet again until 1944, by which time initiatives on policy were largely lost to government.[70] Nor can it be overlooked that the absurd 'balance between the incentive to get well, and the loss of pay if one did', remained an inherent feature of the Industrial Injuries Act, despite all the arguments to the contrary by TUC and BOA–BMA spokesmen.[71]

If the long-term impact of the TUC–BMA alliance is questionable, the short-term impact of the liaison between the orthopaedic lobby within the BMA and the leaders of organized labour is not. More effectively than the BMA as a whole, the orthopaedists managed both to enlist and to retain the support of the leaders of organized labour to advance their own interests. During the Second World War Bevin, Citrine, Smyth and other labour leaders visited the meccas of rehabilitation proclaiming that the

> trades unions were delighted to have been associated with the development of the idea of rehabilitation centres such as had been outlined by Mr. Watson-Jones. The centres already in existence had furnished a practical demonstration of the value of the idea. The Trades Union Congress has recently given its approval to the idea, and has called upon its General Council to demand of the Minister of Health that he bring about an intensive development of this service. . . . The compensation bogey must be destroyed.[72]

In part the liaison between orthopaedists and the TUC and Labour Party was possible because, as in the TUC–BOA–BMA Memorandum, there was a stronger basis for rapport than on the maternity issue. But it was also possible because the orthopaedic lobby took a deliberately accommodating stand. They could afford to, since their interests lay less with government policy towards, and administration of, rehabilitation schemes, than with gaining support for the general principle of specialist-controlled centres. Where need be, particular policy interests could be disguised or held in abeyance.

Partly, too, the positive response to orthopaedic interests by labour leaders rested on the fact that there already existed within one of the most important national industries – coal – an agency through which such support could be fed without causing political and economic commotion. The Miners' Welfare Commission, a semi-governmental body organized on a quasi-democratic basis, occupied anomalous ground between labour and management, capital and the state.[73] Orthopaedic interests had already been generously served by some of the regional Miners' Welfare Funds: that in Nottingham gave £750 to the Harlow Wood Orthopaedic Hospital in 1933, though at that time the hospital was entirely for crippled children; in the same year, the Sheffield Royal Infirmary received £25 000 from the Fund in order to build a fracture unit within the Miners' Welfare ward; and Platt's fracture clinic at the Manchester Royal Infirmary, when it was incorporated into a new orthopaedic and physiotherapy building in 1938, received £13 000 from the Lancashire and Cheshire Miners' Welfare Committee.[74] It was relatively easy, therefore, for the Ministry of Mines and Supplies in 1942 to stimulate the Commission into building rehabilitation centres throughout the coalfields, especially since these might also serve as a sop to the then rampant union militancy among miners.

Finally, the positive response from labour leaders can be related to the ambiguous nature of rehabilitation itself – an ambiguity apparent in the urgent need felt by virtually every writer on the subject to begin by defining the term.[75] Although it was increasingly understood as a 'complex process, not a series of independent contributions by the surgeon, the therapist, the technician, the limb-maker ... and the patient himself',[76] it was not at all clear before the Second World War which, if any, of these groups should be in control. In the absence of a powerful rehabilitation lobby, such as emerged in America, proclaiming the process as a 'third phase of medicine', going beyond 'preventive and curative medicine and surgery',[77] orthopaedic surgeons were able to put their particular interests above those of other less organized professional bodies. Their dominance within the thinking on industrial rehabilitation, which was resentfully noted by Wilson and Levy among others,[78] was well reflected in the plans afoot in 1942 for an experimental rehabilitation centre at the Manchester Docks. The origins of this experiment lay in a meeting between Bevin and the National Dock Council at which it was lamented that no centre existed for rehabilitating 'civilian medical cases such as bronchitis, rheumatism, gastric disorders and general debility, which are apparently the main causes of sick absence amongst dockers'. All that was currently available was 'for orthopaedic cases'. While this latter comment should also be taken as reflecting cracks in the perceived hegemony of orthopaedic interests in this area, it is nevertheless ironic that the Ministry of Health's file on this scheme was headed 'Orthopaedics of Manchester Dockers'.[79]

Clearly, the socioeconomic conditions that worked against the involvement of British orthopaedists in the treatment of industrial injury in the 1920s were overcome in the 1930s. By enrolling organized labour in support of accident and rehabilitation schemes, orthopaedic surgeons furthered both immediate professional self-interests and the longer-term extension of those interests through legitimation of integrated medical services. But before the Second World War, aspirations outstripped realities: the work of the pioneers, Moore, Griffiths, Nicoll, Miller and Watson-Jones, was seen hardly to have 'scratched the surface of the enormous problem of bringing the injured man back into a condition in which his productive power may be utilised to its full capacity'.[80] Even less did these 'comparatively tiny experiments' dent the structure of British medicine. Orthopaedic rehabilitation schemes, like fracture clinics, existed more in theory than in practice; most orthopaedists continued to work mainly with crippled children. Moreover, the child-centred image of the specialism was reinforced by the lavish patronage bestowed on hospitals for crippled children in the 1930s by Lord Nuffield (£125 000 in 1935 alone).[81] No wonder there were those in the Ministry of Health during the war who were under the impression that the 'orthopaedic service' was concerned only with crippled children.[82] Suffice it to conclude here, however, that as the lights over Europe began to flicker once more, the prospects for orthopaedics shone bright. Industry and labour, as courts of appeal in specialty legitimation, seemed about to have their pay-off; for experts in rehabilitation – including the rehabilitation of medicine itself – the future looked good.

11

THE PHONEY WAR

For the orthopaedic services the [Second World] war marked a new beginning. (Richard Titmuss, *Problems of Social Policy*, 1950, p. 476.)

Almost everyone who wrote on British medicine in the wake of the Second World War regarded 'the creation of a framework for a national rehabilitation scheme . . . one of the chief successes of the Government's emergency medical service'.[1] Richard Titmuss, in his official history of the wartime civilian services, was cautious about the comprehensiveness and uniformity of the rehabilitation services by 1945,[2] but his estimation of their overall value was fully in accord with that of Bevin, Beveridge and many other war and postwar politicians and planners. Indeed, his estimate echoed that of the first postwar Minister of Health, Aneurin Bevan, who proclaimed that 'One of the best things that has come out of the war is the development of the Rehabilitation services.'[3]

Rehabilitation, as a part of occupational health, was in fact soon to fall by the wayside in British medicine, just as it did in America.[4] Yet the pre-NHS enthusiasm for it is not hard to understand. In many ways it was a symbol of social corporatism – of equitable welfare medicine serving the ends of both national and individual efficiency. It demonstrated how, through government initiatives, health care could be rationalized and coordinated between and among hospitals and medical professionals, and how, through centralized regional planning and state funding, such services could be made available to all, free at point of delivery.

In view of the increasing commitment to rationalization, corporatism and statism among orthopaedists during the interwar period, and in light of their involvement with the treatment of industrial injury, it might be supposed that the successful development of rehabilitation services during the war was doubly owing to them. It might further be supposed that their part in the story can be read as the chronicle of a specialist group's successful exploitation of the opportunities of the Second World War. But this would be mistaken. Although the wartime

218

rehabilitation services owed something to their efforts, much more was owed to the Ministry of Labour. If one had to find a wartime hero for the story, that hero would be Bevin's junior secretary, George Tomlinson MP (1890–1952), the Chairman of the Inter-Departmental Committee on the Rehabilitation and Resettlement of Disabled Persons.[5] Orthopaedists were in fact unable to exploit the enthusiasm for rehabilitation for their own professional ends. In this area as in others, their main experience of the war was that of being marginalized.

This chapter briefly explores the nature of that marginalizing process, thus opening out the irony of the specialism's inability effectively to capitalize on the restructuring of medicine of which it was so intimately a part. Necessarily this entails narrowing our focus on to the politics of specialization within British orthopaedics, yet in some ways this need for a narrower focus is itself reflective of further significant transformations in the character of medicine as a whole during this period.

Instead of providing an occasion for orthopaedists to consolidate their professional interests, or leading them to exercise greater authority in medical politics generally, the Second World War was an anti-climax. While the official historians of wartime orthopaedics may have been right to proclaim that 'in no other branch of medicine [were] the opportunities so great',[6] a number of factors stood in the way of the realization of those opportunities. Despite the fact that special orthopaedic centres, fracture hospitals, and rehabilitation services came to be needed for civilian as well as military casualties, the specialism was to secure little of the popular prestige that it had through the Great War. Rehabilitation centres such as those organized by Watson-Jones for the RAF were showpieces of modern medicine, but among specialist groups, it was above all the plastic surgeons who captured the public imagination and applause. Indeed, while Archibald McIndoe and his team at the East Grinstead Plastic Centre pioneered their techniques on the badly burned heroes of the RAF, the majority of orthopaedists were compelled to address a problem which, 'up to 1939 . . . was regarded very largely as a feminine complaint, usually the result of wearing unsuitable shoes' – that is, 'feet cases'.[7] Furthermore, as if seeking to add insult to professional injury, the Army discouraged the 'operative treatment for foot troubles in serving soldiers'.[8]

As in the First World War, about 70 per cent of all the injuries in the Second involved open wounds of the extremities.[9] Yet the injuries were not of a type to make this war another 'orthopaedists' war'. Fractures sustained on the battlefield were not only less prevalent, they were less readily segregated from other kinds of injuries. As the official historians explained:

> the increased destructive power of modern bullets and shells; the high speeds of modern transport and combat vehicles on land and sea and in the air; the greatly increased casualty-causing effect of bomb blast; . . . and the hazards associated with special tasks as, for example, parachuting; . . . [meant that] injuries tended to be multiple and it was not uncommon to have to treat as many as a dozen or more separate problems in the same individual.[10]

Thus the division of medical labour tended to be between different groups of specialists, in contrast to the single-specialty 'team work' of the First World War. Within the new arrangement, moreover, the role of orthopaedists became less central and less obvious. Reflecting this loss of special wartime status was the relative disdain of orthopaedists and orthopaedic equipment among the troops. If the experience of the novelist Anthony Burgess is anything to go by, troops now found the Thomas leg splint 'a neolithic masterpiece of tapes and granny knots'.[11]

Nor were orthopaedists particularly well-equipped to deal with the traumatic injuries peculiar to war.[12] They had spent much time debating the *organization* of accidents and rehabilitation services, but far less time discussing the *treatment* of wounds, least of all the treatment of war wounds. An exception here was Josep Trueta, the future professor of orthopaedics at Oxford, whose experience with battle and air-raid casualties during the Spanish Civil War was to render his *The Treatment of War Wounds and Fractures* (1939) a much sought-after publication.[13] At the time of its publication most senior orthopaedic surgeons were still relying on their memory of the treatment of wounds during the Great War, and insofar as such treatments were inappropriate to this conflict, their memories further disadvantaged them. 'When we went to war in September 1939, many of us thought in terms of 1918', reflected an orthopaedist in 1942. 'It was only natural that we should; for twenty years we had put fighting and its problems from our minds. It is not altogether an advantage to a nation to have a background of war at twenty years' distance when it faces another struggle.'[14]

Another of the reasons for the marginalizing of orthopaedics during the Second World War was inadvertently touched upon by this same author when he sought to expose the major therapeutic differences between the two conflicts. His contrast of a soldier wounded in the trenches in 1918 and a patient injured in the great air-raid on Coventry, whose successful orthopaedic treatment resulted in his ability 'to return to his work in the factory', is a reminder of the decidedly unromantic context associated with the bulk of this war's orthopaedics.[15] No matter how much encouragement government might give to those rescuing the 'soldiers of industry', industrial sprains and fractures were not the sort of work to win great public or professional plaudit. Such work was dull and increasingly routine: reported accidents in industry rose from 180 000 to over 300 000 between 1938 and 1943, with the largest increase (400 per cent) among women workers, the majority of whose injuries were to the feet.[16]

Nor was there anything novel or contentious about the organization of orthopaedics during the Second World War. The first of the 19 special orthopaedic centres set up under the Emergency Medical Service (EMS) in England and Wales (there were five others in Scotland) were located in the same places where Jones had established his centres after 1916. In some cases, such as the Wingfield-Morris Hospital in Oxford and the old Alder Hey Infirmary in Liverpool, the orthopaedic centres were housed in the same buildings.[17] And the category of patients intended for referral to the centres were the same as Jones had sought in 1916 through his widened definition of orthopaedics. Although there were to be some minor additions to the list, as well as some significant subtractions, both the Ministry of Health and the War Office agreed that the appropriate cases for the orthopaedic centres were fractures, deformities of the extremities and spine, diseases, derangements and disabilities of joints (including the spine), and injuries of peripheral nerves.[18] As pointed out by the Ministry of Health's Consultant Adviser on Orthopaedics, H. A. T. Fairbank, 'The general plan and the various departments . . . were the same as those made use of in the War of 1914–18. Improvements of undoubted value were made but these were concerned with detail rather than with the general plan.'[19]

Such replication is hardly surprising given that many of those appointed as chiefs-of-staff to the orthopaedic centres were the same as had been appointed by Jones during the First World War and were now long of tooth. T. P. McMurray (1888–1949) headed the team in Liverpool, Girdlestone (1881–1950) administered from Oxford, and Platt from Manchester, while Jones's old surgical colleague and orthopaedic helpmate during the Great War, Sir William Ireland de Courcy Wheeler (1879–1943) was one of the consulting surgeons appointed to the Royal Navy.[20] Rowley Bristow (1883–1947), who was appointed consulting orthopaedic surgeon to the Army in 1940, might have been speaking for them all when he reported in 1943 that 'In so far as is possible, it has been my aim to follow the pattern [Jones] created, in advising the Army in the formation and organization of its orthopaedic service.'[21]

On the basis of this evidence, it is tempting to conclude that the only difference between orthopaedics in the First as compared to the Second World War was the absence of Jones himself. Yet such a conclusion would be difficult to sustain, even if it were realistic to reduce the history of modern orthopaedics to the personality of Jones. The real difference was not Jones's absence, but the want of the political conditions that had enabled Jones to undertake his grandiose organization. Even Harry Platt, who was among the most active agents for the specialism during the Second World War, came nowhere near to obtaining the authority that Jones had held during the First. (Significantly, perhaps, it was Platt who later cautioned his colleagues in the BOA against 'claiming too much for our heroes'.)[22] Opportunities existed during the war for expressing surgical talent and for planning the postwar 'brave new world', but not for the type of empire-building that Jones had undertaken. A major constraint was the separation of the civilian EMS (initially Emergency Hospital Service) from the Army Medical Service, and the independence of the AMS from the services of the Navy and the Air Force. This difference from the First World War, combined with the different nature of the warfare, and the diminished importance of the Army relative to the Air Force and Navy, meant that it was virtually impossible for a Director-General of the AMS to reestablish the kind of exclusive and intimately shared system of power that had existed in 1914–18. Unimaginable were the kinds of privileges acquired by Jones once he had secured his office down the corridor from Keogh in the War Office. Gone, too, was the extensive control over hospital provision and independent financial power of the Red Cross; hence equally unimaginable were the kinds of relations that had existed between Jones and King Manoel and Sir Arthur Stanley. In this war, because of the importance of the EMS, medical power came to reside principally in the civilian Ministry of Health. Consultant advisers to the Ministry informed and oversaw the implementation of Ministry instructions, maintaining close relations with regional advisers, regional consultants on special subjects, and practitioner group advisers (for areas remote from urban hospitals and medical schools).[23] Power, where it was to be had at all, was to be gained through the mandarinate in Whitehall.

Several other features of the political landscape may have conspired against orthopaedic interests. Conflicts between the Ministries of Health and Labour meant that the orthopaedists' alliance with the latter stood them in poor stead with the former. Indeed, the absence of this good standing might explain why the first consultant adviser on orthopaedics to the EMS (appointed by the Ministry of Health in March 1939 'on the nomination of the Presidents of the Royal College of Physicians and Surgeons'), was Gwynne Williams – a medical dean who had played no part in the previous politics of orthopaedics and who was not a member of the BOA.[24] Although Williams was soon replaced by Fairbank, Fairbank was more closely identified with the London medical establishment than with the more politically ambitious Liverpool School. Among the latter, only Platt became involved with the work of the Ministry, though at first merely as part-time assistant to Fairbank.[25] Fortunately, Platt was in official harness before the Ministry of Health began, early in 1940, to regard as threatening the regional hospital planning that Platt had been much involved with through the Nuffield Provincial Hospitals' Trust.[26] Bristow, as noted above, was ultimately delegated to the AMS, while Reginald Watson-Jones secured the appointment of Civilian Consultant Adviser on Orthopaedic Surgery to the RAF. It was in this capacity – independent of the Ministry of Health – that Watson-Jones created the country's first comprehensive rehabilitation service by organizing and integrating the RAF's 16 orthopaedic centres with its four major rehabilitation centres.[27]

Whether the Ministry of Health was acting in a consciously political manner in its initial appointment of Williams remains an open question. As regards the appointment of Fairbank, it seems likely that the Ministry was simply unaware of the subtlety of the Liverpool/London divisions within the BOA, and was influenced more or less unwittingly by the London medical establishment. The latter's influence is perhaps also to be seen in the Army's appointment of the orthopaedic surgeon of St Thomas's Hospital, George Perkins, to be in charge of orthopaedic cases in the British Expeditionary Forces.[28] What is beyond doubt, however, is that ideologues in the BOA believed that the politicians, along with bureaucrats and the military, were paying insufficient heed to their specialist interests. So 'grave' was their concern, especially over 'the apparent lack of organisation for the treatment of the wounded soldiers and of civilians injured in Air Raids', that in 1939 they drew up a schedule of 'possible lines of attack' to enable their views to be heard. One strategy was to obtain a hearing

from Sir Charles Gordon-Watson, the Consulting Surgeon to the
Army at Home. Through him they hoped to gain access to the Director
General of the AMS. Another proposal, worth quoting in full for its
strategic explicitness, was to make 'A direct approach to the Cabinet':

> This line of attack might be conducted through Nuffield, who should
> be asked to write a letter to the Prime Minister voicing the
> dissatisfaction of the senior orthopaedic surgeons with the existing
> preparations for the treatment of the large 'orthopaedic' group of
> potential air-raid casualties and war injuries. This letter should
> remind the Prime Minister of Nuffield's deep interest in the
> prevention and treatment of crippling in patients of all ages. He
> would naturally mention the hospital which he had built at Oxford
> and which had replaced a hospital set up during the last War as a
> military orthopaedic centre under the aegis of Robert Jones. He
> would quote the outstanding achievements of Robert Jones in the
> last War, and refer to the fact that owing to his influence and
> example, by the time America had joined the Allies, the U.S. Army
> Medical Service had already made elaborate and detailed provision
> for military orthopaedics. He might also remind the Prime Minister
> (a) of the recommendations in the recent Interdepartmental Report
> of the Ministry of Health and Home Office on the treatment of
> fractures; and (b) that the personnel of a nation-wide fracture
> organisation directed by orthopaedic surgeons is already available.
> He would urge that both the E.M.S. and A.M.S. should be
> compelled to put the principles enunciated in that report into
> practice without delay.[29]

Both of these proposals were in fact taken up and had positive
outcomes: the AMS acknowledged the seriousness of the situation
through their appointment of Bristow in 1940, while the Ministry of
Health climbed down from its earlier resolve not to establish special
orthopaedic centres until *after* the air-raids had begun.[30] To this extent,
at least, orthopaedists overcame the marginal position of power in
which they found themselves at the outbreak of war.

But in other ways real power and authority eluded them. Primarily
this was because, unlike during the Great War, they were unable to
retain or reestablish unity of control over the various venues and
therapeutic practices for the restoration of the injured. By September
1940, a year after the outbreak of the war and two months after the
onset of enemy bombing, it was obvious to Platt and his colleagues that
the existing organization of orthopaedics was '*increasingly vulnerable to*

the influence of certain disintegrating factors' (my emphasis). Without urgent reform, it was argued, 'a good deal of the orthopaedic material may be lost to the [orthopaedic] centres'.[31] While on the one hand, 'a number of the larger [home] Army Hospitals [were] receiving and treating a considerable proportion of the orthopaedic disabilities of the soldier', on the other, 'the independent orthopaedic scheme of the R.A.F. [was providing] . . . a service which might conceivably have been included in the E.M.S. scheme'. Platt confessed to having 'no information about the orthopaedic arrangements in the Royal Navy' but he had no doubt that they too were 'developing independently'.[32] He also recognized that 'the future policy of the Ministry of Pensions [was] still in the melting pot in regard to (a) responsibility for the treatment of the more serious Service disabilities; and (b) the expansion of Ministry hospitals'. Furthermore, he saw that 'formidable difficulties – administrative, social, and surgical – are likely to be encountered in the future in any attempt to segregate civilian orthopaedic casualties resulting from air-raids of the type experienced during the last two months'. Platt therefore sought to bring forward proposals to overcome what he now regarded as 'the one real and tangible orthopaedic problem which now confronts us' – the 'entirely speculative' future.[33]

But the solutions were mostly piecemeal, reflecting politics of compromise and contingency, rather than those of integrative organizational ideals. This was largely inevitable, since neither the AMS nor the RAF could be expected to sacrifice their independence. The Army remained insistent that 'the serving soldier who is injured and who is likely to be fit to serve again, is better cared for in a military than in a civil hospital'.[34] It was this intransigence that led to the 15 orthopaedic units under Bristow in 1940 being attached to the stationary Army hospitals in Britain. 'It is not a question of surgical skill, nursing, the hospital buildings, equipment, or ancillary services', Bristow sought to explain in light of the then 24 similar orthopaedic centres in the EMS and another seven in the RAF; 'It is no reflection on the E.M.S. [Rather] it is merely the fact that the whole military atmosphere, call it military discipline if you will, is necessarily lacking in the non-service hospital.'[35] In the end, the best that could be arranged between the EMS and the AMS was the division of orthopaedic labour illustrated in Figure 11.1. By 1943 this structure was operating relatively smoothly.

Piecemeal, too, was the rationalization that gradually came about within the EMS orthopaedic service.[36] Here, matters were helped

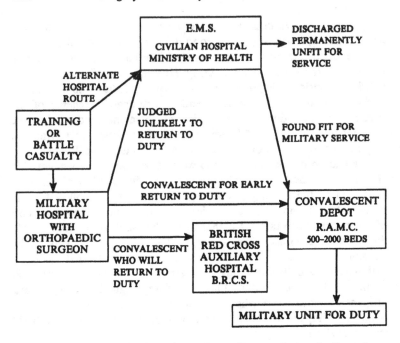

Figure 11.1   Structure for handling orthopaedic cases during the Second
World War

Source: W. Rowley Bristow, 'Some Surgical Lessons of the War', *Journal of Bone and Joint Surgery*, 25 (1943), p. 525. (By courtesy of the *Journal of Bone and Joint Surgery*.)

somewhat by the Inter-Departmental Conference on the Rehabilitation of Persons Injured Through Enemy Action, convened by the Ministry of Health in the summer of 1940. The Conference was concerned with the limited number of hospital beds for orthopaedic cases, as based on pre-air-raid estimates of need, as well as with disparities in the facilities for orthopaedics at different hospitals. It also drew attention to the poor quality of fracture treatment at some hospitals, and to reports that orthopaedic patients were sometimes being left in ill-equipped hospitals.[37] The Conference thus helped to stimulate the grading, and consequent upgrading, of three types of 'fracture hospitals' under the EMS (that is, hospitals with fracture departments). Roughly, the three types – 'A', 'B' and 'C' – corresponded to the Delevingne Committee's classification of hospital provision according to the degree of fracture segregation and degree of orthopaedic unity of control.[38] But with the onset of enemy bombing came the realization that the percentage of serious orthopaedic casualties from air-raids had been greatly overestimated – being closer to 20 per cent as opposed to the estimated 60 to 70 per cent.[39] The effect was to dampen some of the initial sense of urgency. Although new energy for rationalizing and upgrading orthopaedic services was to come from the Ministry of Labour when the manpower shortages in industry began to make themselves felt in 1941, the moment was lost for specifically orthopaedic initiatives. Thereafter, orthopaedists had to manoeuvre mostly within the diffuse territory of rehabilitation – a territory where the Ministry of Health trailed behind the Ministry of Labour and where the orthopaedic scheme of the EMS followed suit. In this domain, moreover, there were now other sets of professional interests to contend with.

In particular, the wartime emphasis on rehabilitation gave impetus to the fields of occupational therapy and physical medicine, both of which had been largely dormant since 1918 when Jones's orthopaedic centres had closed down. It was said of occupational therapy, for instance, that it 'reverted for the most part to the mental hospital service' where it had first been put into practice in the nineteenth century.[40] Physical medicine – involving heliotherapy, thermaltherapy, hydrotherapy and physiotherapy together with gymnastics and 'active' and 'passive' massage – was to be defined with new coherence after the war as 'that branch of medical art which employs physical agents in diagnosis and treatment'; but its major spokesman, Sir Robert Stanton Woods, recorded that in 1939 it was still 'almost universally' unrecognized, its various embryonic parts being lodged in hospital massage departments.[41] Within a year, however, after various appeals from aristocrats, MPs, and the King's physiotherapist, Sir Morton Smart, the Ministry of Health appointed Stanton Woods as chairman of an Advisory Committee on Physical Medicine.[42] At the same they appointed as Honorary Consultant Adviser on Rehabilitation the one-time apprentice to Robert Jones, Rhaiadr Jones.[43] Although the terms of these appointments were vague and somewhat at cross purposes,[44] the immediate tasks involved the inspection of facilities, the assessing of future needs, and overseeing the departments of physiotherapy that the Ministry had decided (in December 1939) to equip at each of the 19 orthopaedic centres.

The most pressing problem facing the development of wartime rehabilitation work was the scarcity of qualified medical and paramedical personnel. This was a problem even before 1943, when physiotherapy, occupational therapy and remedial gymnastics were provided exclusively for armed forces and civil defence patients, air-raid casualties and certain specified evacuees. Thereafter, as civilian industrial fractures were added to the list, the problem of staffing became acute.[45] There is no need here to enter into how the list came to be lengthened through the interventions of Bevin in the Cabinet,[46] nor how the problem of recruitment was dealt with by means of crash courses and changes in age and curriculum requirements.[47] Neither is it necessary to document the means by which these areas of medicine scoured away their respective 'fringe' and 'artsy-crafty' associations.[48] The important point is that, as a result of the demand for rehabilitation services, orthopaedists began to lose control over parts of their practice which had been integral to Jones's conception of it. Shortages of space, in some instances, forced the rehabilitation work to be carried out in places apart from that of the orthopaedic surgeons. Undermined, therefore, was continuity of treatment. Moreover, the practitioners of physical medicine now had before them encouraging American examples of professionalization and specialization.[49]

Orthopaedic surgeons were not unaware of the dangers to their specialism from the independent expansion of physical medicine. But having, as it were, unbound this Prometheus and then welcomed and taken full advantage of the Ministry's drive to expand the provision of rehabilitation within the orthopaedic service, they were in no position to directly challenge its subsequent independent progress. Their own manpower shortages, they realized, necessitated that they avoid taking 'an authoritarian attitude' on the matter.[50] Furthermore, since there had always been orthopaedic practitioners whose main area of expertise was 'manipulation', it was never very clear where the boundaries between orthopaedics and the ambiguously defined field (or fields) of physical medicine/rehabilitation/occupational therapy should be drawn.[51]

But the threat to orthopaedics from the development of rehabilitation ran deeper than merely the loss of certain therapeutic procedures for restoring function. Fundamental was the fear of losing control over the process as a whole. This threat came partly from the would-be 'physiatrists' (as the physical medicine experts were sometimes called) who now took up the call for 'unified medical control' on behalf of their own interests.[52] It also came from Ministry of Labour officials, such as those on the Tomlinson Committee, who insisted that rehabilitation was 'not solely a medical problem'.[53] Thus there was need to state in a BOA 'Memorandum on Rehabilitation' of November 1942 (intended for the Ministries of Health and Labour) that the Association 'would stress its conviction once more that rehabilitation is essentially a part of *treatment*, and that it must be developed and must remain in all its phases under medical [i.e., orthopaedic] control.'[54]

That unity of control mattered more to orthopaedists than loss of control over certain parts of the medical process may be illustrated indirectly, by the apparent absence of concern with the wartime segregation of patients with peripheral nerve injury. From the start of the war, neurologists had been appointed to handle many of these cases.[55] Indeed, in neurology, as in physical medicine, the war had hardly begun before the opportunities for further professionalization were seized. Largely through the influence of George Riddoch, the Consulting Neurologist to the Army on peripheral nerve injury, a Nerve Injuries Committee was set up under the Medical Research Council. This was subsequently reconstituted as an advisory body to the EMS, with Riddoch as chairman. Riddoch soon justified the need for special centres for these cases which led to the selection of three of the orthopaedic centres in England for the purpose, together with two centres in Scotland not within the orthopaedic scheme. There does not appear to have been orthopaedic opposition to this development. It may have helped that Platt's friend, the neurologist Geoffrey Jefferson, who was an honorary member of the BOA, was also on the EMS Advisory Committee on Peripheral Nerve Injury, and that Bristow and Platt had maintained research interests in this field since the First World War. More important, probably, is that this was seen to be an acceptable form of 'segregation within segregation' – the whole of the process remaining under orthopaedic control.[56] Thus independent professionalization in this area tended to enhance rather than threaten orthopaedic unity of control; the segregation of cases within the specialism was within the idealist concept of the subordination of parts to wholes. Indeed, it was clearly seen to conform to practices within orthopaedics, such as the segregation of fracture cases.[57]

A slightly more straightforward illustration of this same concern with unity of control is found in the wartime reformulation of the argument for accident and fracture services. Initially, the war appeared as an opportunity to enforce the recommendations of the Delevingne Report of 1939. Platt, in his capacity as Consultant Adviser on Orthopaedics, inspected EMS provision for fracture treatment and wielded the Report instrumentally, more or less as an official policy document. On his visit to the Royal Salop Infirmary in August 1941,

for example, where he found the arrangements 'unsatisfactory' with no
unity of control or continuity of treatment and no proper coordination
with the massage department, he explicitly referred to the recommendations of the Report, instructing that

(a) Both the in-patient and out-patient fractures should be under
     the control of one member of the hospital visiting staff.
(b) There should be a daily minor fracture clinic conducted by an
     experienced resident. . . .
(c) There should be a weekly major fracture clinic conducted by the
     Surgeon in charge of the Fracture Department.

In conclusion, he reminded the hospital administrators 'of the
recommendation in the Interdepartmental Committee Report (1939)
viz: that the surgeon in charge of an organised fracture department
should be remunerated by a suitable honorarium, as the efficient
control of such a department demanded a good deal of time'.[58]

It is unclear to what extent such means were successful in having the
recommendations of the Delevingne Report implemented, as opposed
to merely making them better known. Available evidence suggests that
at least until 1942, the confused division of labour between the EMS
and the AMS over the treatment of minor fractures was a major
obstacle to systematic reform, while thereafter the nature of the reform
took a different turning, with the emphasis on fracture clinics being
superseded by that on accident services and a general gravitation to the
view (subsequently expressed by William Gissane, the advocate of
separate trauma centres) that the Delevingne Committee, with its stress
on fracture clinics, had 'misunderstood its brief'.[59] In 1942, the same
year in which Girdlestone and his colleagues in Oxford established the
country's first general accident service,[60] the BOA's Sub-committee on
Fractures began to question the appropriateness of its title (which in
1944 was changed to the Fracture and Accident Services Committee).[61]

Capturing and contributing to this shift of emphasis was the
Memorandum on Accident Services drafted by Watson-Jones in
September 1941 and printed and issued by the BOA in February
1943. Projected as 'a natural development of organized fracture
services' – a development therefore to be mainly located within general
hospitals – the accident service was described as one in which every
type of injury was admitted 'but the predominating clinical material
[was] injuries of the locomotor system'. Thus the service was to 'be
under the direction of a surgeon who, in addition to a wide training in
surgery as a whole, has had special training and experience in the

surgery of the locomotor system'. According to early drafts of the memorandum, the Director was not only to have executive clinical control of the whole of its operations, but was also to 'be relieved of lay administrative duties . . . , be paid an adequate salary, and [be] permitted limited private consulting practice within the precincts of the hospital'.[62] Embodied within the call for accident services, therefore, were all the professional gains previously sought through the advocacy of fracture clinics. But in the face of the wartime forces of 'disintegration' operating against the specialism from within and without, there was now additional and urgent reason to extend the claim.[63]

Two strategies in particular were pursued. Most obvious was the preempting of 'traumatology' as a separate and competing specialism. '[I]t would be far from ideal', the memorandum claimed, 'to train a new type of surgeon – the traumatic surgeon. The skill of such a surgeon would be disseminated and scattered over the vast fields of locomotor, abdominal, thoracic, facio-maxillary, ophthalmic, and cerebral surgery for no reason other than that the disability was the result of trauma.'[64] Such fears over the dilution of orthopaedics echoed those of American orthopaedic surgeons in the 1930s. There, discussion on the extent to which 'orthopaedic surgery and traumatic surgery can be mixed' emerged from awareness of the 'attractive and lucrative treatment of traumatic and occupational injuries'.[65] In Britain, however, where there was virtually no private market in trauma services, orthopaedic surgeons had greater reason to fear the power of economic arguments in encouraging the public funding of widely separated accident and trauma centres. The segregated Birmingham Accident Hospital and Rehabilitation Centre, over which Gissane took charge in 1941, may have been established partly by chance, but it also represented a sign of the times which orthopaedists saw reason to resist.[66] As E. A. Nicoll put it at a BOA meeting in 1949, 'if all accident cases in a given area were to be dealt with by a traumatic specialist, cold [i.e., non-traumatic] orthopaedics would become self-contained – which would be unfortunate'[67] – unfortunate especially for those with a professional interest in establishing hospital fracture clinics.

The other strategy embedded in the orthopaedic call for accident services was the subordination of rehabilitation. In early drafts of the Memorandum on Accident Services (subsequently denominated 'Orthopaedic and Accident Services'), rehabilitation was seen as an 'essential feature', but it was accepted that it might have to be conducted in widely separated facilities. Rehabilitation centres, it was said, might 'supplant' existing convalescent homes, though they were not to be 'functionally separated from the Accident Centre'. In the printed version of 1943, greater emphasis was placed on rehabilitation facilities 'within the precincts of the parent hospital', and the stress was more on the facilities than on the personnel to carry out the work. The printed memorandum also identified the wartime rehabilitation centres as 'special', thus suggesting that they were marginal, or that their status was conditional upon the extraordinary conditions of war.[68] It was thus more than appropriate for Platt to deliver an address on 'Rehabilitation as a Hospital Service' at the opening of the rehabilitation department of the Preston Royal Infirmary in March 1946, and make clear that 'if hospitals can carry through their rehabilitation programme to the final act of reconditioning, there will be less need for a large number of residential centres'.[69]

These shifts of emphasis relate to and serve to reflect the changing status and meaning of 'regionalism' before and after the war. By the late 1930s regionalism had become the 'order of the day'; to some, the 'New Order', or the 'New Deal'.[70] As kindled through commissions such as that set up by the British Hospitals' Association under Lord Sankey in 1937, regionalism was essentially another name for integrated planning and rationalization. Hospital planners were now increasingly thinking in terms of groupings geographically larger than existing local government areas. Within these 'natural regions' efficiency and economy were to be served by upgrading and standardizing medical and allied health services. This 'regionalism', which was brought into being during the EMS, differed from that advocated by Jones and Girdlestone in 1919 in their National Scheme for crippled children only insofar as it was not based on new institutions. Thus in 1938, in Ministry of Health discussions on the reorganization of hospital services, 'the orthopaedic service' could be put forward as a 'good example of the possibilities of a regional planning', as opposed to the many 'spasmodic efforts' at the local government level.[71]

A decade later, however, when Girdlestone claimed that the 'regional orthopaedic and accident services' were the extension of existing crippled children's orthopaedic schemes, he was only half right.[72] In part because of the absence of new institutions, NHS 'regionalism' was administrative only; functionally, curative medicine remained centred on the existing teaching hospitals in their urban settings.[73] (Indeed, the Regional Hospital Areas in their 1946 form took the names of towns rather than regions, as had been the case in antecedent plans.)[74]

Structurally, all that was different under the NHS was that local, municipal and county authorities were deprived of their (politically accountable) decision-making power. Consequently – as Dawson and others had desired even before the Great War – hospital consultants were able to assume much greater control over the politics of health.

When Girdlestone in 1949, in the context of NHS 'regionalism', pointed to the 'regional' scheme for crippled children as an 'organisation . . . tried over many years' and which had 'earned the approval of local authorities, voluntary hospitals, and all concerned', he was conflating the older with the newer regionalism. Thereby he sought to legitimate the firm basing of orthopaedic-controlled regional accident services within the major NHS teaching hospitals. If this could be accomplished, then one of the chief professionalizing obstacles associated with rehabilitation services would be overcome – the fact that they had mostly been centred on occupations (coal mining, motor vehicle manufacture and heavy industry, in particular) and located in isolation from the prestigious centres of medical education. Thus, in this sense also, hospital accident services became extensions of fracture clinics, intended to bring together 'under the same administrative and surgical control' all therapies for the restoration of locomotor function.[75]

Much negotiation was to take place and many reports were to be written prior to the 1970s when nearly 80 per cent of Accident/ Emergency departments in major British hospitals were under the control of orthopaedic surgeons.[76] Nonetheless, it was the Second World War that served to concentrate the minds of orthopaedists on the issue. With a sense of great urgency the executive of the BOA in 1944 drew the attention of all its members to the Memorandum on Accident Services, and instructed 'that copies be sent to the British Hospitals' Association, [and] to the Committees of Management and Medical Committees of all hospitals of more than 100 beds'.[77] Insofar as this can be attributed to the obstructions to professional fulfilment on other fronts, the Second World War may indeed have marked a new beginning in British orthopaedics. But it would be more correct to say that the war ushered in a new set of priorities for the specialism, the nature of which was radically to transform its previous social and political significance.

12

AN END TO 'ADOLESCENCE'

With its focus on the elderly and its identification with the highly skilled and radically invasive surgical operation for total hip replacement, orthopaedics today stands poles apart from the specialism of the interwar period. No longer do multitudes of disabled soldiers, crippled children and industrial workers constitute its clinical mainstay,[1] and no longer do rural open-air hospitals and low-status outpatient departments serve as its principal work sites. The 'mature' specialism, with premises at the glamorous surgical centre of hospital medicine, embodies a very different set of social relations and professional interests. Of these, the greater kudos, security and income of its practitioners are signs and symbols.[2]

To trace the postwar history of orthopaedics is beyond the purpose of this study. What is sought in this final chapter, rather, is an understanding of the shifts that were fundamental to the postwar remaking of the specialism. More particularly, our concern is with how those shifts were instigated, and the significance they had for the philosophy and politics of the medical organization discussed in the preceding chapters.

SCIENTIFICITY AND WAR

The scientific and technological transformation of orthopaedics to the specialism that we know today owed little to the war itself. After the cessation of hostilities, orthopaedists openly doubted 'whether there had been [as a result of the war] any great advance in technique or in operative skill'.[3] The hip replacement operation, pioneered in Manchester by Platt's young and ambitious associate, John Charnley (1911–82), was a product of the late 1950s and 1960s. Only since then has orthopaedics taken on the hues of science, with increasing investment in basic research and technology.[4] In 1940 only 15 per cent of the papers published in the *Journal of Bone and Joint Surgery* were concerned with 'investigative research', whereas by 1959 the proportion was closer to 50 per cent.[5] As late as 1954, just after the

Journal of Clinical Orthopaedics was launched and just before the Orthopaedic Research Society held its first meeting 'to encourage research by orthopaedic surgeons',[6] it was still considered 'a little out of routine' for an orthopaedic surgeon to be reviewing literature on antibiotics and chemotherapy.[7]

A further reason why the war cannot be held directly accountable for these changes is, conversely, that the interest in research and development can be traced to long before the war – indeed, to before the Great War.[8] Admittedly this is clearest in America where, throughout the interwar period, there were calls for more 'research work' and 'research workers'.[9] Robert Osgood, in 1921 for instance, stressed the need 'to steal time from the confusion of clinical work' in order to

> follow farther the paths of Ollier, Macewen and Codivalla, and blaze trails beyond, bringing our clinical experience to bear upon the purely experimental investigations, seeking to interpret and better our results by knowledge of the fundamental physiologic processes, the laws of growth, and the minute mechanics of repair.[10]

William S. Baer in his presidential address to the AOA in 1924 was more forthright in calling for orthopaedic research 'either in clinical branches or in the laboratory'.[11] As might be expected from the head of the Department of Orthopaedics at the Johns Hopkins University Medical School, Baer emphasized the need for 'the university spirit – the spirit of truth for truth's sake' and a place where one could 'abstract oneself from the professional duties of the day [so] as to obtain a proper time for contemplation and research'. A means to this end was gained in the mid-1930s in the USA when research fellowships became available through the newly founded American Academy of Orthopaedics.[12]

Although attitudes towards basic and clinical research in British medicine before the Second World War were more equivocal, and the practice less developed, there is no doubt that increasing importance came to be attached to the image of science in medicine. Within orthopaedics, the white coats that Platt introduced into his practice upon his return from America in 1914 were no less symbolically significant than the 1918 constitution of the BOA which proclaimed 'the advancement of the science and art of orthopaedic surgery'. Certainly, BOA members shared little of the disdain for science and technology that Hugh Owen Thomas sometimes expressed.[13] On the contrary, a majority prided themselves on the lengths to which they took their bacteriologically informed aseptic techniques.[14] Widely endorsed was the positivist belief expressed by Jones after the Great War, that it was 'by reason of the advances made in pathological, anatomical, and physiological knowledge' that orthopaedics had been able to 'enlarge its borders on the operative side'.[15] True, little such research was undertaken by orthopaedists themselves. The experimental work conducted by Hey Groves prior to the Great War on the repair of fractures in animals is notable for being so unusual;[16] moreover, it is a telling reminder of priorities in British orthopaedics in the interwar period that Hey Groves's research ceased once he joined Jones in the politics of orthopaedic organization. In general, the social and philosophical interests of British orthopaedists were not conducive to the reductive concerns of 'pure' scientific research. Yet it would be wrong simply to juxtapose these concerns. As we have indicated, the pursuits of rationalization, coordination, efficient controls over processes, and 'team work' were recognizable parts of an increasingly important appearing science of management. It should also be borne in mind that in few other fields of medicine in Britain – let alone in other specialisms – were there well-developed traditions of 'pure', clinical or experimental research. In spite, or because, of efforts by the Medical Research Council to foster clinical research, developments were piecemeal until after the Second World War when an effective university research base was established.[17]

However, shortly before the Second World war orthopaedic leaders themselves began to distance their specialism from its 'less than scientific' past. Rowley Bristow, for instance, in a keynote address of 1937, opined that 'In the past, orthopaedic surgeons have been called upon to play a leading part in the actual details of organisation and administration of what may be called the social aspect of orthopaedics.' Bristow now called for more basic research, justifying it on his perception of the specialism's altered circumstances. 'The problem of the treatment of the crippled child [was now] largely solved, or at least well in hand', he argued and (quoting H.A.T. Fairbank) ' "it seems . . . likely that eventually all fracture clinics will be under the care of the orthopaedic surgeon of the hospital" '.[18] Thus for purposes of professional self-preservation it was already beginning to seem that

the vision of the future generation [of orthopaedists] must be . . . to the attainment of a second objective – viz., the furtherance of investigation and active research into the basic problems which form the background of orthopaedics – problems in the applied anatomy, physiology, and pathology of the locomotor system.

Obviously, then, the Second World War was not responsible for initiating research in orthopaedics; nor is it clear that it did much to stimulate those interests. The war, rather, by changing the political reality of medicine as a whole, indirectly encouraged orthopaedists to shift their professional sights away from the wider politics of medical organization. By 1948 the opportunities for such politics were largely gone, and orthopaedists were more or less compelled to redefine the nature of their specialty in the narrower terms of research and hospital medicine.

This process of professional redefinition began with the consolidation of the position of most consultants as a privileged elite within British medicine during the Emergency Medical Service and later under the NHS. Hospital consultants, more than others in medicine, were to gain what Aneurin Bevan declared should be the doctors' 'full participation in the administration of their own profession'.[19] Not only were they now freed from the bossy lay governors of voluntary hospitals and the local politicians in charge of municipal facilities and services, but they also gained some control over the regional allocation of central government resources. Inasmuch as these had been the goals of the reforming ideologues in orthopaedics since the Great War, their interests were well met. Only later would it become apparent that these same events had reshaped the political arena in such a way as to siderail certain other key elements in their interwar thinking. At the time, the seeming benefits of the new medical order were sufficient to bring to an end the interwar calls for rationalization and reform. In order to consolidate their position within the new hospital structure, and to combat the process of marginalization outlined in the previous chapter, orthopaedists needed to adopt a new image – one based on the authority accorded to science.

More precisely, scientific and technological research became a strategy vital to professional survival, for it was a means of securing a footing on the top rung of the medical hierarchy – the university medical schools. To argue the case for research and development – whether basic, clinical or technological – was simultaneously to argue the need for an appropriate place to conduct that work, and increasingly that place was seen to be the university medical school. Again, this understanding and the strategy arising from it pre-dated the war. Bristow in his 1937 address maintained that research could only come about when orthopaedic surgery is accorded its rightful academic status in the medical schools and universities, and when research is generously endowed. We must use our enthusiasm and our powers of persuasion to bring about this ideal state of affairs. ... I would like to see a bigger representation of orthopaedic surgeons on the various examining boards.

But what was then coming to seem merely professionally worthwhile was now appearing as imperative. The *Report of the Inter-Departmental Committee on Medical Schools* (1944), under the chairmanship of the banker Sir William Goodenough, left no doubt that a specialism without a scientific profile was doomed to the proverbial 'dustbin of history'. The Goodenough Committee found that in most teaching hospitals in Britain the accommodation for teaching and research fell well below ordinary requirements when compared to America and Europe. The future of British medicine, it maintained, depended on implementing essentially the Johns Hopkins model of medical education – the integration of medical schools into universities, and the appointment of full-time professors for teaching and research under the control of the University Grants Committee.[20] It is not surprising, therefore, that increasingly the lament of British orthopaedists echoed that of their American colleagues: that one of 'the greatest drawbacks to basic fundamental research in orthopaedic surgery is the paucity of adequately remunerated chairs of orthopaedic surgery'.[21]

The 'social', low-tech profile of orthopaedics was thus forced to change, and the history of British orthopaedics had either to be modified or ditched. By 1948 – the year that Charnley and a few other bright sparks set sail for America to visit leading orthopaedic centres – it had become conventional to denigrate a past which, it now seemed, had foolishly inverted its constitution and 'put art before science'. If 'young men of ability and character' were to be recruited to the specialism, argued the president of the BOA, its past was best left behind.[22]

Rapid 'progress' was already under way and, in general, as the rhetoric of science burgeoned, so the old conservative surgery waned along with the holistic philosophy of which it was a part. By 1947, Charnley was 'strongly supporting radical measures' in the treatment of backache, and after his visit to America he drastically revised his views on the conservative treatment of tuberculous joints.[23] Having declined an invitation to work in the USA (because he believed the potential for human experimentation was much greater in British hospitals), he also began to drive a wedge between orthopaedics and both physical medicine and osteopathy. This 'robust attitude', which Charnley's biographer suggests was characteristic of the new generation in orthopaedics, also entailed drawing distinctions between patients with 'real' organic disease and those suspected of having only 'psychosomatic' complaints. Among this new generation of orthopaedists, there were few calls for heeding patients' individual socioeconomic and psychological circumstances. On the contrary, the seeming lack of scientific 'rigour' surrounding such concerns, or surrounding such procedures as massage, made these preoccupations seem suspect. During the early years of the Cold War the long-departed champion of massage therapy, J. M. M. Lucas-Championnière, was resurrected only in order to be denounced as 'no true scientist' and banished as a 'surgical Karl Marx'![24] Significantly, it was also in 1947 that the orthopaedic physician James Cyriax published his *Textbook of Orthopaedic Medicine* in order to bridge what he felt was the growing gap between surgery on the one hand and 'do-nothingism' on the other. The continued demand for Cyriax's text itself testifies how the gap has remained open, as also does Cyriax's call in the preface to the 1978 edition for an independent specialty status for orthopaedic medicine.

Yet it would be naive to suppose that the ideologies and conceptions that had so recently informed the specialism's past were suddenly discarded. Ideas linger, in part because those who have held them do not disappear, or do not disappear all at one historical moment. Nor, in embracing one kind of outlook, are all others necessarily forsaken. While, in retrospect, the increased involvement of orthopaedic surgeons with research and technology can be seen to parallel the demise of their preoccupations with the organizing principles of unity, team work and integration, this historical trajectory was by no means so self-evident at the time. By turning to the other major concern of British orthopaedists in the period between the setting up of the EMS in 1939 and the initial operation of the NHS in 1948 – the securing of a commanding place for the specialism in medical curricula – one can trace what was conceived by some to be the consummation of integrative ideals.

PURSUING THE MEDICAL CURRICULUM

Interest in research in orthopaedics as a means to gaining entry into university medical schools was usually accompanied by concern with securing a place for the specialism in medical curricula. But these two aims were also separable, or at least capable of being prioritized. Clearly, this was how it appeared to Platt in 1945 when he wrote: 'I do not think orthopaedic surgeons will fail to make contributions both massive and audacious to technological advances, but I foresee the danger that our contributions to general ideas may be negligible, unless we can ensure . . . recruitment to our specialty.'[25] From his perspective, the place of the specialism in the structure of medical education was more fundamental than the pursuit of research and development, for upon it appeared to depend not only the status and security of orthopaedics within medicine's new academic echelons, but – once again – its very survival.

Statistical sources, no less than anecdotal ones, suggest that Platt was right. According to one set of retrospective figures, for 1938–9, there were less than half as many orthopaedic surgeons in Britain as there were otolaryngologists and less than a third the number of ophthalmologists. Collectively, orthopaedists comprised only 4.1 per cent of all consultants and specialists in medicine and surgery. A decade later this percentage had hardly changed (4.3); indeed, it was still under 5 per cent in 1964.[26] An examination of the 814 Fellows of the Royal College of Surgeons who died between 1930 and 1951 (that is, those likely to have been trained mainly between 1885 and 1905) reveals that 53 (6.5 per cent) said they had a specialist interest in orthopaedics.[27] All such statistics must be treated with caution, however, for both before and after the NHS, specialist status was self-determined, rather than certificated as in the USA, and there was no register of specialists (though a proposal for one was made in 1944).[28] Thus the figure for the number of orthopaedic consultants working within the NHS in 1949 (227) scarcely tallies with the total of BOA members in 1948 (Table 12.1). But the most obvious explanation for this disparity is that not all BOA members were employed as consultant specialists under the NHS. Among 'Associate' members, especially, were many junior-level recruits.

Perhaps as worrying to Platt may have been the signs of possible stagnation in the specialism. Although the membership of the BOA had swollen from 24 in 1918 to nearly 300 by 1940, since 1934 only nine members had risen from the ranks of 'Associate' to 'Active' – the latter usually indicating full commitment as an orthopaedic specialist to hospital consultancy work and private practice.[29] This lack of expansion may reflect elitism within the BOA; but, mainly, it seems to indicate – like the low percentage of orthopaedists relative to other

Table 12.1   British Orthopaedic Association members, 1918–48

Year	'Active'	'Associate'	Total
1918	24	–	24
1920	35	–	35
1922	54	25	79
1928	94	67	161
1934	98	106	204
1940	98	195*	293
1948	155	286	441

* 69 (35%) of these became 'Active Members' by 1948.
Source: BOA Annual Reports.

specialisms – the perceived lack of opportunities in the field for lucrative private practice, and the continued unglamorous outpatient department image. Only five of the 28 general teaching hospitals in Britain in 1938–9 had slightly over 10 per cent of their beds given over to teaching orthopaedics.[30] Consequently, as Platt was made aware on his tours of hospital inspection for the EMS, orthopaedic work tended to be conducted by persons either improperly trained in orthopaedics or uncommitted, or both.[31] Of course the demands of the war made the manpower problem even worse; by 1942, 'it was manifest that the supply of properly qualified men was inadequate both in the Army and in the E.M.S.' The same member of the BOA who pointed this out, further echoed what was also becoming widely apparent – 'that it was of little value to plan post-war accident services [as Watson-Jones was doing] unless a supply of men to man these services was forthcoming'.[32] Although 'crash' courses for fracture treatment during the war were a help, in the long term more permanent inroads into medical curricula were demanded.

As we saw in Chapter 6, Robert Jones had been concerned with this problem during and after the Great War. But his efforts to turn the Shepherd's Bush Infirmary in Hammersmith into a postgraduate school to teach courses in orthopaedics similar to the six-month ones that he had run during the war were unsuccessful. When, in 1935, this same former Poor Law infirmary became the Royal Postgraduate Medical School, postgraduate training in orthopaedics was finally secured, though for the first 50 years of the school there was no professor of orthopaedics.[33] Scarcely more successful was the postgraduate course that Jones and his colleagues established in Liverpool.

Before 1925, when the then young demonstrator, Watson-Jones, enrolled on the course and carried away the Gold Medal, there were only two other students. Not until 1945, when the Liverpool Department of Orthopaedics was reconstituted as a separate department in the Faculty of Medicine, were there more than a dozen students at any one time. The fact that it remained the only course of its kind in the world is further indicative of its general standing. In Britain, because a fellowship in one of the Royal Colleges served as the prerequisite to hospital consultancy, a diploma in orthopaedics was of no particular value to a would-be orthopaedic consultant. Both before and after 1948 (when the number of students enrolled on the course rose to 52 before sliding back), as many as 70 per cent of those on the course were from outside the British Isles.[34]

Chief among the complaints raised against the Liverpool degree – especially by Platt in the 1940s – was that it reinforced a regard of orthopaedics as a narrow specialism suited only for postgraduate study. Such a view had not been intended by Jones, who consistently argued that 'an orthopaedic surgeon must have a thorough working knowledge of general surgery' above all else.[35] By the time Platt came to utter his opinions on education, the grounds for the undergraduate teaching of orthopaedics as a branch of general surgery had not diminished, but had strengthened on two counts.

The first reason – ironically – was the success of orthopaedists in carving out their own hospital space, as a result of which fewer undergraduates gained access to fracture and trauma cases. Hitherto, there had been clinical sessions in orthopaedics in the outpatient departments of several of the larger teaching hospitals.[36] Although attendance at these clinical sessions or at orthopaedic lectures was not compulsory, most undergraduates received some instruction and experience in orthopaedics during their period as hospital dressers. Now – just as the opponents of orthopaedic specialization in the 1920s had forecast – they were less likely to do so, especially in places such as Manchester, Leeds, Bristol, Liverpool, Oxford and some of the London schools where orthopaedic specialization had gone furthest. At Liverpool, this problem was aired in December 1938, and it was decided to make it mandatory for students to spend three months as dressers in orthopaedic wards during their final year (including one month in the casualty department).[37] At St Bartholomew's Hospital in 1943 the Policy Committee of the Medical College recommended 'that students should be attached to the Orthopaedic Department for a period of 3 months and attend both in-patient and out-patient sections' (though as late as 1956, this policy had still not been implemented).[38] Prior to the NHS there was little incentive for students to devote themselves to it, since only a few orthopaedists could hope to obtain a good living from private practice.[39] In a place the size of Aberdeen, it remained highly debatable until the NHS whether there would 'ever be a living here for a specialized orthopaedic surgeon'.[40]

The other reason why it seemed more than ever necessary to stress that orthopaedics was a branch of general surgery was, paradoxically, general surgery's demise in the face of the growing strength of specialisms with non-generalist orientations. Neurosurgery, urology, thoracic and cardiac surgery, plastic surgery, among other divisions, not only fragmented the general surgery that had been known in the nineteenth century, but they also challenged the post-1880s dominance of abdominal surgery within that general surgery. Abdominal surgeons sought to cling to the power and authority that they had acquired, but the practical basis for this was seriously eroded by the time of Moynihan's death in 1936. Thereafter, as Platt maintained in a lecture in 1945 on 'Orthopaedics in Medical Education', it was far more difficult for abdominal surgeons to claim the 'automatic right . . . to assume leadership in the affairs of surgery and to occupy every university chair in surgery'.[41] The 'abdominal wave', Platt insisted, had 'spent its force' and the abdominal surgeon 'in his capacity to contribute to techniques and to general and fundamental notions . . . is now only one among many'. Senior orthopaedists might delight in the fact, for in dispelling what Jones had always envied as the 'lure of abdominal surgery',[42] authority and leadership in surgery, it was hoped, would pass to them. They were left the representatives of what 'is undoubtedly the largest field so far abstracted from surgery as a whole'.[43]

The orthopaedists' concern to secure a place in medical curricula had gained momentum in the late 1930s as the spectre of war grew more menacing and the manpower shortages of the last war began to be recalled. To Watson-Jones and the others who constituted the BMA's Orthopaedic Group Committee (which met for the first time in October 1938 in anticipation of 'an emergency') it seemed self-evident that 'the present very unsatisfactory state of the teaching of orthopaedic surgery should be one of the first objects to be considered and dealt with by the Committee'.[44] Watson-Jones, who had already raised the matter at a BOA executive meeting, had apparently only been waiting for the formation of this expressly 'medico-political body' in order to bring forward his preliminary memorandum. Unfortunately, there are no details of the memorandum either in the files of the BMA or the BOA, and it may be that Watson-Jones's wartime involvements in the RAF, private practice, and other matters – notably his drafting of the BOA memorandum on accident services – precluded his continuance with it. It is unlikely to have been much different, however, from the Memorandum on Education drafted by Platt in May 1942 at the request of the BOA executive in their anticipation of the hearings later that year of the Goodenough Committee on Medical Education. Platt was ideally suited for the task (which may be why Watson-Jones left him to it): along with Girdlestone and Ernest Rock Carling, he had been a charter member of the Advisory Committee of the Nuffield Provincial Hospitals' Trust, of which Goodenough was the chairman. He was also a close colleague of Sir John Stopford, the vice-chancellor of Manchester University and formerly its Dean of Medicine, who was the vice-chairman of the Goodenough Committee.[45]

In essence, Platt's memorandum called for formalizing on a national scale the kind of educational programme that he and Stopford had been developing in Manchester since the mid-1930s. In this, orthopaedics formed a systematic and examinable part of undergraduate teaching and clinical work in general surgery with 'not less than twenty-four lectures' and not less than two months of intensive clinical experience under a tutor in orthopaedics.[46] As worked out later in somewhat more detail:

> During the first two clinical years all students should spend part of their surgical dressership on the orthopaedic unit and should attend a comprehensive course of systematic lectures covering both 'trauma' and 'cold orthopaedics'. In the 6th, the final clinical year, teaching should be available in the general orthopaedic outpatient clinics and major fracture clinics and in revision classes and seminars, where small groups can be brought into contact with the orthopaedic clinical tutors. During this year, one or more large-scale demonstrations should be staged in a country hospital, where the long-term therapeutic and sociological policies of orthopaedics can be brought vividly to the notice of the student, now in his most receptive mood, on the eve of qualification.[47]

Such instruction (cum-indoctrination) was not solely intended to lead students into orthopaedic specialization; primarily, it was intended to situate orthopaedics in general medicine so as to no longer render problematic either the status of orthopaedic surgery or recruitment to it. Against the view 'of some clinical teachers' (and others 'who should know better') that orthopaedics was only a postgraduate subject, Platt insisted that 'orthopaedic teaching must permeate the undergraduate course' and become 'an integral part of clinical medicine and surgery'.[48] Far from wanting merely the superaddition of orthopaedics to the general medical curriculum, Platt was calling for the curriculum to be constituted around it! At the same time, although he did not deny the value of postgraduate training courses and classes, he doubted the wisdom of diploma courses in orthopaedics. That offered by Liverpool he regarded as a 'back-door approach to specialism'.[49]

Petty rivalries between Manchester and Liverpool might be detected in the latter remark, but this should not obscure that at the root of Platt's rhetoric lay the strong legacy in orthopaedics of integrative principles. Basic to his thinking were those features which had been central to the specialism's recent past: the union of curative and preventive practices; the indivisibility of medicine and surgery; holistic understanding of the relations between mind and body and between socioeconomic and psychosocial processes; and, not least, the harmonious networking of regional and national medical services for crippled children and accident victims. Platt's claims for orthopaedics in medical education were thus staked on a higher intellectual ground than those of some of his colleagues who would have been satisfied simply with a succession to general surgery by default,[50] or those who sought only to claim that 'the well-trained orthopaedic surgeon is no less competent to supervise the education of undergraduates than any other surgeon'.[51]

In a context of ever greater fragmentation of both the practice and the bodily object of medicine, Platt was holding up orthopaedics as the last bastion against what he later styled 'surgical apartheid'.[52] Although in the 1960s he had to confess that 'no one individual can today hope to be able to practice and teach the totality of surgery',[53] in the 1940s he still hoped that 'the present trend will be reversed and the specialisms reabsorbed [or] . . . de-differentiate[d]'.[54] Resonant was the prophecy of Osgood in the 1920s, of 'a time when the specialty will have found its soul by losing it in general medicine'.[55] At the very least, echoed Platt, 'each surgical specialty must continue to advance the general craftsmanship of surgery, and – of even greater significance – it must contribute to general ideas'.[56]

Such ideals were by no means anachronistic in the decade between the setting up of the EMS and the start of the NHS. Despite the public enthusiasm for medical specialists, the intellectual status of specialization was still open to considerable debate. The socialist and influential spokesman for science, J. D. Bernal, for instance, pondered the value of specialization in his *The Social Function of Science* (1939) and concluded that 'the monopolising of a certain little corner of knowledge, not making it easy for others to understand, so as to enjoy that delicious sense of personal possession of knowledge, is the ultimate crime of the scientist'. Four years later, the gynaecologist and former member of Lord Dawson's Consultative Council to the Ministry of Health, Victor Bonney, in his Hunterian Oration, argued in language strikingly close to Platt's that 'too restricted attention to one subject cramps the outlook, narrows the mind, destroys the sense of proportion, deforms the speciality, and tends in the end to transform it into a cult'. These and other sources of criticism were cited in an article in the *Lancet* of July 1944 in which Sir Sheldon Dudley, Medical Director-General of the Royal Navy, took the orthopaedic surgeons themselves to task for the narrowness of their gaze. 'It is impossible in the Navy', Dudley claimed, 'to get some orthopaedic surgeons to realise that venereal diseases and tuberculosis are more important problems than fractures, when judged by their effect on the fighting efficiency of the Fleet.'[57] Leaving aside what this suggests – a marked difference between the principled pronouncements of the leaders of the specialism and the practices of its rank and file practitioners (though the latter in this case may only have been temporary recruits from general surgery) – the viewpoints of Dudley and Platt were not greatly different. Although Dudley wanted 'the ordinary doctor' to be trained to diagnose and treat 'all the easy, common and emergency cases that are within each specialty', he and Platt were in agreement on the blinkering effects of specialization. So, too, were the postwar spokesmen for paediatrics, the rhetoric of whom was suffused with the merits of generalism over specialism.[58] Perhaps still more significant is that the Goodenough Report also emphasized the need for a coherent unification of medical knowledge within 'an atmosphere of specialization'.[59] Indeed, the basic orientation of the Report was towards a holistic integration of the components of medical education, and its Committee was strongly committed to fostering the educational framework necessary for the development of a 'social medicine' in which GPs would combine curative and preventive practices.[60] In its section on orthopaedics, the Goodenough Report endorsed the view of Platt that 'possibly no other type of surgical condition offers so many opportunities for inculcating the important principle of considering every aspect of surgical illness'.[61]

To the extent that the Goodenough Report gave positive encouragement to Platt's ambition to move 'the surgery of the extremities' closer to the centre of medical education, it is possible to say that after the Second World War (though not *because* of it), modern orthopaedics finally gained some of the ground that Jones and his men had striven for since the Great War. The problem of recruiting orthopaedic surgeons in Britain was to persist,[62] but Platt could rest satisfied in 1950, claiming that

> orthopaedic surgeons are now fully alive to the dangers of teaching undergraduates a 'specialism', and are turning more and more to the exposition of general principles. With the integration of orthopaedics in the curriculum of medicine and surgery as a whole, the antithesis of 'general' *vis à vis* 'Special' surgery becomes meaningless.[63]

Yet, in many ways this was wishful thinking – at best, a compromise with the ideals of the past that it purported to cement. However professionally advantageous it may have been to secure what Bristow had called the 'rightful academic status [of orthopaedics] in the medical schools and universities', that step undermined the principles of unity that had been central to the original conception of modern orthopaedics. The pursuit of a niche in the medical curriculum, and the advocacy of research and development, helped to legitimize the university medical school as *the* centre of medicine, and to underwrite its fragmentary and wholly curative approach to patients' bodies.

This had not been the intention of Platt and Bristow's generation in seeking academic status for orthopaedics. Platt, while deeply involved with the Manchester Joint Hospitals' Advisory Board in the late 1930s, characteristically spoke of the universities as but one of the 'constituent bodies . . . central [to] future hospital policy' in any large urban centre; the most notable other constituent being the local authorities and the voluntary hospitals. All, including GPs and local MOsH, were seen as equally important to a 'unified orthopaedic service', curative and preventive.[64] But in the course of the wartime medical planning, and more especially the planning for the NHS, the place and role of university medical schools was subtly elevated. No longer were they merely attachments to the major teaching hospitals and, therefore, part of the hubs around which regional medical services could be planned. Instead, they became the hub itself, central to the structure of medicine and the hegemony of hospital consultants within it.[65] As research interests and technology access came to legitimate that position and the place of full- or part-time salaried specialist consultants within it, the principles and practices of coordination between different groups and institutions fell by the wayside.

Of course in seeking 'academic status', orthopaedists did not consciously renounce their interest in the integration of regional health services. Because the plans for the NHS were largely dominated by concerns with upgrading hospitals within regional structures,[66] many of the shortcomings of the service from the point of view of integrated delivery only became clear several years later. With regard to rehabilitation, it was some twelve years after Watson-Jones had proclaimed that ideals had become facts as a result of the war,[67] that the *Report of the Committee of Inquiry on the Rehabilitation, Training and Resettlement of Disabled Persons* (1956) made plain that the NHS had effectively destroyed all hopes for integrated rehabilitation services by leaving intact the old division between industrial health and the rest of medicine. Whilst lamenting that 'the medical profession as a whole has failed to realize the scope and potentialities of effective rehabilitation', the Report also observed how the links between hospital services and industry had floundered.[68] Similarly, a major orthopaedic review of accident services a few years later made clear that 'One of the problems of the National Health Service has been a division of responsibility . . . [between] the hospital, general practice and the public health services.'[69] Thus well into the 1970s the BOA would be compelled to reassert the need for a comprehensive accident service of well-integrated sectors.

These and other divisions in the NHS, when set next to the principles of continuity and unity celebrated by interwar orthopaedics, make it easy to understand how Platt could come to the conclusion in 1976 that in some respects the wartime and postwar legislation had had a 'deadening effect on creative thinking in the planning of hospital clinical services'.[70] While on the one hand the NHS made a safe haven for a group of practitioners whose survival had hitherto been precarious in the private patient market, on the other hand, it subtly stripped the specialism of its former meaning, role, and intellectual purpose in the organization of modern medicine. The NHS further drew a veil over that meaning and example by encouraging the specialism to pursue scientific research and development. To the extent that the latter was conducted successfully, and the image of the specialism remade, orthopaedics became a conventional member of the club of surgical specialties – a fact perhaps reflected in the election of Platt to the presidency of the Royal College of Surgeons in 1954.

For orthopaedics, then, the NHS, like the EMS of the Second World War, did not represent the culmination of all the efforts at rational planning and organization that had gone before. But neither did it signify the wholesale negation of the past; after all, the NHS was no less the outcome of the same material forces, ideologies and events that had structured orthopaedics and its approach to the body. In the final analysis, the NHS in the history of the modern specialism stands as a political marker to the beginning of the end of a notable and revealing 'adolescence'.

NOTES

The place of publication is London unless otherwise indicated. The first reference to any printed source within a chapter is given in full; subsequent citations within the same chapter are normally abbreviated to author's surname and short title.

1 Introduction

1. See J. Noble and C. S. B. Galasko (eds), *Recent Developments in Orthopaedic Surgery* (Manchester, 1987); idem, *Current Trends in Orthopaedic Surgery* (Manchester, 1988).
2. See D. S. Barrett, 'Are Orthopaedic Surgeons Gorillas?', *BMJ*, 24–31 Dec. 1989, pp. 1638–9, and Elizabeth Morgan's discussion of her time in an orthopaedic department in an American hospital in her *The Making of a Woman Surgeon* (New York, 1980), pp. 172ff.
3. See Paul Vickers, 'Orthopaedics', in I. Craft *et al.* (eds) *Specialized Futures: essays in honour of Sir George Godber* (1975), pp. 1–29; and Royal College of Surgeons of England, Commission on the Provision of Surgical Services, *Report of the Working Party on The Management of Patients With Major Injuries* (November 1988).
4. See, for example, the special issue of *JBJS*, 32B (Nov. 1950), 'Half a Century of Progress in Orthopaedic Surgery, 1900–1950', pp. 451–740. Cf. N. E. Vanzan Marchini (ed.), *L'ortopedia nella storia e nel costume* (Udine, 1989), which brings to light the many interconnecting social, scientific, artistic and cultural features that shaped the development of orthopaedics in Italy.
5. For an overview, see Gerald Larkin, *Occupational Monopoly and Modern Medicine* (1983), ch. 1: 'Perspectives on Professional Growth'.
6. Rosen, *The Specialization of Medicine with Particular Reference to Ophthalmology* (New York, 1944), p. 4.
7. Sydney Halpern, *American Pediatrics: the social dynamics of professionalism 1880–1980* (Berkeley, 1988), pp. 3–4. See also Andrew Abbott, *The System of Professions: an essay on the division of expert labor* (Chicago, 1988).
8. Against the latter approach, so far as professions are concerned, see Eliot Freidson, *Profession of Medicine: a study of the sociology of applied knowledge* (Chicago, 1970). Freidson's advice, to let the sociological theory of professionalization flow from empirical studies, rather than vice versa, is acted on in Rita Schepers's *De Opkomst van het Medisch Beroep in Belgie: de evolutie van de wetgeving en de beroepsorganisaties in de 19th eeuw* (Amsterdam, 1989).
9. See, for example, Glenn Gritzer and Arnold Arluke, *The Making of Rehabilitation: a political economy of medical specialization, 1890–1980* (Berkeley, 1985). For examples of other ways of approaching the history

of specialization in medicine, see Stanley J. Reiser, 'Technology, Specialization, and the Allied Health Professions', *J. Allied Health*, 12 (1983), pp. 177–82; Lester King, 'Medicine in the USA: XXI: specialization', *JAMA*, 9 Mar. 1984, pp. 1333–8; A. B. Davis, 'Twentieth Century American Medicine and the Rise of Specialization: the case of anesthesiology', in F. J. Coppa and R. Harmond (eds), *Technology in the Twentieth Century* (Iowa, 1983), pp. 73–88; B. E. Blustein, 'New York Neurologists and the Specialization of American Medicine', *Bull. Hist. Med*, 53 (1979), pp. 170–83; and Joel Howell, 'The Changing Face of Twentieth-Century American Cardiology', *Annals of Internal Medicine*, 105 (Nov. 1986), pp. 772–82.

10. For the USA, see Charles Rosenberg, *The Care of Strangers: the rise of America's hospital system* (New York, 1987), esp. p. 210; Paul Starr, *The Social Transformation of American Medicine* (New York, 1982), p. 165 et passim; and Rosemary Stevens, *American Medicine and the Public Interest* (New Haven, 1971).

11. See Ivan Waddington, *The Medical Profession in the Industrial Revolution* (Dublin, 1984).

12. For a detailed elaboration of key aspects of this change within a specific provincial context, see Steve Sturdy, 'The Political Economy of Scientific Medicine: science, education and transformation of medical practice in Sheffield, 1890–1922', *Med. Hist.*, 36 (1992), pp. 125–59, esp. pp. 136–7.

13. For the most part, the institutionalization in British hospitals of 'consultant specialists', so called, was post-NHS. Non-consulting, salaried 'surgical specialists' were called for during the First World War and were sometimes appointed to municipal hospitals in the interwar period. See below Chs 8 and 9.

14. On the specialty boards, see Stevens, *American Medicine*.

15. See Frank Honigsbaum, *The Division in British Medicine: a history of the separation of general practice from hospital care, 1911–1968* (1979).

Among several more or less social-contextual studies of specialisms in Britain recently to appear are: Ornella Moscucci, *The Science of Woman: gynaecology and gender in England, 1800–1929* (Cambridge, 1990); W. F. Bynum, C. Lawrence and V. Nutton (eds), *The Emergence of Modern Cardiology, Med. Hist.*, Suppl. no. 5 (1985); Lindsay Granshaw, *St. Mark's Hospital, London: a social history of a specialist hospital* (1985); and Jennifer Beinart, *A History of the Nuffield Department of Anaesthetics, Oxford 1937–1987* (Oxford, 1987). The latter three are to some extent triumphalist accounts and it is significant that the latter two are both the result of commissions from the specialist institutions themselves, and, indeed, are as much about those institutions as about specialty formation.

There are also historically informed sociological studies of bodies of specialty medical knowledge. Among these is Larkin's *Occupational Monopoly*, which draws heavily on American studies of professionalization to illustrate how the boundaries were drawn around the 'marginal' medical occupations of ophthalmic opticians, radiographers, physiotherapists and chiropodists. Studies such as these, on boundary formation, owe debts to the sociology of scientific knowledge as pursued in relation to the study of the emergence of scientific disciplines. For example, G. Lemaine and Roy Macleod (eds), *Perspectives on the Emergence of Scientific Disciplines* (The Hague/Paris, 1976). See here, in particular, Michael Worboys, 'The Emergence of Tropical Medicine: a study of a scientific speciality'. Sharing this sociological perspective, but extending into the role of economic forces is David Cantor's 'The Contradictions of Specialization: rheumatism and the decline of the spa in inter-war Britain', in Roy Porter (ed.), *The Medical History of Waters and Spas, Med. Hist.*, Suppl. no. 10 (1990), pp. 127–44.

16. Rosemary Stevens, *Medical Practice in Modern England: the impact of specialisation and state medicine* (New Haven, 1966), p. 11.

17. She reminds us at the outset of her book, that 'in 1939 there was only one specialist to every six or seven GPs, and they were highly concentrated: more than one third of them were in London alone', *Medical Practice in Modern England*, p. 3. However, American comparisons may be exaggerated. As late as 1940, 70 per cent of doctors in the USA were still in general practice. See David Rothman, *Strangers at the Bedside* (New York, 1991), pp. 113–14.

18. *Medical Practice in Modern England*, p. 4.

19. Stevens's view of scientific knowledge as a strong determinant in medical relations is shared by Daniel Fox in his *Health Policies, Health Politics: the British and American experience, 1911–1965* (Princeton, 1986). However, compare Daniel Fox, 'The National Health Service and the Second World War: the elaboration of consensus', in Harold L. Smith (ed.), *War and Social Change: British society in the Second World War* (Manchester, 1986), pp. 32–57 at pp. 32 and 49. For Honigsbaum, too, it appears that 'the advance of knowledge is mainly responsible' for the fact that 'doctors nearly everywhere are divided into two main classes, general practitioners and specialists'. *Division in British Medicine*, p. 1.

20. Sir Harry Platt, interview, 1959. 'But now', he recollected, 'the limelight has shifted to others – amongst them the plastic surgeons and the cardiac surgeon – and we are finding that the field of orthopaedics is a contracting universe.'

21. 'Corporatism' is used throughout this study to refer to non-individualist, non-class oriented professionalizing activities, which have been informed by and, are themselves expressions of, managerial notions of efficiency. This usage, which relates to practice, rather than to political theories of the state, has largely been informed by Harold Perkin, *The Rise of Professional Society: England since 1880* (1989), ch. 7: 'Toward a Corporate Society', esp. p. 290.

22. Such is the claim, however, of Ted Bogacz with respect to neurology in his 'War Neurosis and Cultural Change in England, 1914–22: the work of the War Office Committee of Enquiry into "Shell-Shock"', *J. Contemp. Hist.*, 24 (1989), pp. 227–56 at p. 227.

23. However, see Roger Cooter (ed.), *In the Name of the Child: health and welfare, 1880–1940* (1992).

24. The point is underlined by Eduard Seidler's 'An Historical Survey of Children's Hospitals', in Lindsay Granshaw and Roy Porter (eds), *The Hospital in History* (1989), pp. 181–97.

25. Among general texts, almost alone in even mentioning the subject of industrial injury as a part of public health is Anthony S. Wohl's *Endangered Lives: public health in Victorian Britain* (1983). For the few specialist writings, see below Chs 5 and 7.
26. Among the exceptions are J.V. Pickstone, *Medicine and Industrial Society: a history of hospital development in Manchester and its region, 1752–1946* (Manchester, 1985), ch. 7, and David Green, *Working-Class Patients and the Medical Establishment: self-help in Britain from the mid-nineteenth century to 1948* (Aldershot, 1985).
27. Noel Whiteside in her review of Peter Hennock, *British Social Reform and German Precedents: the case of social insurance, 1880–1914* (Oxford, 1987) in *Social Hist. Med.*, 1 (1988), pp. 243–4. But see P.W.J. Bartrip, *Workmen's Compensation in Twentieth-century Britain: law, history and social policy* (Aldershot, 1987), and Paul Weindling (ed.), *The Social History of Occupational Health* (1985). As with so many of these subjects, more research has been conducted in North America than in Britain or Europe. See, for example, David Rosner and Gerald Markowitz (eds), *Dying for Work: workers' safety and health in twentieth-century America* (Bloomington, Indiana, 1987); Eric Tucker, *Administering Danger in the Workplace: the law and politics of occupational health and safety in Ontario, 1850–1914* (Toronto, 1990); and Daniel M. Berman, *Death on the Job: occupational health and safety struggles in the United States* (New York, 1978).
28. See Susan Reverby and David Rosner, 'Beyond the Great Doctors' in Reverby and Rosner (eds), *Health Care in America: essays in social history* (Philadelphia, 1979), pp. 3–16; and, for exemplification, Peter Wright and A. Treacher (eds), *The Problem of Medical Knowledge: examining the social construction of medicine* (Edinburgh, 1982).
29. Bruno Latour, *The Pasteurization of France*, trans. Alan Sheridan and John Law (Cambridge, Mass., 1988), p. 218.
30. Of the little work that has been conducted on medicine from patients' perspectives, most has focused on the seventeenth and eighteenth centuries. See, Lucinda M. Beier, *Sufferers and Healers: the experience of illness in seventeenth-century England* (1988); Dorothy Porter and Roy Porter, *In Sickness and in Health: the British experience, 1650–1850* (1988); Dorothy Porter and Roy Porter, *Patient's Progress: doctors and doctoring in eighteenth-century England* (Cambridge, 1989); and Mary Fissell, *Patients, Power, and the Poor in Eighteenth-Century Bristol* (Cambridge/New York, 1991). An exception is F.B. Smith's *The People's Health, 1830–1910* (1979).
31. Christopher Lawrence, 'Democratic, Divine and Heroic: the history and historiography of surgery', in Lawrence (ed.), *Medical Theory, Surgical Practice: studies in the history of surgery* (1992), pp. 1–47, esp. pp. 13–15.
32. An important exception is Jane Lewis, *What Price Community Medicine? The philosophy, practice and politics of public health since 1919* (Brighton, 1986).
33. *Report of the Working Party on the Management of Patients with Major Injuries*, p. 6., and see Susan P. Parker, 'Injuries in America: a national disaster', in Russell C. Maulitz (ed.), *Unnatural Causes: the three leading*

killer diseases in America (New Brunswick, NJ, 1989), pp. 135–45 at p. 135.

34. Rosemary Stevens, 'The Curious Career of Internal Medicine: functional ambivalence, social success', in Russell Maulitz and D. Long (eds), *Grand Rounds: one hundred years of internal medicine* (Philadelphia, 1988), pp. 339–64 at p. 360.

35. Reginald Elmslie at a meeting of the Royal Society of Medicine, 1944, quoted in Sir Sheldon Dudley, 'Naval Experience in Relation to National Health Service', *Lancet*, 29 July 1944, p. 134.

36. Harry Platt (quoting Tennyson), 'Education', *Lancet*, 17 Nov. 1943, p. 643 (my italics).

2 The Medical Context of Bones

1. For the history of bone and joint surgery prior to 'orthopaedics' see Edgar M. Bick, *Source Book of Orthopaedics* (Baltimore, 1937); Bruno Valentin, *Geschichte der Orthopadie* (Stuttgart, 1961), esp. pp. 158–85; and David Le Vay, *The History of Orthopaedics: an account of the study and practice of orthopaedics from the earliest times to the modern era* (Carnforth, Lancashire, 1990).

2. Printed for A. Millar, 1743, 2 vols. Facsimile reprint Philadelphia and Montreal, 1961.

3. Among Andry's most famous contests was that with the surgeon Jean-Louis Petit, the author of a treatise on injuries and diseases of bone. See R. Beverly Raney, 'Andry and the *Orthopaedia*', *JBJS*, 31A (1949), pp. 675–82. See also H. Winnett Orr, 'Nicolas Andry, founder of the orthopaedic specialty', *Clinical Orthopaedics*, 4 (1954), pp. 3–9; the entry on Andry in the *Dictionaire de Biographie Francaise* (1936), vol. 2, pp. 1013–16; and Toby Gelfand, 'Empiricism and Eighteenth-Century French Surgery', *Bull. Hist. Med.*, 44 (1970), pp. 40–53.

4. M. Foucault, *Surveiller et punir: naissance de la prison* (Paris, 1975).

5. See under these men's names in Valentin, *Orthopädie*; and Le Vay, *History of Orthopaedics*.

6. The third edition of Robley Dunglison's *Medical Lexicon: a new dictionary of medical science* (Philadelphia, 1842) contains 'Orthopaedia – The part of medicine whose object is to prevent and correct deformity in the bodies of children'. However the ninth edition (1853) adds 'Often used, however, with a more extensive signification, to embrace the correction or prevention of deformities at all ages. *Orthosomatics, Orthosomatice* (from . . . "right" and . . . "body") has been proposed as a preferable term.' See also J. Cohen, 'Orthopaedics' in J. Walton, P. B. Beeson and R. B. Scott (eds), *Oxford Companion to Medicine* (Oxford, 1986), p. 954.

7. See *Testimonials to George Combe* (Edinburgh, 1836), p. xvi. According to the *Oxford English Dictionary* 'mental orthopaedics' was a term used by the *Popular Science Monthly* in 1900 in reference to the use of hypnotism.

8. *Travels in Europe and the East in the Years 1834–41* (1842), pp. 54–7.

Mott, whose intention at this time was to establish 'an American Orthopaedic Institution', had just spent three years studying this surgery under the Parisian, Jules Guérin. He therefore regarded the French as preeminent in the field. On the impact of the French school of orthopaedics on American surgeons, see William S. Bigelow, *Memoirs of Henry Jacob Bigelow* (Boston, 1900), pp. 28–9; and Buckminster Brown (founder of the Boston Orthopaedic Institution), 'Orthopaedic Surgery in Europe [communicated from Paris, 19 May 1846]', *Boston Med. & Surg. J.*, 34 (1846), pp. 428–33.

9. See Charles Macalister, *The Origins and History of the Liverpool Royal Southern Hospital, with personal reminiscences* (Liverpool, 1936), p. 58.

10. See J. A. Cholmeley, *The History of the Royal National Orthopaedic Hospital* (1985). The Royal Orthopaedic and Spinal Hospital, Birmingham, dates from 1817, but it only took this title in 1888. It was originally known as 'The Institution for the Relief of Hernia, Club Feet, Spinal Diseases, and all Bodily Deformities' and was mainly concerned with the supply of trusses and braces. Until 1871 it had only outpatient facilities; by the First World War it had 30 beds. See J. Ernest Jones, *A History of the Hospitals and other Charities of Birmingham* (Birmingham, 1909).

11. Cholmeley, *RNOH*, pp. 4–5.

12. See, for example, *Med. Times & Gaz.*, 13 Feb 1869, p. 182. See also the discussion on the meaning of the word in *Med. Communications*, 9 (1981), pp. 93–9.

13. As a cripple, Little was similar to several others who took an interest in, or took up the practice of, orthopaedics – among them, Agnes Hunt, Harry Platt, G. R. Girdlestone, Alexander Mitchell, Walter Mercer, Elliot Brackett and Arthur Stanley. Dr John Bull Brown, who opened the first American 'Orthopaedic Institute' in Boston in 1838, became interested in orthopaedics after his son developed tuberculosis of the spine. The son, Buckminster Brown (1819–90) subsequently studied under Little and Stromeyer in the 1840s. On W. J. Little, see A. Rocyn Jones, 'Pioneers of Orthopaedics. VI – William John Little', *Cripples' J.*, no. 22 (Oct. 1929), pp. 325–330; idem, *JBJS*, 31B (1949), pp. 123–6; Cholmeley, *RNOH*, chs 1 and 2; and Jay Schleichkorn, *'The Sometime Physician', William John Little* (Farmingdale, New York, 1937).

14. On the latter's traditional role, see, for example, Timothy Sheldrake, Sen., *A Practical Essay on the Club Foot* (1798; 2nd edn, 1806), and *Animal Mechanics Applied to the Prevention and Cure of Spinal Curvature and other Personal Deformities* (1832). Sheldrake, a mechanic, was truss-maker to the Westminster Hospital. His son followed him in the same occupation. Henry Heather Bigg (1826–81) and his grandson, H. R. H. Bigg (1853–1911), also plied this trade. H. H. Bigg, the author of *Orthopraxy. The mechanical treatment of deformities, debilities, and deficiencies of the human frame, a manual* (1865; 3rd edn, 1877), described himself as 'Anatomical Mechanist to the Queen' and to various London hospitals. T. P. Salt of Birmingham, author of *A Treatise on Deformities and Debilities of the Lower Extremities and the Mechanical Treatment* (1866), described himself as an 'anatomical and orthopaedic mechanic'.

15. W. J. Little, 'Orthopaedic Surgery' in T. Holmes (ed.), *A System of Surgery*, vol.3 (1862), pp. 557–614 at p. 557.
16. *Lancet*, 11 July–15 August 1857: pp. 28, 133, 161. Samuel Hahnemann, the inventor of homoeopathy, was among those who went beyond the non-surgical or 'bloodless' orthopaedics of those whose manipulative practices developed out of traditional bonesetting. In 1831 he argued strongly against the use of machines in the treatment of scoliosis (spinal curvature) in favour of homoeopathic doses of physic. See 'Hahnemann as Orthopedic Surgeon', *Allg. Hom. Ztg.*, 53 (1857), p. 107, reprinted in Richard Haehl, *Samuel Hahneman, His Life and Work*, trans. M. C. Wheeler, ed. J. H. Clarke and F. J. Wheeler, vol. 2 (1927), p. 399.
17. Little, 'Orthopaedic Surgery', 1862, p. 561.
18. See handbill reproduced in Cholmeley, *RNOH*, p. 52. Until the advent of the NHS, the RNOH's full title was '. . . Hospital for the Treatment and Cure of all Crippled and Deformed Patients' (ibid., p. 75).
19. *62nd Annual Report for 1900*, p. 24. For confirming evidence see Cholmeley, *RNOH*, pp. 31, 67, 77.
20. An analysis of the 3000 cases seen by ROH surgeons William Adams and E. F. Lonsdale in 1852, showed 1600 cases of knock knee and bow leg, 500 cases of foot deformity, 450 cases of spinal deformity and 250 contracted joints. Cholmeley, *RNOH*, p. 15.
21. Adolf Lorenz, *My Life and Work: the search for a missing glove* (New York, 1936), p. 78.
22. See Colcott Fox, 'On Medical Education and Specialism', *BMJ*, 4 Oct. 1890, pp. 784–6.
23. 'Hospital', *Ency. Brit.* (11th edn 1911) vol. 13/14, p. 795.
24. See E. Muirhead Little, 'Specialism and General Surgery', *JOS*, 1 (1919), pp. 63–6 at p. 64. See also the editorial in *Lancet*, 3 Oct. 1891, p. 774; and L. Granshaw, ' "Fame and Fortune by Means of Bricks and Mortar": the medical profession and specialist hospitals in Britain, 1800–1948', in Granshaw and Roy Porter (eds), *The Hospital in History* (1989), pp. 199–220, esp. p. 212.
25. Henry Heather Bigg, *Orthopraxy*, cited in A. R. Shands, Jr, 'William Ludwig Detmold, the first American orthopaedic surgeon', *Current Practice in Orthopaedic Surgery*, 3 (1966), pp. 3–17 at p. 15.
26. See Richard Barwell, assistant surgeon to Charing Cross Hospital, *On the Cure of Club-Foot Without Cutting Tendons* (1863), a second edition of which was published in 1865.
27. Reprinted in his *Remarks on the Treatment of Infantile Congenital Club Foot* (1876).
28. Cholmeley, *RNOH*, p. 75.
29. References to the use of anaesthetics are rare throughout orthopaedic literature. Unique is William Mayo's comment in 1907 on the work of Robert Jones at the Nelson Street Clinic: 'In osteotomy, clubfoot, and similar operations the patients are allowed to go home after recovering from the anaesthetic. All operations are done under ether anaesthetic.' The fact that this is seen as worthy of comment may suggest that it was unusual at the time. However, the comment may be referring to ether rather than chloroform – mortality from the latter having become a

controversial issue in the 1890s and 1900s. Mayo, 'Present-Day Surgery in England and Scotland', *J. Minn. State Med. Assn*, 1 Dec. 1907, offprint, p. 6.

30. C. B. Keetley, quoted in *Trans. BOS*, 3 (1899), p. 35. For the orthopaedists' perception of their contribution to Listerism and Lister's place in the history of their specialism, see H. Winnett Orr, 'The Contribution of Orthopaedic Surgery to the Lister Antiseptic Method', *JBJS*, 19 (1937), pp. 575–83; and A. Rocyn Jones, 'Lister', *JBJS*, 30B (1948), pp. 196–9.

31. On Lister's amputations between 1871 and 1877, see W. Watson Cheyne, *Antiseptic Surgery: its principles, practice, history, and results* (1881), pp. 372ff.

32. Granshaw, 'The Development and Reception of Antisepsis in Britain, 1867–1890', in J. V. Pickstone (ed.), *Medical Innovations in Historical Perspective* (1992), pp. 17–46; Nicholas J. Fox, 'Scientific Theory Choice and Social Structure: The Case of Joseph Lister's Antisepsis, Humoral Theory and Asepsis', *Hist. Sci.*, 26 (1988), pp. 367–97; and Christopher Lawrence and Richard Dixey, 'Practising on Principle: Joseph Lister and the germ theories of disease', in C. Lawrence (ed.), *Medical Theory, Surgical Practice: studies in the history of surgery* (1992), pp. 153–215.

33. C. G. Wheelhouse, 'A Review of the Progress of Surgery', *BMJ*, 10 Aug. 1878, pp. 204–11 at p. 206. Wheelhouse was senior surgeon to the Leeds General Infirmary and a leading figure in the BMA.

34. Among the pioneers of subcutaneous osteotomy was the general surgeon, William Adams (1820–1900). Adams was Little's successor at the ROH, but most of his work on subcutaneous osteotomy was conducted at the Great Northern Hospital. See A. Rocyn Jones, 'William Adams', *JBJS*, 33B (1951), pp. 124–9.

35. Compare Little's 'Orthopaedic Surgery' in Holmes, *Surgery*, vol. 3, 1862, p. 562, with the same entry in vol. 2, 1883, p. 229.

36. Macewen adopted the procedure from the German advocate of Listerian principles, Richard von Volkmann (1830–89), on whom see Valentin, *Orthopädie*, p. 104. On Macewen (1848–1924), the author of *The Growth of Bone* (1912), see A. K. Bowman, *The Life and Teaching of Sir William Macewen* (1942).

37. Sampson Gamgee, 'The Unity of Surgical Principles in Wound and Fracture Treatment', *Lancet*, 18–25 Nov. 1882, pp. 840–1, 885–7 at p. 886. While Listerism laid stress on the *seed* of infection, the alternative tradition placed stress on the *soil* or general environment. Only in the 1880s did soil and seed come together through immunology to offer a new reading of germ theory. I am grateful to Mick Worboys for helpful discussions on this subject.

38. See Owsei Temkin, 'Surgery and the Rise of Modern Medical Thought' in his *The Double Face of Janus* (Baltimore, 1977), pp. 487–96. The practice, as developed in France, also had philosophical and professional meanings. See Gelfand, 'Empiricism', p. 51. For America, see Gert H. Brieger, 'From Conservative to Radical Surgery in Late Nineteenth-Century America' in Lawrence (ed.), *Medical Theory, Surgical Practice*, pp. 216–31.

39. See Max Neuburger, *The Doctrine of the Healing Power of Nature Throughout the Course of Time*, trans. L. J. Boyd (New York, [1942]).

40. Syme was also among those who took up subcutaneous tenotomy at the suggestion of Stromeyer, though he became highly critical of those 'doing an incalculable amount of mischief to the public and the profession' practising tenotomy under the name of 'Orthopaedists'. See *BMJ*, 22 Apr. 1853, p. 351; Robert Paterson, *Memorials of the Life of James Syme* (Edinburgh, 1874); and H. Winnett Orr, *On the Contributions of Hugh Owen Thomas of Liverpool, Sir Robert Jones of Liverpool and London, John Ridlon, MD, of New York and Chicago to Modern Orthopedic Surgery* (Springfield, Illinois, 1949), p. 135n. James Gamgee Beaney's *Conservative Surgery* (Melbourne, 1859) was dedicated to Syme.

41. F. C. Skey, *Operative Surgery* (1850), p. ix.

42. On Fergusson, see *DNB*; H. Willoughby Lyle, *King's and Some King's Men* (1935), p. 119; D'Arcy Power, *British Masters of Medicine* (1936); and G. Gordon-Taylor, 'Sir William Fergusson', *Med. Hist.*, 5 (1961), pp. 1–14. See also Fergusson, *Lectures on the Progress of Anatomy and Surgery During the Present Century* (1867).

43. See Mathias Roth, *Notes on the Movement-Cure, or Rational Medical Gymnastics* (1850). Significantly, Roth was a homoeopath. See T. J. Surridge, 'Dr Mathias Roth: Swedish medical gymnastics and homoeopathy', in David McNair and N. A. Parry (eds), *Readings in the History of Physical Education* (Hamburg, 1981), pp. 95–102. The preceding quotations are from W. Hale White, MD, 'Massage and the Weir Mitchell Treatment', *Guy's Hosp. Repts.*, 45 (1888), pp. 267–80. Liverpool's Royal Southern Hospital installed gymnastic equipment in 1858, to 'accelerate the perfect use of the patient's limbs, and shorten the period of their recovery'; and in 1906 a Massage and Exercise Department was opened (Macalister, *Liverpool Royal Southern*, pp. 69–72). In 1874 the National Orthopaedic Hospital in London appointed an honorary professor of gymnastics, and in 1910 a school of massage was started. (Cholmeley, *RNOH*, pp. 74, 93.) Guy's Hospital seems fairly representative of the larger voluntary hospitals in establishing a massage department in 1888 and in incorporating Swedish methods of gymnastics in the 1900s: Cameron, *Guy's Hospital*, p. 393.

44. See *JBJS*, 21 (1939), p. 811; and see J. L. Thornton, 'Orthopaedic Surgeons at St.Bartholomew's Hospital, London', *St Barts. Hosp. J.*, 59 (1955), pp. 195–204 at p. 198. Marsh was a corresponding member of the American Orthopaedic Association.

45. See Callender, 'Seven Years of Hospital Practice', *St Barts Hosp. Repts*, 14 (1878), pp. 183–95. See, also, F. C. Skey, 'On Fractures', in his *Operative Surgery*, pp. 137–73; O. H. and S. D. Wangensteen, *The Rise of Surgery* (Folkestone, 1978), pp. 48–51; John Woodward, *To Do The Sick No Harm: a study of the British voluntary hospital system to 1875* (1974), p. 165; and R. B. Fisher, *Joseph Lister, 1827–1912* (1977), p. 124. See, also, Jessie Dobson, 'Pioneers of Osteogeny: George William Callender [1830–79]', *JBJS*, 31B (1949), pp. 127–9.

46. John Hilton, *On the Influence of Mechanical and Physiological Rest in the Treatment of Accident and Surgical Diseases, and the Diagnostic Value of*

Pain: a course of lectures, delivered at the Royal College of Surgeons of England in the years 1860, 1861, and 1862 (1863). The second edition was entitled *Rest and Pain* (1877).

47. H. Marsh, 'On Manipulation; or, the use of forcible movement as a means of surgical treatment', *St Barts Hosp. J.*, 14 (1878), pp. 205-19; idem, 'Bone-setting', *BMJ*, 27 May 1911, pp. 1231-9.

48. *BMJ*, 5 Jan. 1867, pp. 1-14, reprinted in S.Paget, *Clinical Lectures and Essays* (1875), and S. Paget (ed.), *Selected Essays and Addresses* (1902). For historical discussion and further references, see Roger Cooter, 'Bones of Contention? Orthodox medicine and the mystery of the bone-setter's craft', in W. F. Bynum and Roy Porter (eds), *Medical Fringe and Medical Orthodoxy, 1750-1850* (1987), pp. 158-73.

49. Marsh, *Diseases of the Joints* (1886; German trans. 1888). New editions appeared in 1895 and 1910.

50. See, for example, J. Russell Reynolds, *On the Relation of Practical Medicine to Philosophical Method, and Popular Opinion: being the annual oration delivered before the North London Medical Society on February 10th, 1858* (1858), p. 17. On nineteenth-century relations between orthodoxy and fringe medicine, see Bynum and Porter, *Medical Fringe*; and Roger Cooter (ed.), *Studies in the History of Alternative Medicine* (1988).

51. Quoted in Frances Power Cobbe's powerful antivivisectionist article, 'The Medical Profession and Its Morality', *Modern Rev.*, 2 (1881), pp. 296-328 at p. 312. On William Fergusson and the antivivisection movement, see Richard French, *Antivivisection and Medical Science* (Princeton, 1975), pp. 57-9.

52. See, for example, Albert G. Walter, *Conservative Surgery in its General and Successful Adaptation in Cases of Severe Traumatic Injuries of the Limbs, with a report of cases* (Philadelphia, 1867). By 1875 Walter's conservative surgery was widely accepted by the members of the staff of the Western Pennsylvania Hospital in treating numerous industrial accident cases. See T. L. Hazlett and William W. Hummel, *Industrial Medicine in Western Pennsylvania, 1850-1950* (Pittsburgh, 1957), pp. 25-33.

53. Brieger, 'From Conservative to Radical Surgery', pp. 216, 226.

54. *BMJ*, 3 July 1917, pp. 35-8. Cf. Lloyd Stevenson, 'Science Down the Drain: on the hostility of certain sanitarians to animal experimentation, bacteriology and immunology', *Bull. Hist. Med.*, 29 (1955), pp. 1-26.

55. See Fox, 'Lister's Antisepsis', Lawrence and Dixey, 'Lister', and Granshaw, 'Reception of Antisepsis'.

56. In a paper of 1908, for instance, on the treatment of tuberculous hip joints in 900 children at the Alexandra Hip Hospital in London, it was shown by Sir Anthony Bowlby that far better results could be obtained through environmental and dietary measures than through operative surgery. See 'Contributions of the late Sir Henry Gauvain to Orthopaedic Surgery', *JBJS*, 30B (1948), p. 385.

57. Lorenz, *My Life and Work*, esp. ch. 7, 'Calamity Creates a "Dry" Surgeon', and ch. 9, 'Developing Bloodless Operations'. On the British reception of Lorenz, see the editorial in the *Lancet*, 7 Feb. 1903, pp.

381–2. Lorenz's manipulative techniques for hip dislocation were taken up at the RNOH, see Cholmeley, *RNOH*, p. 108.

58. 'The International Medical Congress and the Progress of Medicine', *Westminster Rev.*, 116 (1881), pp. 403–39 at p. 428.

59. Information on the life and work of Thomas is drawn from T.P. McMurray, 'The Life of Hugh Owen Thomas', *Liverpool Medico-Chirurg. J.*, 43 (1935), pp. 3–41; D. McCrae Aitken, *Hugh Owen Thomas: his principles and practice* (Oxford, 1935); Frederick Watson, *Hugh Owen Thomas: a personal study* (1935); H. Williams, 'Hugh Owen Thomas' in his *Doctors Differ* (1946), pp. 95–123; Orr, *Contributions*; David Le Vay, *The Life of Hugh Owen Thomas* (Edinburgh, 1956); and T. Goronwy, 'From Bonesetter to Orthopaedic Surgeon', *Ann. Roy. Coll. Surg. Eng.*, 55 (1974), pp. 134–42, 190–8, together with material from his own writings, and the note book, case book and letter book in the Royal College of Surgeons.

60. See Orr, *Contributions*, pp. 48, 67, and Le Vay, *Thomas*, p. 72.

61. Quoted in Le Vay, *Thomas*, p. 197.

62. Thomas, *Principles of the Treatment of Diseased Joints* (Liverpool, 1883), pp. 63–4. See also, A. G. Timbrell Fisher, 'Hugh Owen Thomas', in his *Treatment by Manipulation in General and Consulting Practice* (5th edn, 1948), pp. 9–11.

63. *Lectures and Essays*, quoted in McMurray, 'Thomas', p. 19.

64. This was mainly put about by Robert Jones, notably in his delivery of the first triennial Thomas Memorial Lecture, 25 April 1922, as quoted in Timbrell Fisher, 'Thomas', p. 8. Another advocate of Thomas's ideas, John Ridlon (on whom see below) wrote to Thomas's biographer, Frederick Watson, 'I visited MacEwen [sic] in Glasgow, Wright in Manchester, William Adams in London (and many others) and not one of them had a kindly word to say of Thomas.' Quoted in Watson, *Thomas, a personal study*, p. 69.

65. McMurray, 'Thomas', p. 17.

66. Wheelhouse, 'Progress of Surgery', p. 210. Thomas's reputation, especially for his splints, owed much to Rushton Parker, a London trained surgeon who became Professor of Surgery at the Liverpool Royal Infirmary. Parker first recognized the novelty of Thomas's splints in 1875 when he accompanied his father, Dr Edward Parker (a Liverpool Police Surgeon) on a visit to a patient in Thomas's care. See LeVay, *Life of Thomas*, p. 58.

67. In July 1879 the manager of the Leeds General Infirmary reported that Thomas's splints had been distributed to patients at a cost £20 19s 0d. Board Minutes, Leeds General Infirmary, contained in a letter to Professor McFarland, 1 Jan. 1962, held in the Hugh Owen Thomas archives, Liverpool Medical Institution. For praise of Thomas's splints at the International Medical Congress, London 1881, see Rushton Parker, 'On the Treatment of Fractured Femur' in his *Surgical Cases and Essays* (Liverpool, 1882), pp. 31–7. Sir John Eric Erichsen, *The Science ad Art of Surgery* (8th edn, 1884), vol. 2, pp. 389–91, 458 (Erichsen was the president of the Surgical Section at the 1881

International Medical Congress); Thomas Bryant, *A Manual for the Practice of Surgery* (4th edn, 1884), vol. 2, pp. 496, 502.

68. Ridlon (1852–1936) visited Thomas again in June 1890, while en route to Berlin to organize an Orthopaedic Section of the International Medical Congress. See Orr, *Contributions*, p. 15. See also, Ridlon, 'Personal Remembrances of Hugh Owen Thomas', *JBJS*, 17 (1935), pp. 506–9. On the AOA, see below, Ch. 3.

69. This led Thomas to retitle his *Review of the Past and Present Treatment of Disease of the Hip, Knee and Ankle Joints* (Liverpool, 1878) *An Argument with the Censor at St. Luke's Hospital, New York* (Liverpool, 1889).

70. See for example the debate between Thomas and Frederick Treves over the nature and treatment of intestinal obstruction, recounted in Stephen Trombley, *Sir Frederick Treves: the extra-ordinary Edwardian* (1989), pp. 63–5. Even those in favour of Thomas's views often hesitated to mention his name in print. See, for example, Gamgee, 'Wound and Fracture Treatment', p. 887. Among American orthopaedists approved of by Thomas – besides Ridlon – was Louis Bauer of St Louis (from whose institution Thomas received an honorary doctorate).

71. See Orr, *Contributions*, p. 116.

72. For a full list and assessment of Thomas's publications, see Le Vay, *Thomas*, pp. 60–126.

73. See, for example, Thomas, *Hip, Knee, Ankle*, 2nd edn, 1876, pp. 217, 232–3.

74. In the wake of Jones's triumphs during the First World War, and probably at his prompting, the Liverpool Medical Institution decided in January 1918 to promote a memorial to Thomas as the 'surgeon famous throughout the world for his pioneer work in orthopaedics, and as the inventor of the splints which bear his name'. *BMJ*, 25 Jan. 1919, p. 112. The memorial took the form of a triennial oration, the first of which was given by Jones in September 1920. See Frederick Watson, *The Life of Sir Robert Jones* (1934), p. 239.

75. For character references on Jones, see Mayo, 'Present Day Surgery', p. 6; B. Moynihan, *The Robert Jones Birthday Volume* (1928), p. vii; and Leo Mayer, 'Reflections on Some Interesting Personalities in Orthopaedic Surgery During the First Quarter of the Century', *JBJS*, 37A (1955), p. 380.

3 Politics and Professionalization

1. In 1908, the London Hospital appointed T. H. Openshaw; in 1912, St Bartholomew's appointed R. C. Elmslie; and in 1913, Guy's appointed W. H. Trethowan; while at the century-old Birmingham Royal Orthopaedic and Spinal Hospital, one of Jones's former assistants in Liverpool, Naughton Dunn, was appointed. See H. Osmond-Clarke, 'Half a Century of Orthopaedic Progress in Great Britain', *JBJS*, 32B (1950), p. 630; and St. J. D. Buxton, 'Sir Thomas Fairbank', *JBJS*, 38B (1956), pp. 4–21.

2. See Harry Platt, 'Orthopaedics in Continental Europe, 1900–1950', *JBJS*, 32B (1950), pp. 570–86; Leo Mayer, 'Orthopaedic Surgery in the United States of America', *JBJS*, 32B (1950), pp. 461–569.

3. E. Muirhead Little (quoting Sir James Paget, the president of the 1881 Congress), 'Specialism and General Surgery', *JOS*, 1 (1919), pp. 63–6 at p. 64.

4. Sir Montague Burrows, *Cripples' J.*, 3, no. 11 (Jan. 1927), p. 277. See also James R. Learmonth, 'Surgery and Preventive Medicine' (1949) reprinted in his *The Thoughtful Surgeon*, ed. D. M. Douglas (Glasgow, 1969), pp. 99ff.; H. Winnett Orr, *On the Contributions of Thomas, Jones and Ridlon* (Springfield, Ohio, 1949); and Frederick Watson, *The Life of Sir Robert Jones* (1934).

5. Reginald Watson-Jones, obituary on Jones in *JBJS*, 15 (1933), pp. 541–3.

6. Watson, *Life of Jones*, pp. 86ff.; William Mayo, 'Present-day Surgery in England and Scotland: from notes made on a recent short visit', reprinted from *J. Minn. State Med. Assn.*, 1 Dec. 1907, p. 6; Jones and Oliver Lodge, 'The Discovery of a Bullet Lost in the Wrist by Means of the Roentgen Rays', *Lancet*, 22 Feb. 1896, pp. 476–7; and C. Thurston Holland, 'X-rays in 1896', *Liverpool Medico-Chirur. J.*, 45 (1937), pp. 61–77. Jones was cautious about the value of X-rays, however; see his comments in 'Present Position of the Treatment of Fractures', *BMJ*, 7 Dec. 1912, pp. 1589–94 at p. 1594; and idem., 'Discussion on the Treatment of Fractures, with special reference to its organization and teaching', *BMJ*, 22 Aug. 1925, pp. 319–22 at p. 319.

7. See Harvey Cushing, 'The Society of Clinical Surgery in Retrospect', *Annals of Surgery*, 169 (1969), pp. 1–9 at p. 8 (a paper of 1922 in which he recalls his visit to Jones's clinic in 1910). Among other American visitors, besides John Ridlon, were William Mayo, George Packard, J. B. Murphy, and the orthopaedists, Virgil Gibney, Robert Osgood, Leo Mayer, Joel Goldthwait, M. S. Danforth, Arthur Gillette and Robert Lovett. Adolf Lorenz visited the clinic in 1903.

8. See below, Chs 4 and 8, and on the Ship Canal, Ch. 5.

9. Leo Mayer, 'Reflections on Some Interesting Personalities in Orthopaedic Surgery During the First Quarter of the Century', *JBJS*, 37A (1955), pp. 374–83 at p. 379, where he recalls his visit of 1913.

10. Charles Macalister, *The Origin and History of the Liverpool Royal Southern Hospital with Personal Reminiscences* (Liverpool, 1936), pp. 61–2.

11. See Anson Rabinbach, *The Human Motor: energy, fatigue, and the origins of modernity* (Berkeley, 1992).

12. 'Orthopaedist, one who practises orthopaedia': Robley Dunglison, *Medical Lexicon* (9th edn, Philadelphia, 1853).

13. For a full list, see 'A Chronological List of Sir Robert Jones's Contributions to Surgical Literature', *JBJS*, 39B (1957), pp. 212–17.

14. Watson, *Life of Jones*, p. 239.

15. Mayo, 'Present-day Surgery', p. 6.

16. Watson, *Life of Jones*, p. 85, quoting Macalister.

17. Thornton Brown, *The American Orthopaedic Association: a centennial history* (n.p., [1986]), p. 4.

18. Royal Whitman, 'A Review of the Evolution of the Orthopaedic Branch of Surgery in New York City', *JBJS*, 29 (1947), pp. 250–3; Leo Mayer, 'Orthopaedic Surgery in the United States of America', *JBJS*, 32B (1950), pp. 461–569; and T. Brown, *AOA*.

19. Royal Whitman, 'The Emancipation of Orthopaedic Surgery', *Proc. Roy. Soc. Med.*, 36 (1943), pp. 327–9.

20. At the annual meeting of the AOA in 1893, for instance, only one paper out of 31 dealt with invasive surgical procedures. See Leroy C. Abbott's presidential address to the AOA, 1947: *JBJS*, 29 (1947), pp. 840–50 at p. 841. Fred Albee, who developed bone-grafting techniques just prior to the war, recalled that around 1905, when he began work in the department of orthopaedic surgery at the New York Post Graduate Medical School Clinic, 'there was practically no open technical operative surgery being done'. Albee, *A Surgeon's Fight to Rebuild Men: an autobiography* (1950), p. 47.

21. Virgil P. Gibney, 'Orthopaedic Surgery: its definition and scope' (a reply to AOA charter member Newton M. Shaffer's 'What is Orthopaedic Surgery'), *N.Y Med. J.*, 7 Nov. 1891, reprinted in *Gibney of the Ruptured & Crippled*, ed. A. R. Shands (New York, 1969), pp. 129–34.

22. A.M Phelps, quoted in E.A. Chandler's presidential address to the AOA, 1953: *JBJS*, 35A (1953), p. 1.

23. Royal Whitman, quoted in Chandler, ibid., p. 1.

24. Jones, 'Orthopaedic Surgery in its Relation to the War', *Recalled to Life*, 1 (1917), p. 51.

25. For a list of foreign members of the AOA elected in 1890, see Orr, *Contributions*, p. 186n.

26. Presumably, Tubby had a hand in the founding of the department, though on this or anything else to do with the department nothing is said in John Langdon-Davis's *The Westminster Hospital, 1719–1948*, 2 vols (1952). Tubby's interest in orthopaedics emerged after his appointment at the National Orthopaedic Hospital in 1891, and after studying tendon transplant under Fritz Lange in Berlin and Otto Vulpius in Heidelberg. See obituaries in *BMJ*, 1 Mar. 1930, pp. 419–20; *Lancet*, 1 Mar. 1930, pp. 485–6; and A. Rocyn Jones, *JBJS*, 35B (1935), pp. 139–43. Tubby was also to become the visiting surgeon to the Sevenoaks Children's Hospital for Hip Diseases.

27. Most notably, Jones and Tubby, *Modern Methods in the Surgery of Paralysis* (1903).

28. *Trans.BOS*, 1 (1896), p. 48. Prior to this, Jones, in proposing the idea of a British Orthopaedic Society, had written to William Adams, asking if he would act as its president. Adams gave his blessing to the idea, but declined the offer of the presidency on the grounds of ill-health. Watson, *Life of Jones*, p. 93.

29. Quoted in Watson, *Life of Jones*, p. 94.

30. Ibid., p. 93.

31. Such was initially the case for W. H. Trethowan at Guy's Hospital. See H. C. Cameron, *Mr. Guy's Hospital, 1726–1948* (1954), p. 377. Another example is Reginald Elmslie, even though he spent 10 months training at the ROH before his appointment to the orthopaedic department at

Barts. The failure to find vacancies in the prestigious voluntaries for chosen fields of pursuit led some budding surgeons to change specialties. A.M. Henry Gray for example switched from gynaecology to dermatology because he could find no opening at the London teaching hospitals. See *Lives of the Fellows of the Royal College of Surgeons* (1970), pp. 135–7.

32. Stephen Trombley, *Sir Frederick Treves: the extra-ordinary Edwardian* (1989), p. 19.

33. City Orthopaedic Hospital, Minute Book, 8 Mar. 1905. The City's comments were made during the discussions over the amalgamation of the three London orthopaedic hospitals precisely because the amalgamation was not going to alter the educational situation.

34. Cited in Mayer, 'Orthopaedic Surgery in America', p. 487. In one of his reports as Surgeon-in-Chief to the New York Orthopaedic Dispensary and Hospital, Shaffer maintained that 'in true orthopaedic surgery, operative work, per se, has no real status'. Quoted in George M. Goodwin, *Russell A. Hibbs: pioneer in orthopaedic surgery, 1869–1932* (New York, 1935), p. 40.

35. 'A Discussion on the Operative Treatment of Club-Foot' in the section on Surgery at the Annual Meeting of the BMA, Glasgow, August 1888, *BMJ*, 27 Oct. 1888, pp. 919–27. Smith was the translator of F. Busch's *General Orthopaedics, Gymnastics, and Massage* (1886). In 1903 he emerged as one of the leading defenders of Lorenz's 'bloodless' operation for congenital dislocation of the hip. See *Lancet*, 17 Jan.1903, pp. 173–3, 194; *Lancet*, 14 Feb.1903, p. 209; and *BMJ*, 29 Aug.1903, pp. 462–3.

36. Orr, *Contributions*, p. 195. Jones gave full backing to Thomas's views in publications such as 'On the So-called "Abuse of Rest"', *Liverpool Medico-Chirurg. J.*, 6 (1886), pp. 18–24, and 'A Protest Against the Routine Excision of Joints', ibid., 8 (1888), pp. 420–39.

37. Mayo, 'Present-day Surgery', p. 5.

38. See remarks in Sir William Heneage Ogilvie, 'Clinical Surgery' (1956), reprinted in his *Surgery: orthodox and heterodox* (Oxford, 1948), pp. 75 and 44.

39. *BMJ*, 6 March 1926, p. 451.

40. 'The International Medical Congress and the Progress of Medicine', *Westminster Rev.*, 116 (1881), pp. 403–39 at p. 432. See also Morrell Mackenzie, 'Specialism in Medicine', *Fortnightly Rev.*, 37 (1885), pp. 772–87; and George Rosen, *Specialization of Medicine* (New York, 1944), p. 63.

41. George Rosen, *The Structure of American Medical Practice, 1875–1941* (Philadelphia, 1983), esp. ch. 1: 'Competition in the Medical Market'.

42. J. Jackson Clarke, assistant surgeon to the North-West London and City Orthopaedic Hospitals, 'Orthopaedic Surgery', *Practitioner*, 6 (1897), pp. 318–24 at p. 319 (my emphasis).

43. Glenn Gritzer and Arnold Arluke, *The Making of Rehabilitation* (Berkeley, 1985), p. 42. On Goldthwait (1866–1961), see *Cripples' J.*, 5 (July 1929), pp. 243–50.

44. Quoted in Mayer, 'Orthopaedic Surgery in America', p. 469.

45. Minute Book, 8 Mar. 1905.
46. Hence there were to be few appointments of orthopaedic surgeons to children's hospitals. At Britain's oldest and most prestigious children's hospital, at Great Ormond Street, London, it was not until 1946 that Eric Lloyd (1892–54) was officially designated an 'orthopaedic surgeon', though he had been performing the same work there for over twenty years. See *JBJS*, 37B (1955), pp. 161–3. The orthopaedist H. A. T. Fairbank was appointed a *general* surgeon to the Great Ormond Street Hospital in 1906. Of course, for some of those working in the area of bone and joint surgery in the children's hospitals the label 'orthopaedist' was inappropriate because of specialist interests in more than one field. A. S. Blundell Bankart (1879–1951), for instance, one of 'Jones's men' during the First World War, simultaneously practised orthopaedic surgery and neurosurgery in the surgery of children, and held appointments before the war at both the RNOH and the Maida Vale Hospital for Nervous Diseases. See *JBJS*, 33B (1951), pp. 278–80. The Australian, Sir Robert B. Wade (1874–1954), Honorary Surgeon to the Royal Alexandra Hospital in Sydney from 1912, was established as a 'paediatric surgeon' before becoming a children's orthopaedist. See *JBJS*, 36B (1954), pp. 662–3.
47. I am grateful to Sheila Reid for this information. Her work on children's hospitals in Britain has guided me to many of the sources in this section.
48. See, for example, Edinburgh's Royal Hospital for Sick Children, *Annual Report for 1893*, p. 467. For a useful general introduction to the history of children's hospitals in Britain and the rise of paediatrics, see 'Background to the Care of Children in Hospital', in Nuffield Foundation, *Children in Hospital: studies in planning*, (1963), pp. 1–11.
49. Summary contained in *Annual Report for 1900*, p. 221. For Edinburgh see Douglas Guthrie (ed.), *The Royal Edinburgh Hospital for Sick Children, 1860–1960* (Edinburgh, 1960), p. 72.
50. The situation with respect to orthopaedic hospitals in pre-war America was slightly different; see, for example, Charles Sinkler, 'The Philadelphia Orthopedic Hospital and Infirmary for Nervous Diseases', in Frederick P. Henry (ed.), *Founders' Week Memorial Volume* (Philadelphia, 1909), p. 795; and H. Platt, 'The Special Orthopaedic Hospital, past and future', *J. Roy. Coll. Surg. Edin.*, 21 (1976), pp. 67–74 at p. 73.
51. Morris Vogel, *The Invention of the Modern Hospital: Boston, 1870–1930* (Chicago, 1985), pp. 64ff.; and see Clement A. Smith, *The Children's Hospital of Boston* (Boston, 1983), ch. 12. Bone and joint surgery similarly became the focus of attention at the Children's Hospital of Philadelphia: see F. A. Packard's article on the hospital (written *c.* 1897) in Henry (ed.), *Founders' Week Memorial Volume*, pp. 770–6.
52. See John F. Fulton, *Harvey Cushing, a biography* (Springfield, Illinois, 1946), p. 92; and A. McGehee Harvey, 'Orthopedic Surgery at Johns Hopkins', *Johns Hopkins Med. J.*, 150 (1982), pp. 221–45. The latter's orthopaedic department was headed by the ambitious young surgeon William Stevenson Baer. On Bradford, who established the orthopaedic service at the Boston Children's and, from 1903, was the first full

professor of orthopaedics at the Harvard Medical School, see obituary in *JBJS*, 8 (1926), pp. 461–5.

53. Harry Platt, 'Orthopaedic Surgery in Boston' *Med. Chron.*, 58 (1914), pp. 473–9.

54. *Diseases of the Bones; their pathology, diagnosis and treatment* (Manchester, 1887). On Thomas Jones, see William Brockbank, *The Honorary Medical Staff of the Manchester Royal Infirmary, 1830–1948* (Manchester, 1965), pp. 83–4.

55. See *Medical Directory* and his obituary in the *Lancet*, 10 Apr. 1920, p. 840. Both Wright and Thomas Jones had been students at Guy's Hospital where they were inspired by the Listerian surgeon Henry Howse, who had a special interest in the excision of joints in children.

56. See Brockbank, *Manchester Royal Infirmary*, pp. 158–9; and *Contemporary Biographies*, p. 149, and William Waugh, *John Charnley: the man and the hip* (1990), pp. 14–15. On Pendlebury, see also Henry J. Eason, 'The Manchester Children's Hospital', *The Child*, 2 (1911), pp. 142–9. Yet another 'orthopaedist' to emerge from training at Pendlebury was the general surgeon William Percy Montgomery (1867–1911) who was appointed assistant honorary surgeon and lecturer on orthopaedics at the Manchester Royal Infirmary in 1900. Brockbank, op. cit., p. 126.

57. See Helen Clapesattle, *The Doctors Mayo* (Garden City, New York, 1943), pp. 508–9; the obituary on Stiles by John Fraser in *Year Book of the Royal Society of Edinburgh* (Edinburgh, 1946), pp. 30–3, and that by W. J. Stuart, in Guthrie (ed.), *Royal Edinburgh Hospital*, pp. 45–9. Stiles was also responsible for the fundamental research on bone and joint tuberculosis carried out by John Fraser. See John Fraser, *Tuberculosis of the Bones and Joints in Children* (1914). On Henderson (who became president of the AOA in 1934), see *Sketch of the History of the Mayo Clinic and the Mayo Foundation* (Philadelphia, 1926), pp. 42–3 (which also gives an account of the rise of bone and joint operations from 1890 to 1924); and his obituary in *JBJS*, 36A (1954), pp. 1087–8.

58. On the history of the Edinburgh Chair of Orthopaedics, see Walter Mercer, 'Inaugural Lecture', *Edin. Med. J.*, 56 (1949), pp. 173–86. Stiles's wartime involvement in orthopaedics is mentioned in his 'Practice and Training of Orthopaedic Surgery', *Lancet*, 19 Apr. 1919, pp. 671–2.

59. Clapesattle, *Doctors Mayo*, p. 452. Stiles spent working holidays at the Mayo Clinic in 1906 and 1911. See Stiles, 'Surgical Training: reminiscences and suggestions', *BMJ*, 18 Oct. 1919, pp. 502–4.

60. Stuart, *Edinburgh Children's Hospital*, p. 46.

61. See Platt, 'Boston'.

62. At the Johns Hopkins University Medical School in Baltimore the Division of Orthopedic Surgery was the first of the surgical specialisms to have a completely part-time staff. See Harvey, 'Orthopedic Surgery at Johns Hopkins', p. 201.

63. Cited in Mayer, 'Orthopaedic Surgery in America', p. 480. For the context of specialization in American medical education in this period, see William Rothstein, *American Medical Schools and the Practice of Medicine: a history* (New York/Oxford, 1987), pp. 101–9.

64. The exceptions were Jones, who was appointed Lecturer in Orthopaedic Surgery for five years at the University of Liverpool in 1909 (Watson, *Life of Jones*, p. 134), and Tubby, who was Lecturer on Clinical and Orthopaedic Surgery at the Westminster Hospital Medical School some time after his appointment at the Westminster in 1895. In his *The Advance of Orthopaedic Surgery* (1924), Tubby remarked on the 'striking contrast' between the teaching of orthopaedics in America and Britain, and described the British situation as 'deplorable' (p. 11). For the situation in Germany, which was not much better than in Britain, see Beat Ruttimann, *Wilhelm Schulthess (1855–1917) und die Schweizer Orthopädie Seiner Zeit* (Zurich, 1983), pp. 167ff.; and Han-Heinz Eulner, 'Orthopädie', in his *Die Entwicklung der Medizinischen Spezialfacher an den Universitaten des Deutschen Sprachgebiets* (Stuttgart, 1970), pp. 387–96. For other European countries, see Platt, 'Orthopaedics in Continental Europe', pp. 570–86.

65. See John R. Paul, *A History of Poliomyelitis* (New Haven, 1971), esp. p. 61 on the pioneer orthopaedist and physiotherapist, Charles F. Taylor of New York.

66. Jones, 'A Note on Operative Treatment of Flail Paralytic Elbow', *Pediatrics* [New York], 14 (1902), p. 344, and Jones, 'A Note on the Surgical Treatment of Spastic Infantile Paralysis', *Pediatrics*, 14 (1902), p. 298.

67. A. McGehee Harvey, *Science at the Bedside: clinical research in American medicine, 1905–1945* (Baltimore, 1981), p. 159. See also, Sydney A. Halpern, *American Pediatrics: the social dynamics of professionalism, 1880–1980* (Berkeley, 1988), p. 51ff.

68. Note gynaecology's similar claim against the opponents of specialization, cited in Ornella Moscucci, *The Science of Woman: gynaecology and gender in England, 1800–1929* (Cambridge, 1990), pp. 79–81 *et passim*.

69. See 'List of Office Bearers, Past Members of Council', *Reports of the SSDC*, 6 (1905–8), pp. xvii–xxxvii.

70. See George Carpenter, 'Introductory', *Br. J. Childr. Dis.*, 1 (1904), pp. 1–3, and Carpenter, 'The History of the Society for the Study of Diseases in Children', *Reports of the SSDC*, 8 (1907–8), pp. xxix–lxi. On professionalization in American paediatrics, see A. Jacobi's introduction to George Payne, *The Child in Human History* (1916); Harold Faber and R. McIntosh, *History of the American Pediatric Society, 1887–1965* (New York, 1966); and Halpern, *American Pediatrics*.

71. In 1901 the SSDC had 237 members, 82 of whom were provincial. See *Br. J. Childr. Dis.*, 1 (1904), p. iii.

72. Ibid., p. iii and Charles Macalister, *The Origin and History of the Royal Liverpool Country Hospital for Children at Heswall from the time of its inception (1895–1898) and foundation (1899) to the year 1930* (Campden, Glous., 1930).

73. Cf. Frederick Churchill, 'Should Diseases of Children be Made a Speciality at Our General Hospitals?', *Lancet*, 18 Nov. 1882, pp. 870–1.

74. *Trans. BOS*, 1 (1896), p. 2.

75. See 'The Children's Section at the British Medical Association at Oxford – Where Was It?', *Br. J. Childr. Dis.*, 1 (1904), pp. 410–11.

76. Platt recollected that in 1913, during the proceedings of the International Medical Congress, 'it was evident to the onlooker that a coolness had developed between these two distinguished men'. 'British Orthopaedic Association: First Founders' Lecture', reprinted in Platt, *Selected Papers* (Edinburgh, 1963), p. 118. Jones's name last appears on the membership list of the SSDC in 1907.

77. See Maurice Davidson, *The Royal Society of Medicine: the realization of an ideal* (1955). There exists as yet no adequate study of the politics of the Royal Society of Medicine in these years, but there are some hints in Lindsay Granshaw, *St. Mark's Hospital London* (1985), p. 180.

78. The idea for an orthopaedic section at the International Congresses came from Newton Shaffer who, with Harry M. Sherman and John Ridlon, organized a propaganda committee. Ridlon, as secretary, claims to have written to every orthopaedic surgeon in the world. See the letter quoted in Orr, *Contributions*, p. 204.

79. For tactical reasons E. M. Little was also appointed the first president of the BOA. The RSM subsection on orthopaedics was granted full sectional status in 1922. See Little's presidential address (4 Nov. 1913) in *Proc. Roy. Soc. Med.*, 7, pt 3 (1913–14), pp. 43–6, and A. Rocyn Jones, 'The Evolution of Orthopaedic Surgery in Great Britain – President's Address', *Proc. Roy. Soc. Med.*, 31, pt 1 (1937–8), pp. 19–26 at p. 25. Platt was elected president of the section in 1931. He later reflected that the section isolated orthopaedists both physically and ideologically from the Association of Surgeons, organized by Moynihan in 1920.

80. On Moynihan, who shared a summer cottage in Norway with William Mayo, see Donald Bateman, *Berkeley Moynihan, surgeon* (1940); on Horsley, see J. B. Lyons, *The Citizen Surgeon, a biography of Sir Victor Horsley* (1966).

81. Horsley was typical of most of the cohort before the war in devoting about 60 per cent of his time to neurosurgery and 40 per cent to general surgery. Lyons, *Horsley*, p. 49.

82. Mayo, 'Present-day Surgery', p. 7.

83. Quoted in Lyons, *Horsley*, p. 121.

84. *Ordered to Care: the dilemma of American nursing, 1850–1945* (Cambridge, 1987), p. 143.

85. Rabinbach, *Human Motor*, p. 6.

86. *BMJ*, 27 Oct. 1923, pp. 739–45 at p. 739.

87. See Peter English, *Shock, Physiological Surgery, and George Washington Crile: medical innovation in the progressive era* (Westport, Conn., 1980).

88. On the latter see the essays by Christopher Lawrence and Joel Howell in W. F. Bynum, *et al.* (eds), *The Emergence of Modern Cardiology, Med. Hist.*, Suppl. no. 5 (1985), pp. 1–52. The 'new psychology' of Adolf Meyer and others at the Johns Hopkins Hospital and at Michigan and Boston, beginning around 1906, was also based on dynamic functional theories. See Margo Horn, *Before It's Too Late: the child guidance movement in the United States, 1922–1945* (Philadelphia, 1989), p. 17.

89. Sir William Heneage Ogilvie, 'Orthodoxy and Heterodoxy in Surgery', *Lancet*, 25 Apr. 1931, pp. 897–902 at p. 897.

90. See Rosemary Stevens, 'The Curious Career of Internal Medicine: functional ambivalence, social success', in Russell Maulitz and D. Long (eds), *Grand Rounds* (Philadelphia, 1988), pp. 339–64 at p. 343.

91. See 'The Association of Fellows of the Royal College of Surgeons', *Lancet*, 13 Jan. 1883, pp. 79–80. On Tait, see J. A. Shepherd, *Lawson Tait, the rebellious surgeon (1845–1899)* (Lawrence, Kansas, 1980).

92. Mayo, 'Present-day Surgery', p. 2. See also Lawson Tait on 'our metropolitan rivals' in his 'Address in Surgery', *BMJ*, 2 Aug. 1890, pp. 267–73 at p. 269.

93. See English, *Crile*, pp. 69–70.

94. Crile wrote on fracture treatment in the 1890s. In 1913 Murphy's private clinic was observed to be 'the most popular in America' for the treatment of diseases of the bones and joints. See Robert Milne, 'Orthopaedic Surgery in America', *London Hosp. Gaz.* (1913), pp. 275–83 at p. 278. On Murphy, see Loyal Davis, *Surgeon Extraordinary, the life of J. B. Murphy* (1938), p. 263 and Moynihan, 'John B. Murphy, Surgeon' in Moynihan, *Selected Writings*, edited by A. White Franklin (1967), pp. 38–69.

95. See Cushing, 'Society of Clinical Surgery'; Franklin H. Martin, *The Joy of Living: an autobiography* (New York, 1933), vol. 1, chs 40 and 41; George Grey Turner, 'A Visit to America in War Time', *Univ. Durham Coll. Med. Gaz.*, 19 (15 Mar. 1919), pp. 17–23 at p. 18. Stiles is included in the undated photograph of the members of the Society of Clinical Surgery prefacing Cushing's article.

96. Loyal Davis, *Fellowship of Surgeons, a history of the American College of Surgeons* (Springfield, Illinois, 1960).

97. Martin, *Autobiography*, vol. 1, p. 404; J. M. T. Finney, *A Surgeon's Life: the autobiography of J. M. T. Finney* (New York, 1940), pp. 129–32; E. A. Codman, 'Autobiographical Preface' to his *The Shoulder* (Boston, 1934), pp. 39–41.

98. See Hedley Atkins, 'Travelling Clubs' in his *The Surgeon's Craft* (Manchester, 1965), pp. 192–5. Among other members of the Club, which first met in Leeds, were Hogarth Pringle, Alexis Thompson, Ernest Finch and George Grey Turner. See the latter's obituary in *Br. J. Surg.*, 52 (1965), pp. 641–6 at p. 643.

99. E. Hey Groves, 'A Surgical Adventure: an autobiographical sketch', reprinted from *Bristol Medico-Chirurg. J.*, 50 (1933), p. 21; and Platt, 'British Orthopaedic Association', pp. 116–25.

100. The political-minded consultants, Ernest Finch of Sheffield and Harry Platt of Manchester, came on the Board after the war. On Platt, see below Ch. 9. Finch held the Chair of Surgery at the Sheffield Medical School where he was responsible for the creation of the departments of orthopaedics, neurology and thoracic surgery.

101. When he first heard about the *British Journal of Surgery*, Macewen claimed that 'there was little in British surgery of any value and few men capable of writing anything worth while'. He changed his mind after Moynihan commissioned a biographical article on him for the *Journal*. See Bateman, *Moynihan*, p. 283. Macewen was also an honorary member of the AOA from 1890.

102. He was brought in partly by Martin, who manoeuvred him into giving the opening address to the American College of Surgeons 'to elevate the standard of surgery, [and] to educate the public'. See Martin, *Autobiography*, vol. 1, pp. 324 and 435–40; and Bateman, *Moynihan*, pp. 280–1.

103. Among the British who presented the Great Mace to the American College of Surgeons after the war were the BOA members, H. A. T. Fairbank, Robert Jones, T. H. Openshaw, A. H. Tubby and Sir William de Courcy Wheeler, along with Moynihan, Anthony Bowlby, Arthur Chance, C. S. H. Frankau, Henry Gray, Arbuthnot Lane, John Lynn-Thomas, Sir George Makins, Stiles, William Thorburn, S. Maynard Smith and A. Webb-Johnson, all of whom either wrote on fracture treatment during the war or were involved with Jones's orthopaedic centres.

104. Moynihan's Association of Surgeons of Great Britain and Ireland was not founded until 1920. See H. Platt, 'The Foundation of the Association of Surgeons of Great Britain and Ireland', *Br. J. Surg.*, 69 (1982), pp. 561–3. Moynihan was also behind the postwar Surgical Union which first met in Leeds in 1921. See Peter Boreham, *Surgical Journeys: a history of the Surgical Union which became the 1921 Surgical Travelling Club of Great Britain* (Braunton, Devon, 1990).

105. For bibliographies on these subjects, see *JBJS*, 32B (1950), pp. 550–69; and Tubby, *Advance of Orthopaedic Surgery*, pp. 118–22.

106. According to Mayer, it was the work on orthoplasty of the German general surgeon, E. Payr, that paved the way for the subsequent work in this field of W. S. Baer, Murphy and others. Mayer, 'Reflections', p. 378. Arbuthnot Lane, general surgeon of Guy's Hospital, who toured and demonstrated his 'open' treatment of fractures in America in 1910, was another major influence. See Edgar M. Bick, *Source Book of Orthopaedics* (Baltimore, 1937), p. 311. As the *Lancet* remarked in review of Robert Osgood's *The Evolution of Orthopaedic Surgery* (St Louis, 1925), 'if Paget and Brodie are to be claimed as orthopaedic surgeons, where is the line to be drawn?', 14 Aug. 1926, p. 336.

107. Significantly, perhaps, it was Tubby, the most conspicuous of the 'modernists' in British orthopaedics to be left out of Jones's war and postwar organization of the specialism, who retrospectively stressed the importance of technical factors in his *Advance of Orthopaedic Surgery*. Although Tubby joined the BOA in 1918, he played no active role. During the war he was posted to Alexandria. See Tubby, *A Consulting Surgeon in the Near East* (1920).

4   The Cause of the Crippled Child

1. See Deborah Dwork, *War is Good for Babies and Other Young Children: a history of the infant and child welfare movement in England, 1898–1918* (1987); Jane Lewis, *The Politics of Motherhood: child and maternal welfare in England, 1900–1939* (1980); George K. Behlmer, *Child Abuse and Moral Reform in England, 1870–1908* (Stanford, 1982); Harry

Hendrick, *Images of Youth: age, class and the male youth problem, 1880–1920* (Oxford, 1990); and idem, 'Constructions and Reconstructions of British Childhood: an interpretative survey, 1800 to the present' in A. Prout and A. James (eds), *Constructing and Reconstructing Childhood* (Basingstoke, 1990), pp. 35–59.

2. See K. W. Jones, 'Sentiment and Science: the late nineteenth century pediatrician as mother's advisor', *J. Soc. Hist.*, 17 (1983–4), pp. 79–96; Sydney Halpern, *American Pediatrics: the social dynamics of professionalism, 1880–1980* (Berkeley, 1988); and Rosemary Stevens, *American Medicine and the Public Interest* (New Haven, 1971), p. 200.

3. See Nikolas Rose, *The Psychological Complex: psychology, politics and society in England, 1869–1939* (1985); Margo Horn, *Before It's Too Late: the child guidance movement in the United States, 1922–1945* (Philadelphia, 1989); Gillian Sutherland, *Ability, Merit and Measurement: mental testing and English education, 1880–1940* (Oxford, 1984); and Adrian Wooldridge, 'Child Study and Educational Psychology in England, 1880–1950', DPhil thesis, Oxford, 1985. Cf. Barbara Ehrenreich and Deirdre English, *For Her Own Good: 150 years of the experts' advice to women* (1979), ch. 6: 'The Century of the Child'.

4. See Central Council for the Care of Crippled Children, *Directory of Orthopaedic Institutions, Voluntary Organisations and Official Schemes for the Welfare of Cripples* (1935); *Burdett's Hospitals and Charities Annual* (1930); Charity Organization Society, *Charities Annual and Register*, 1902–1930; and D. G. Pritchard, *Education and the Handicapped, 1760–1960* (1963). Unless otherwise indicated, information on the societies for the physically handicapped referred to in this chapter is drawn from these sources.

5. See Clark Nardinelli, *Child Labour and the Industrial Revolution* (Indiana, 1990). Children nevertheless continued to be victims of industrial accidents. See C. P. Hampson, *Salford Through the Ages* (Salford, 1930; reprinted 1972), pp. 195–200; and A. Clerke, *The Effects of the Factory System* (1899), pp. 108–13.

6. [Ebba de Ramsay], *A Few More Steps Onward! last year's experience of a cripples' home in Sweden, by a Swedish lady* (1878), p. 6. See also the letter from the *Manchester Guardian* quoted in Frederick Engels, *The Condition of the Working Class in England* (1845) in Marx and Engels, *On Britain* (2nd edn, Moscow, 1962), p. 315.

7. De Ramsay, *Onward*, p. 6. Cf. Montaigne, who relates that in the sixteenth century cripples were sought out for their supposed sexual prowess. 'Of Cripples' in *The Essays of Montaigne*, trans. by E. J. Trechmann (1935), vol. 2, pp. 497–508. I am grateful to Roy Porter for this reference.

8. Stanley Smith, 'The Alexandra Hospital for Children with Hip Disease', *The Child*, 3 (1912–13), pp. 728–31; 'London Letter', *Arch. Pediatrics*, 2 (1885), pp. 579–80; and H. B. Lee, 'The End of the Alexandra Hospital', *St Barts Hosp. J.*, 62 (1958), pp. 10–11. The hospital treated around 150 inpatients in 1884.

9. See Ruth Hodgkinson, *The Origins of the National Health Service, the medical services of the New Poor Law, 1834–1871* (1967), pp. 543ff.; and

F. B. Smith, *The People's Health, 1830–1910* (1979), p. 387 *et passim*. Returns from an 1861 survey of persons in workhouses for more than five years revealed that for Liverpool, Manchester and Oldham, 'crippled' inmates constituted 4.26 per cent, 7.19 per cent and 9.64 per cent respectively. Others doubtless would have been included in the categories for the 'infirm', 'sick', 'paralyzed' and 'aged'. My thanks to Mary Fissell for this data.

10. Vivian A. Zelizer, *Pricing the Priceless Child: the changing social value of children* (New York, 1985). See also Carolyn Steedman, 'Bodies, Figures and Physiology: Margaret McMillan and the late nineteenth-century remaking of working class childhood' in Roger Cooter (ed.), *In the Name of the Child: health and welfare, 1880–1940* (1992), pp. 19–44. The transformation also coincided with the fall in infant mortality and the decline in the national birth rate. See Anders Brandstom and Lars-Goran Tedebrand (eds), *Society, Health and Population during the Demographic Transition* (Stockholm, 1988); and Anna Davin, 'Imperialism and Motherhood', *History Workshop J.*, 5 (1978), pp. 9–65.

11. Peter D. Mohr, 'Philanthropy and the Crippled Child: the Band of Kindness and the Crippled Children's Help Society in Manchester and Salford, 1882–1948', MSc thesis, Manchester University, 1991.

12. See C. S. Loch, 'Charity and Charitism', *Ency. Brit.* (11th edn, 1910), vol. 5, pp. 860–91. For the other side of the story, see G. Wagner, *Barnardo* (1979) and J. W. Bready, *Doctor Barnardo* (1935). For an excellent local study of these competing styles, see Mohr, 'Philanthropy'.

13. See Charles L. Mowat, *The Charity Organization Society, 1869–1913: its ideas and work* (1961); Helen Bosanquet, *Social Work in London, 1869 to 1912: a history of the Charity Organization Society* (1914); and Gareth Stedman Jones, *Outcast London: a study in the relationship between classes in Victorian society* (Harmondsworth, 1976). In its efforts to eradicate mendicity and put philanthropy on a 'rational' basis, the COS scrutinized parents seeking to send their children gratis to cripples' homes. But in so doing, it only further exposed and compounded the problems of cripples: of 1863 'confirmed mendicants' accosted by the COS over fifteen months in 1908–9, 701 were found to be in some way physically defective, while 447 (24 per cent) were found to be 'crippled'. See *Med. Record*, 1 May 1909, p. 781; for the repression of begging in homes for cripples – 'one of the major sources of mendicity' – see *Charity Organization Reporter*, 23 May 1872, p. 96.

14. See Duchess of Sutherland (founder and patron of the Potteries Cripples' Guild), 'The Work of the Potteries Cripples' Guild', *Charity Organization Rev.*, 16 (1904), pp. 81–5; idem, 'The Children of the Potteries', *Pall Mall Mag.*, 32 (1904), pp. 1–10; and Edgar Allen (President of the International Society for Crippled Children), 'The Evolution of the Problem of the Cripple', *Cripples' J.*, 5 (1929), pp. 330–8 at p. 334.

15. See Treloar and H. J. Gauvain, 'The Treatment of Cripples at Alton', *The Child*, 1 (1910), pp. 178–87; H. Gauvain, 'The Lord Mayor Treloar Cripples' Hospital and College', *Cripples' J.*, 2 (Oct. 1924), pp. 64–70; H. Gauvain and E. M. Holmes, 'The Evolution of Hospital Schools',

Lancet, 13 Apr. 1929, pp. 790–1; and C.W. Lawrence, *William Purdie Treloar* (1925).

16. Cited in Pritchard, *Education and the Handicapped,* p. 161, who adds, 'Certainly, twenty-eight boys who had passed through the hospital and craft school were sufficiently well made to fight and fall in the First World War.'

17. Cited in H. Osmond-Clarke, 'Half a Century of Orthopaedic Progress in Great Britain', *JBJS,* 32B (1950), p. 620.

18. On the late nineteenth-century shift in attitude to the state, see Gertrude Williams, *The State and the Standard of Living* (1936), ch. 1; Behlmer, *Child Abuse,* esp. ch. 7: 'Voluntary Effort and the State'; Geoffrey Searle, *The Quest for National Efficiency: a study in British politics and social thought, 1899–1914* (Oxford, 1971); and B.B. Gilbert, 'Health and Politics: the British Physical Deterioration Report of 1904', *Bull. Hist. Med.,* 39 (1965), pp. 143–53. For a perspective on degeneracy, see Daniel Pick, *Faces of Degeneration: a European disorder, c. 1848–1918* (Cambridge, 1989).

19. Madeline Rooff, *Voluntary Societies and Social Policy* (1957), p. 11; Charles R. Henderson, *Introduction to the Study of the Dependent, Defective and Delinquent Classes, and of Their Social Treatment* (Boston, 1901), pp. 138ff.; and Pritchard, *Education and the Handicapped,* pp. 153ff. Burdett, *Hospitals and Charities Annual for 1930* (pp. 719–25) lists 68 surviving settlements, 40 in London, 24 in the provinces and 4 in Scotland.

20. See A.J. Kidd, 'Charity Organization and the Unemployed in Manchester, *c.* 1870–1914', *Social Hist.,* 9 (1984), pp. 45–66 at p. 62.

21. The shift from degeneracy to social environment was signalled in the evidence presented to the *Report of the Inter-Departmental Committee on Physical Deterioration* (1904), vol. II: List of Witnesses and Minutes of Evidence, Cd 2210. For doctors and the state, see Jeanne L. Brand, *Doctors and the State: the British medical profession and government action in public health, 1870–1912* (Baltimore, 1965); Elizabeth Macadam, *The New Philanthropy* (1934); and T.S. Simey, *Principles of Social Administration* (1934).

22. See 'Guildhall Conference on Invalid Children', *Charity Organization Rev.,* 16 (1904), pp. 61–6, and below.

23. Cited in 'The Sick Children of the Poor', *The Times,* 29 Nov. 1889, p. 12.

24. See Board of Education, *Annual Report of the CMO for 1913,* p. 122, and for 1919, p. 9; Bosanquet, *Social Work in London,* pp. 249, 230; COS, *Charities Annual and Register for 1910,* pp. clxxviii–clxxx, cclxvii–cclxix, 48; H. Warrington (Chairman), 'Invalid Children's Aid Association', *The Child,* 1 (1910), pp. 195–6; A Heddle, *Forty Five Years: a brief history of the Woolwich and District ICAA, 1892–1937* (1937); and 'The Guildhall Conference on Invalid Children [organized by the ICAA]', *Charity Organization Rev.,* 16 (1904), pp. 61–6.

25. Ivy A. Ireland, *Margaret Beavan of Liverpool, her character and work* (Liverpool, 1938). On the Leasowe Hospital (renamed the Margaret Beavan Hospital in 1932), see *Annual Reports* (Birkenhead Central Reference Library). Charles Macalister was a consulting physician to the

Hospital, and Jones's assistant, George P. Newbolt, was the first chairman of the Medical Board. George Newman regarded Leasowe in 1917 as 'the high-water mark of the scientific treatment of the bones and joints'. See Leasowe, *Annual Report for 1917*, p. 8. On the work of the ICAA in Liverpool in 1910, see *Lancet*, 12 Feb. 1910, pp. 466–7; see also Ursula Townsend's letter to Dr. Alfred Eichholz (Chief Medical Inspector to the Board of Education), 20 June 1920, in PRO:ED/50/153.

26. See *DNB* entry. On Chailey, see Nora A. Smith, 'Guilds of Play and of Brave Poor Things', *Outlook*, 90 (1908), pp. 78–82; and Mrs Kimmins, 'The Cripple Colony at Chailey in Connection with the Guild of Brave Poor Things', *Trans. 2nd Intern. Congress of School Hygiene*, 2 (1907), pp. 753–7, and idem, *Heritage Craft School and Hospital, Chailey, 1903–1948* (1948).

27. See Mrs Janet Trevelyan (née Ward), *The Life of Mrs Humphry Ward* (1923); Enid Huws Jones, *Mrs Humphry Ward* (New York, 1973), esp. pp. 126–31; and John Sutherland, *Mrs Humphry Ward: eminent Victorian, pre-eminent Edwardian* (Oxford, 1990). On the Joint Advisory Committee, see PRO:ED/50/152.

28. Robert Bremner (ed.), *Childhood and Youth in America: a documentary history*, vol. 1: *1600–1865* (Harvard, 1970), p. 758.

29. Cf. *Charities Annual and Register*, 1902, p. 52, and 1930, p. 67.

30. See, for example, Miss Fowler (COS), 'The Care and Relief of Crippled and Invalid Children', pamphlet reprinted from the *Charity Organization Society Rev.*, 1896, pp. 2–31.

31. Gauvain and Holmes, 'Hospital Schools', p. 791. Compare, however, the medical attention provided in Dr Barnardo's special hospitals for handicapped children, dating from the late 1880s. See June Rose, *For the Sake of the Children: inside Dr Barnardo's, 120 years of caring for children* (1987), ch 7.

32. On the 'medicalization' of Chailey that Jones undertook in 1927, see the letter from him to Girdlestone, 13 July 1927, in Girdlestone Papers, BOA. G. Murray Levick was then appointed Medical Director and Reginald Elmslie and W. Rowley Bristow were among the consulting orthopaedic surgeons.

33. See obituary in *JBJS*, 27 (1945), pp. 342–4.

34. This was the view of Miss Haines, a social worker active in London's East End in the early 1900s. Cited by Dr Alfred Eichholz at the 'Conference on Education in Hospital Schools and Schools for Physically Defective Children', *Cripples' J.*, 1 (1925), p. 303. See also the evidence of Rev. J. S. Lidgett (Warden of Bermondsey Settlement and Chairman of the Committee of St Olave's Poor Law Union Infirmary) to the Committee on Defective and Epileptic Children (Education Department), *Report of the Departmental Committee on Defective and Epileptic Children*, vol. II: Minutes of Evidence etc. (1898), Cd 8747. 'The great want', he felt, 'is a medical institution where medical treatment can be combined with specially adapted methods of education' (pp.107–8).

35. E. D. Telford, *The Problem of the Crippled Child* (Manchester, 1910), p. 7.

36. Jones and Charles Macalister were typical in becoming active in the Liverpool Society for the Care of Invalid Children shortly after its founding by Margaret Beavan in 1905. See Thomas Bickerton, *Medical History of Liverpool from the earliest days to the year 1920* (1936), pp. 234–5.

37. *Annual Reports* and a typescript history are contained in the Manchester Medical Collection, John Rylands Library, University of Manchester. The Dispensary, originally known as the Salford Medical Mission, was purchased by Grimké (1817–86) in 1876 from the Salford Ragged and Industrial Schools which, since 1854, had cared for and trained some 200 pauper children. A statement in 1902 refers to the special treatment for the paralyzed and disabled and describes the object of the Mission as providing medical advice and help for the poor of the immediate neighbourhood by means of a general dispensary for all forms of sickness, including the alleviation of the physical and mental deficiencies of crippled children. As a residential school, it was also unusual in returning children to their parents during weekends. At the death of Mrs Emma Grimké in 1906, William Mather, the future mayor of Manchester, Vice-President of the Manchester Cripple Children's Help Associaton, and founder in 1902 in Salford of the first open-air school for debilitated children, took over as administrator of the Mission and invited subscriptions to place the institution on a public basis. The Dispensary was sufficiently important for Sir William Treloar to visit it in 1909 and Dr Montessori in 1919. G. A. Wright was its honorary surgeon. See, also, Board of Education, *Annual Report of the CMO for 1919*, p. 119; and Mohr, 'Philanthropy', pp. 40–9.

38. *Annual Report for 1882*, pp. 7, 13.

39. Joseph Bell, 'Five Years' Surgery', *Edin. Hosp. Reports*, 1 (1893), pp. 466–74 at p. 467.

40. See Alfred Tubby, 'Is the Urban Hospital Treatment of External or Surgical Tuberculosis Justifiable?', *Practitioner*, 71 (Sept. 1903), pp. 313–19.

41. 'The Sick Children of the Poor', *The Times*, 29 Nov. 1889, p. 12.

42. Charles Macalister, *The Origins and History of the Royal Liverpool Country Hospital For Children at Heswall from the time of its inception (1895–1898) and foundation (1899) to the year 1930* (Campden, Glous., 1930).

43. Ibid., p. 7.

44. See his *The Treatment of Epilepsy* (Edinburgh, 1899); obituary in *Lancet*, 29 Mar. 1919, p. 530; Charles Macalister, *Royal Southern Hospital* (Liverpool, 1936), pp. 58, 139–40; Jean Barclay, 'Langho Epileptic Colony: 1906–1984: a contextual study of the origins, transformations and demise of Manchester's "Colony for Sane Pauper Epileptics"', PhD thesis, Manchester University, 1988; idem, '*The First Epileptic Home in England': a centenary history of the Maghull Homes, 1888–1988* (Glasgow, 1990). Alexander was one of the two medical men consulted by the COS in their report on epilepsy – the other being Dr Francis Warner. He also had more than a passing interest in work that would

later come within the remit of orthopaedics: in 1881 he was awarded the Jacksonian prize for his essay on the 'Pathology and Surgical Treatment of Diseases of the Hip-Joint', and, in 1883, he won the Sir Astley Cooper prize at Guy's Hospital for his essay on the 'Pathology and Pathological Relation of Chronic Rheumatoid Arthritis'. I am grateful to Jean Barclay and the late Nora Kemp for information on Alexander.

45. See Bickerton, *Medical History of Liverpool*, p. 126.

46. The celebratory literature suggests otherwise, and also attributes anti-urban and open-air motives. See H. G. Carlisle, MD, 'Orthopaedic Centres VIII: The Royal Liverpool Children's Hospital, Heswall', *Cripples' J.*, 2 (Oct. 1925), pp. 127–34. See, also, F. Watson, *The Life of Sir Robert Jones* (1934), pp. 111–12ff.

47. For a brief history and sources, see Adrian Allan, *List of the Archives, etc., of West Kirby Residential School, 1881–1980*, University Archives, Liverpool.

48. See MS Minutes of the Executive Board of the Royal Liverpool Children's Hospital, 9 July 1900, and *Annual Reports of the Liverpool Country Hospital for Chronic Diseases of Children*, 1900–1919. It may not be coincidental that William Alexander's private residence was at Heswall.

49. Quoted in *Liverpool Daily Post and Mercury*, 6 Nov. 1899. And see MS Minutes of the Executive Board of the Royal Liverpool Children's Hospital, 16 Oct. 1899.

50. Surgical cases were selected from Jones's clinics at the Royal Southern Hospital and Nelson Street. Medical cases came from the Stanley Hospital until 1900, when the Hospital established an outpatient department for children. Thereafter, most cases came from the Royal Southern. See Macalister, *Heswall*, p. 41. For an example of an 'unsuitable' case being sent by the Liverpool Royal Infirmary, see MS Minutes of the Executive Board of the Royal Liverpool Children's Hospital, 29 Jan. 1900.

51. See *Annual Reports*; Adrian Allan, *List of Archives, etc., of the Royal Liverpool Children's Hospital*, University Archives, Liverpool; Board of Education, *Annual Report of the CMO for 1910*, p. 81; and 'Work of the Liverpool Country Hospital in the Curative and Educational Treatment of Crippled Children', *Lancet*, 10 Aug. 1907, pp. 397–8.

52. Tubby, 'Urban Hospital', esp. at pp. 318–19.

53. *LCC Report*, 1907, p. 48, cited in James Kerr, 'Orthopaedic Treatment', in his *The Fundamentals of School Health* (1926), p. 184.

54. See W. S. Craig, *John Thomson: pioneer and father of Scottish paediatrics, 1856–1926* (Edinburgh, 1968), esp. p. 22; Janet Penrose Trevelyan, *Evening Play Centres for Children, the story of their origin and growth, with a preface by Mrs Humphry Ward* (1920); 'Edinburgh Cripple and Invalid Children's Aid Society', *The Child*, 4 (1913), pp. 196–8; and Thomson, 'On Home Care and Treatment of Physically Defective Infants and Young Children', *Br. J. Childr. Dis.*, 21 (1924), pp. 263–8.

55. City of Birmingham Education Committee, *Report of a Special Sub-committee of Enquiry concerning Physically-Defective Adults and Children Presented to the Education Committee 27th October, 1911*. The

subcommittee had been formed in December 1909 to 'consider if it was possible to organise any further help for the cripples living within the City'. The Birmingham Cripples' Union had been active since 1899, and in 1902 opened the first school for physically defective children. See M. Champness, 'Birmingham and District Cripples' Union', *The Child*, 2 (1911–12), pp. 506–10.

56. See Janet Hill, 'Mary Dendy, Her Life and Her Work for the Permanent Care of the Feebleminded In the Manchester Area', Honours Paper, Department of Biology and Geology, Manchester University, 1984. For critical background, see Joanna Ryan and Frank Taylor, *The Politics of Mental Handicap* (Harmondsworth, 1980), ch. 5.

57. On Ashby's reports and the medical interest in his work, see the editorial 'The Education of Crippled and Mentally Defective Children', *Lancet*, 7 Apr. 1906, p. 980. On Swinton House, see E. D. Telford, *Problem of the Crippled Child*; idem, 'The Residential School for Crippled Children', *The Child*, 2 (1911), pp. 121–6; Board of Education, *Annual Report of the CMO for 1909*, pp. 142–3; L.C. Smith, 'The Swinton School for Cripples', *Westminster Rev.*, 152 (1908), pp. 223–7; and Mrs Ward's criticism in the printed Minutes of the Joint Advisory Committee, Education Sub-Committee, 22 Apr. 1919, pp. 4–5, in PRO:ED/50/152.

58. Francis Warner, MD, FRCP, 'Results of an Inquiry as to the Physical and Mental Condition of Fifty Thousand Children seen in One Hundred and Six Schools', *J. Roy. Stat. Soc.*, 56 (1893), pp. 71–100; and the follow-up study, ibid., 59 (1896), pp. 125–68.

59. See A. H. Hogarth, *Medical Inspection of Schools* (1909); and George Newman, *The Building of a Nation's Health* (1939). For background on the medical inspection of schoolchildren, see also Kerr, *Fundamentals*, and William Mackenzie and E. Matthew, *The Medical Inspection of School Children* (Edinburgh, 1904).

60. Charles Webster, 'The Health of the School Child During the Depression', in Nicholas Parry and David McNair (eds), *The Fitness of the Nation: physical and health education in the nineteenth and twentieth centuries* (Leicester, 1983), p. 74.

61. However, before the war, the SMS did more to encourage local authorities and voluntary agencies to meet the needs of the blind, deaf and the mentally handicapped than the physically handicapped. See Gauvain and Holmes, 'Hospital Schools', p. 791. This prioritization was in line with the greater emphasis on the mentally handicapped in reports such as that by Warner (1893, 1896), the COS (1893) and the Departmental Committee on Defective and Epileptic Children (1898).

62. E. M. Little (compiler), *History of the BMA, 1832–1932* (1932), p. 135. The LCC rejected the subcommittee's proposal 'in favour of arrangements with voluntary hospitals'. See also Brand, *Doctors and the State*, p. 187. For Elmslie's involvement, see R. Elmslie, 'The Classes of Physically Defective Children for Whom Instruction in Special Schools is Necessary', *Trans. 2nd Intern. Congress on School Hygiene*, 2 (1907), pp. 762–7; idem., *The Care of Invalid and Crippled Children in School* (1909); and idem., 'School Provision for Physically Defective Children', *School Hygiene* (1910), abstracted in *Br. J. Childr. Dis.*, 8 (1911), pp.

232–3. See also *The Medical Officers of Schools Association: a record of 50 years' work, 1884–1934* (Hertford, 1934), pp. 39, 42, 44, 91, 98.

63. See the editorial in *Br. J. Childr. Dis.*, 4 (1907), pp. 213–16, and *BMJ*, 18 Jan. 1908, p. 165. See also the criticism of 'Lady Bountifuls' in Douglas McMurtrie, 'Social and Moral Considerations Related to the Medical and Surgical Care of Crippled Children', *N.Y. Med. J.*, 30 Dec. 1911, pp. 1330–2.

64. See Board of Education, *Annual Report of the CMO for 1910*, p. 81.

65. E. Muirhead Little and Mary Little, 'A Hospital School for Cripples', *Trans. 2nd Intern. Congress on School Hygiene*, 2 (1907), pp. 757–62.

66. See Macalister, *Heswall*, p. 44; 'Liverpool Country Hospital for Chronic Diseases of Children (Heswall)', *Lancet*, 15 June 1907, p. 1675; and Board of Education, *Annual Report of the CMO for 1913*, pp. 75–6.

67. See 'Grants in Aid', Board of Education, *Annual Report of the CMO for 1913*, p. 80. In 1909 the LCC sent a circular to hospitals in London seeking their cooperation in the treatment of schoolchildren (GLRO:PH/SHS/2); for impoverished hospitals, such as the Poplar in London, this was an important source of income.

68. Kerr, *Fundamentals*, p. 181. Political and Economic Planning, *Report on the British Health Services* (1937), p. 300, cited estimates from surveys carried out in 1933 of 'potential cripples, that is mostly children, who needed immediate orthopaedic treatment' in three regions:

| | Per 1000 at each age | | |
	Under 5	5–16	Over 16
Devon area	10.4	13.5	4.0
Shropshire area	10.3	13.0	2.9
Birmingham	6.8	6.2	3.1

69. The criticism came from Christopher Addison and George Newman. See Minutes of the Joint Advisory Committee, 22 Apr. 1918, and Newman's attached note, in PRO:ED/50/152. See also 'The Meaning of The Term Crippled', *Med. Record* [New York], 94 (1918), pp. 939–41.

70. *Annual Report of the CMO for 1919*, p. 103. Compare the figures for Shropshire given in Robert Jones and G. R. Girdlestone, 'The Cure of Crippled Children: proposed national scheme', *BMJ*, 11 Oct. 1919, pp. 457–60.

71. See Gauvain and Holmes, 'Hospital Schools', p. 791.

72. Board of Education, *Annual Report of the CMO for 1911*, pp. 169ff.

73. Ireland, *Beavan*, pp. 39–40; Board of Education, *Annual Report of the CMO for 1913*, pp. 75–6; Leasowe, *Annual Report for 1917*, p. 8.

74. Cited in Carlisle, 'Heswall', p. 132.

75. Board of Education, *Annual Report of the CMO for 1913*, pp. 74ff., and 191. Of the 1161 children in 'PD' schools at this time, 345 had been sent by local education authorities, 45 by National Insurance Committees, 106 by voluntary bodies, 203 by private individuals, 43 were committed under Sections 45 and 62 of the Children's Act, 1908 (and paid for by the Home Office), 16 from other sources, but 403 (34 per cent) were sent by Boards of Guardians. Board of Trade statistics reveal that nationally

LEAs had spent only £394 000 more on PD schools in 1931 than in 1921. The number of PD schools rose over this period from 500 to 607, and the number of pupils from 36,459 to 48,934.

76. See John V. Pickstone, *Medicine and Industrial Society* (Manchester, 1985), ch. 10; cf. Rosemary Stevens, *Medical Practice in Modern England* (New Haven, 1966), p. 36.

77. Letter, marked 'private', to Sir Amherst Selby-Bigge, of the Board of Education, 21 Dec. 1919 in PRO:ED/50/153. See also the impressions of the American orthopaedic surgeon H. Winnett Orr while stationed at Cardiff during the war: *JOS*, 2 (1920), pp. 147–9.

78. Newman, memo dated 14 Nov. 1919 in PRO:ED/50/152.

79. See R. B. Osgood, 'Progress in Treatment of the Crippled Child', *Cripples' J.*, 4 (1927–28), pp. 91–101 and 221–31, at p. 98; 'Needs of School for Deformed Children', *Boston Med. & Surg. J.*, 30 Nov. 1911, p. 849; and obituary on E. H. Bradford, *JBJS*, 8 (1926), pp. 461–5. For useful international comparisons, see E. M. Goldsmith, 'Crippled Children, Education of', in Paul Munroe (ed.), *Cyclopaedia of Education* (New York, 1911–13), vol. 2, pp. 230–4.

80. Gillette, 'State Care of Indigent Crippled Children', *Trans. Amer. Orthop. Assn.*, 12 (1899), pp. 249–53; 'State Care of Indigent Crippled and Deformed Children', *Proc. Nat. Conf. Charities and Correction*, 1904, pp. 285–94; idem., 'Care of Cripples', *AJOS*, 6 (1909), pp. 723–6; E. Reeves, 'Minnesota State Hospital for Crippled Children', *Amer. Baby* (July 1911), pp. 37–9; and Osgood, 'Progress', p. 224. On Gillette (1864–1921) see obituary in *JOS*, 3 (1921), pp. 159–60, 246–8.

81. Henry H. Kessler, *The Crippled and the Disabled: rehabilitation of the physically handicapped in the United States* (New York, 1935), pp. 35ff. See also 'Recent Progress in the State of Crippled and Deformed Children', *N.Y.Med.J.*, 30 Jan. 1904 (editorial), pp. 217–18; H. Winnett Orr, 'Duty of the State in the Care of Crippled and Deformed Children', *Detroit Med. J.*, 5 (1905–6), pp. 195–8; idem., 'Reason for the State Aid of the Crippled and Deformed; Some of the Problems Involved', *AJOS*, 9 (1911), pp. 218–23; and D. C. McMurtrie, 'The Advantages of, And The Need for State Care of Crippled Children', *Ohio State Med. J.*, 8 (1912), pp. 207–9.

82. See N. Shaffer (founder of the New York State Reconstruction Home at Haverstraw for the care of crippled children), 'What New York State is doing for its Crippled and Deformed Children at West Haverstraw State Hospital', *N.Y. State Med. J.*, 8 (1908), pp. 126–9.

83. 'What Shall We Do With Our Cripples? Report of meeting of the Section on Orthopedic Surgery of the New York Academy of Medicine', *JAMA*, 3 Apr. 1909, pp. 1134–6; and *Med. Record* [New York], 1 May 1909, pp. 780–2. Speakers included De Forest Willard, Newton Shaffer and Virgil P. Gibney. See also, among other writings by orthopaedic surgeons at this time: N. Shaffer, 'The Care and Treatment of the Dependent and Neglected, Crippled and Deformed Children', *Proceedings, Conf. on the Care of Dependent Children* (1909), pp. 80–2; John Ridlon, 'A General Consideration of the Needs of Crippled Children, Their Treatment and the Results to be Expected', *Illinois Med. J.*, 16 (1909), pp. 409–16;

Charlton Wallace, 'Education of the Crippled Child', *Arch. Pediatrics*, 27 (1910), pp. 345–52; and Van Buren Thorne, 'Cripples Made Straight by Marvellous Surgery [on the work of Dr E.G. Abbott of the Children's Hospital, Portland, Maine]', *New York Times*, 3 Dec. 1911, magazine section, pp. 3–5.

84. In his introduction to George Payne, *The Child in Human History* (1916), p. iv.

85. And see Henry E. Abt (Director Bureau of Information of the International Society for Crippled Children), *The Care, Cure, and Education of the Crippled Child ... a complete bibliography ... and directory* (Elyria, Ohio, 1924).

86. See 'Care of Cripples [in Germany]', *JAMA*, 19 Sept. 1908, p. 1024; *Lancet*, 15 May 1909, p. 1425; *BMJ*, 1 May 1909, p. 1092; and Beat Rüttimann, *Wilhelm Schulthess (1855–1917) und die Schweizer Orthopädie Seiner Zeit* (Zurich, 1983), pp. 167–74. See also 'International Exposition of Cripple Aid [on the cripples' section at the International Hygiene Exposition at Dresden]', *JAMA*, 28 Jan. 1911, p. 284; and 'The Care of Crippled Children in Germany', *N.Y. Med. J.*, 99 (1914), pp. 370–7.

87. Biesalski, 'The Development of the Provision for the Crippled in Germany and its Present Standing', *Rehab. Rev.*, 1 (1927), pp. 184–7, at p. 184. See also Biesalski, *Krüppelfursorge* (Berlin, 1913); idem., 'Wesen und Verbreitung des Krüppeltums in Deutschland', *Z. Orthop. Chir.*, 20 (1908), pp. 323–43. On Biesalski, see *JBJS*, 12 (1930), pp. 448–50 and 37A (1955), p. 377; and M. Feja, 'Das Krüppelkind Ein Beitrog zur Geschichte der Kösperbehindertenfürsorge', MD dissertation, Albert-Ludwig University, Freiburg, 1975, for access to which I am grateful to Professor Eduard Seidler.

88. See Kessler, *Crippled and Disabled*, p. 34.

89. Biesalski, 'Crippled in Germany', p. 184, who adds that Germany now (1925) has at least 78 such institutions with some 11 000 beds. For slightly different definitions of what constitutes a 'cripple', see the Report of the Birmingham Educational Committee (1911), and that used by the Massachusetts State Survey Commission of 1905, cited in Kessler, *Crippled and Disabled*, p. 51.

90. Biesalski, 'Crippled in Germany', p. 186.

91. Agnes Hunt, *This is My Life* (1938). The book was written to amuse Jones during his last illness. See also anon., *The Heritage of Oswestry: the origin and development of the Robert Jones and Agnes Hunt Orthopaedic Hospital, Oswestry, 1900–1975* (Oswestry, 1975); and Dorothy Wrenn, 'The Lyrical Adventure, the story of Agnes Hunt', MS, *c.* 1970, copy in Shropshire Local History Library.

92. Arthur Stanley, preface to Hunt, *Life*, p. vi. For Thomas's work at Rhyl, see his *Hip, Knee and Ankle* (2nd edn, 1876), p. 169n., and A.J. Carter, 'Hugh Owen Thomas: the cripple's champion', *BMJ*, 21–28 Dec. 1991, pp. 1578–81.

93. Hunt, 'Reminiscences', quoted in Wrenn, 'Lyrical Adventure', p. 65.

94. Hunt, *Life*, p. 7.

95. Hunt, 'Chailey and Papworth, a personal impression', *Cripples' J.*, 1 (1924), pp. 27–9.
96. T. Fraser, in Hunt (ed.), *The Story of Baschurch* (Oswestry, n.d.), p. 41; and see Hunt, *Life*, pp. 129ff.
97. See, for example, *Med. Annual* (1905), p. 464.
98. On the origins of open-air therapy for consumption, see *BMJ*, 29 Apr. 1905, pp. 960–1; *Lancet*, 21 Jan. and 17 June 1899, pp. 193–4, 1650–1; and Sir Humphry Rolleston, 'British Pioneers in the Modern Treatment of Tuberculosis', *Canadian Med. Assn. J.*, 29 (1933), pp. 113–21.
99. Walter S. Ufford, *Fresh Air Charity in the United States* (New York, 1897). See, also, Roger Cooter, 'Open-Air Therapy and the Rise of Open-Air Hospitals', *Bull. Soc. Social Hist. Med.*, no. 35 (Dec. 1984), pp. 44–6.
100. See Linda Bryder, *Under the Magic Mountain* (Oxford, 1988) and F. B. Smith, *The Retreat of Tuberculosis, 1850–1950* (1988).
101. See, for example, Arthur Ransome, *The Principles of 'Open-Air' Treatment of Phthisis and of Sanatorium Construction* (1903); and Guy Hinsdale, 'Atmospheric Air in Relation to Tuberculosis', *Smithsonian Miscellaneous Collections*, 63 (1914), pp. 1–136.
102. See Marjorie Cruickshank, 'The Open-air School Movement in English Education', *Paedagogica Historica*, 17 (1977), pp. 62–74; and Linda Bryder, '"Wonderlands of Buttercup, Clover and Daisies": the open-air school movement, 1907–1939', in Cooter (ed.), *Name of the Child*, pp. 72–95.
103. See *Annual Reports of Baschurch*, Shropshire County Council Record Office, 1387/1–46.
104. Watson, *Life of Jones*, p. 116ff.
105. See G. R. Girdlestone, 'The Modern Treatment of Tuberculosis of the Bones and Joints', *JBJS*, 6 (1924), pp. 519–37 at p. 532.
106. Josep Trueta, *Gathone Robert Girdlestone* (Oxford, 1971), p. 41. Girdlestone confessed himself 'infected with orthopaedics' after having been 'touched' by Jones. See 'Orthopaedic Influence on the Treatment of Fractures', *Lancet*, 13 Nov. 1943, pp. 593–5.
107. See *Annual Report for Baschurch, 1908–9*. Jones was to be one of Williams's few life-long intimates. See Watson, *Life of Jones*, p. 270.
108. See Jones's contribution to *The Story of Baschurch*, pp. 31–4, and his 'And Some Others' (n.p., 1916) (a published fund-raising letter), Shropshire Country Record Office 1357/34.
109. Heswall, *Annual Report for 1910*; and Macalister, *Heswall*, p. 36. One of the open-air wards was named after Thomas, another after Hunt.
110. Hunt, *Cripples' J.*, 6 (1930), p. 480.
111. From the 1890s, while working as a district nurse, Hunt had been acutely aware of the problem of the relapse of cripples through insufficient after-care. See, *Life*, pp. 145–6, and on this subject Stevens, *American Medicine*, p. 135, and J. David Hirst, 'The Growth of Treatment Through the School Medical Service, 1908–18', *Med. Hist.*, 33 (1989), pp. 318–42 at p. 328.
112. *Annual Report for 1914*; other such details in this section are taken from the *Annual Reports* for the respective years.

113. In 1922 an inpatient cost £136 10s 0d p. a. and an outpatient £2 12s 4½d. See *Annual Report of the Shropshire Orthopaedic Hospital for 1922*, cited in Board of Education, *Annual Report of the CMO for 1922*, p. 97.

5 Happenings by Accident

1. That relatively more people were sustaining injuries is not easy to substantiate. John Benson has argued on the basis of mortality statistics that coal mining was actually becoming a safer industry during the second half of the nineteenth century: *British Coalminers in the Nineteenth Century: a social history* (Dublin, 1980), p. 43. But since what constitutes an 'accident' is socially determined, statistics on 'accidents' must be treated as subjective attempts at the construction of evidence. See Peter Bartrip and P. T. Fenn, 'The Measurement of Safety: factory accident statistics in Victorian and Edwardian Britain', *Historical Research*, 63 (1990), pp. 58–72. On the social construction of accidents, see Karl Figlio, 'How Does Illness Mediate Social Relations? Workmen's compensation and medico-legal practices, 1890–1940', in P. Wright and A. Treacher (eds), *The Problem of Medical Knowledge* (Edinburgh, 1982), pp. 174–224; and idem., 'What is an Accident?', in Paul Weindling (ed.), *The Social History of Occupational Health* (1985), pp. 180–206.

2. See John Rule, *The Experience of Labour in Eighteenth-Century Industry* (1981), p. 91.

3. John Roberton, 'On the need of additional as well as improved hospital accommodation for surgical patients in manufacturing and mining districts, but especially in Manchester', *Trans. Manchester Stat. Soc.*, 1860 (reprinted pamphlet).

4. Benjamin Moore, *The Dawn of the Health Age* (1911). Similarly lacking reference to the victims of accidents is Sidney and Beatrice Webb's *The Prevention of Destitution* (1911), and Lady Waldorf Astor's *The Health of the People* (1917). Following suit, neither F. B. Smith's *The People's Health, 1830–1910* (1979) nor Anthony Wohl's *Endangered Lives: public health in Victorian Britain* (1983) contain sections on the adult victims of accidents.

5. 'Treatment of Fractures', *BMJ*, 24 May 1913, pp. 1118–9. BMA, 'Report of the Committee on Treatment of Simple Fractures', *BMJ*, 30 Nov. 1912, pp. 1505–41.

6. See, for example, Timothy Holmes (ed.), *A System of Surgery*, 4 vols (1862; 3rd edn, 1883); Frederick Treves (ed.), *A Manual of Surgery*, 3 vols (1886); C. C. Choyce (ed.), *A System of Surgery* (2nd edn, 1923). See also William Williams Keen (ed.), *Surgery, its principles and practice* (Philadelphia, 1907) wherein Robert Lovett writes the section on orthopaedics, while a general surgeon writes on fractures.

7. James K. Young (professor of orthopaedic surgery, Philadelphia Polyclinic) and J.M. Spellissy, *Cyclopaedia of the Diseases of Children, medical and surgical*, ed. W. A. Edwards (1899), pp. 135–48. The *Cyclopaedia* was the first standard work for American paediatrics. The chapter on 'Fractures and Dislocations' was by John H. Packard.

8. Most notably at the Rochester General Hospital, New York, where, in 1907, the general surgeons agreed to have all fractures, excepting those of the cranium and ribs, assigned to the orthopaedic service. This was recalled by James W. Sever in the discussion on Leo Mayer's 'Reflections', *JBJS*, 37A (1955), p. 383.

9. At the London Hospital around the turn of the century, only about 3 per cent of the over 15 000 annual accident cases were referred to the Orthopaedic Department (established 1908) headed by Thomas Openshaw. Typically the London's *Annual Report* for 1909, in commenting on the expansion of the Orthopaedic Department, referred only to Openshaw's 'excellent work among crippled children'.

10. *Trans. BOS*, 1 (1896), p. 33.

11. The debate centred on the relative merits of two radically different types of treatment: the surgically invasive open-plating techniques pursued preeminently by Arbuthnot Lane of Guy's Hospital, and the wholly conservative massage treatment pursued with equal preeminence by the French surgeon J. M. M. Lucas-Championnière. The latter's techniques had been on the ascendant in France since the publication of his *Traitement des fractures par le massage et la mobilisation* (1895) in which it was claimed that by these means, and without resort even to the usual apparatus for reduction and traction, perfect anatomical and functional results could be obtained. The main British exponent of these views was James Mennell, whose *The Treatment of Fractures by Mobilisation and Massage* (1911) – the subject of an editorial in the *BMJ* of 16 March 1912 (pp. 627–8) – contained an introduction by Lucas-Championnière.

    Jones was appointed a member of the BMA's Committee of Investigation into Fracture Treatment, but declined the offer (see Minutes of the Committee, 29 March 1911, p. 2), probably in order to avoid giving tacit approval to its recommendations. His own position, which he stated in reply to the BMA Report, was a partial synthesis of the opposing views. See *BMJ*, 7 Dec. 1912, pp. 1589–94, and Jones, 'An Orthopaedic View of the Treatment of Fractures', *JOS*, 11 (Oct. 1913), pp. 314–35. Ironically, the latter's title was the first in which 'orthopaedics' was linked to 'fractures'.

12. Indeed, the staging of the debate over simple fractures was largely the result of efforts on the part of Harold Stiles and Victor Horsley. The debate gained Arbuthnot Lane the sympathy of the 'younger men' in British surgery. See W. E. Tanner, *Sir W. Arbuthnot Lane: his life and work* (1946), pp. 102–3, and BMA Annual Meeting, 'Section of Surgery', *BMJ*, 30 July 1910, p. 292. The first volume (1913) of Moynihan and Hey Groves's *British Journal of Surgery* included a report of a surgical visit to Guy's Hospital to witness Lane's treatment of fractures in his 'marvellously organised' clinic (pp. 314–16).

13. R. Harris, 'The Future of Orthopaedic Surgery', the president's address, 1948, *JBJS*, 30A (1948), p. 806.

14. See G. H. Darwin, *The Different Methods of Lifting and Carrying the Sick and Injured* (Manchester, 1888), p. 7. According to John Furley, discussing the 1880s, 'Frequently members of the medical profession . . .

would say, ". . . we know nothing about the transport of injured people, as we always have our patients brought to us at the hospital or elsewhere".' Furley, *In Peace and War: autobiographical sketches* (1905), p. 278.

15. Thomas Bond, 'Railway Injuries' in Christopher Heath's *Dictionary of Practical Surgery* (3rd edn, 1889), vol. 2, pp. 294–306. See also Wolfgang Schivelbusch, *The Railway Journey: the industrialization of time and space in the 19th century* (Leamington Spa, 1986), p. 134n., who notes the appearance of a literature by medical experts on railway accidents after 1864. On street accidents, see below.

16. *BMJ*, 2 May 1868, p. 443.

17. *BMJ*, 13 June 1885, p. 1210.

18. Roberton, 'Hospital Accommodation'.

19. See Joan Mottram, 'The Life and Work of John Roberton (1797–1876) of Manchester, Obstetrician and Social Reformer', MSc thesis, Manchester University, 1986.

20. The Royal Infirmary's Central Branch in Roby Street, which was completed in 1914 and closed in 1943, consisted of an outpatient department, an accident room with four observation beds, and an operating theatre and X-ray room. The branch had four wards of 10 beds in each and was largely for surgical cases. See William Waugh, *John Charnley, the man and the hip* (1990), p. 14.

21. Hilary Marland, *Medicine and Society in Wakefield and Huddersfield, 1780–1870* (Cambridge, 1987), p. 130.

22. A. Thom Thomson, 'Proposed New Infirmary at Oldham, to editor', *Lancet*, 7 Nov. 1868, p. 609.

23. See *Workmen's Compensation Committee: minutes of evidence*, 1920, Cmd.816, p. 303.

24. In addition to the sources provided in Chapter 2, see John Poland, 'Statistics of Amputations at Guy's Hospital for 42 Years', *Guy's Hosp. Reps.*, 3rd ser., 45 (1888), pp. 239–65; C. Schimmelbusch, *The Aseptic Treatment of Wounds* (1894); and Henry Burdett, *The Relative Mortality after Amputations of Large and Small Hospitals, and the influence of the antiseptic (Listerian) system upon such mortality* (1882).

25. Frederic Skey, 'On Fractures' in his *Operative Surgery* (1850), pp. 137–73 at p. 137.

26. For an example of the traditional routine for fracture treatment at Guy's Hospital, see T. B. Layton, *Sir William Arbuthnot Lane: an enquiry into the mind and influence of a surgeon* (Edinburgh, 1956), pp. 21, 73; and, for St Bartholomew's Hospital, see the letter of James Taylor, quoted in John West, *The Taylors of Lancashire: bonesetters and doctors, 1750–1890* (Worsley, 1977), pp. 66–7.

27. Nelson Hardy, 'Casualty Out-Patients', *BMJ*, 13 June 1874, pp. 777–8. For T. B. Macaulay's remark on fracture treatment, which was *en passant* and unsubstantiated, see his *The History of England*, vol. 1 (1849), p. 424.

28. See, for example, Andrew Fullerton (Belfast), in the 'Discussion on the Operative Treatment of Simple Fractures', *BMJ*, 8 Oct. 1910, pp. 1054–5.

29. At the Glasgow Royal Infirmary in 1860, for example, fracture cases constituted 15 per cent of admissions – the highest percentage (see David Hamilton, 'Surgery and World War I', paper delivered to the Wellcome Symposium on Modern Medicine and War, 26 Feb. 1988). Similarly at the Kent and Canterbury Hospital, patient records from 1893 reveal that fracture cases outnumbered all others. F. Marcus Hall, *et al.*, *The Kent and Canterbury Hospital, 1790–1987* (Canterbury, 1987), p. 66.

30. For estimates of the duration for the repair of different types of fractures under ideal conditions, see Robert Jones, 'A Few Common Errors in the Treatment of Fractures', *Prov. Med. J.*, 7 (1888), pp. 244–6. For the relative incidence of different kinds of fractures, see Lewis Stimson, *A Practical Treatise on Fractures* (5th edn, 1908), pp. 19–20.

31. See *BMJ*, 8 Oct. 1910, p. 1059.

32. *Report of the Accident Relief Society for 1851* contained with the MS on the Society in the archives of the London Hospital. The Society's motto was 'The Hospital receives the Injured man, but does not provide for his Family'.

33. See Seebohm Rowntree, *Poverty, a study of town life* (1900), p. 168.

34. W. C. Mallalieu, 'Joseph Chamberlain and Workmen's Compensation', *J. Econ. Hist.*, 10 (1950), pp. 45–57; and Gilbert Stone, *A History of Labour* (1921), pp. 367–84.

35. Peter Bartrip and S. B. Burman, *The Wounded Soldiers of Industry: industrial compensation policy, 1833–1897* (Oxford, 1983), pp. 65 and 212ff.

36. See David Green, *Working-Class Patients and the Medical Establishment: self-help in Britain from the mid-nineteenth century to 1948* (1985).

37. Ronnie Cole-Mackintosh, *A Century of Service to Mankind: a history of the St. John Ambulance Brigade* (1986), p. ix. See also Nigel Corbet-Fletcher, *The St. John Ambulance Association, its history and its part in the ambulance movement* (1929), and J. Clifford, *For the Service of Mankind: Furley, Lechmere and Duncan, St. John Ambulance Founders* (1971).

38. See evidence from MOsH to the parliamentary *Report as to the Practice of Medicine and Surgery by Unqualified Persons in the United Kingdom* (1910), Cmd 5422.

39. See Henry Burdett, *Hospital Sunday and Hospital Saturday. Their origin, progress, and development* (1884). See also Thomas B. Stead (Leeds), 'Working-Class Support to Hospitals, &c.', *Charity Organization Rev.*, 5 (March 1889), pp. 97–101; (Cave), *Interim Report of the Voluntary Hospitals Committee*, 1921, Cmd 1206; and Brian Abel-Smith, *The Hospitals, 1800–1948* (1964), pp. 135ff.

40. See Viscount Knutsford (Sidney Holland), *In Black and White* (1926), pp. 132–8.

41. 'Account of Accidents Brought to the London Hospital, 1842–1908', *Annual Report for the London Hospital*, 1909.

42. J. V. Pickstone, *Medicine in Industrial Society: a history of hospital development in Manchester and its region, 1752–1946* (Manchester, 1985), p. 113.

43. Cited in Marland, *Medicine in Wakefield and Huddersfield*, pp. 436–7.

44. Thomas Bickerton, *Medical History of Liverpool* (1936), p. 116.
45. *Burdett's Hospital Annual for 1890*, p. 64.
46. John Poland, *Records of the Miller Hospital and Royal Kent Dispensary* (Greenwich, 1893), p. 258.
47. Jo Manton, *Sister Dora: the life of Dorothy Pattison* (1971).
48. Pickstone, *Medicine and Industrial Society*, pp. 142 and 150.
49. Manton, *Sister Dora*, pp. 251, 341, 221; and Tait, 'Address in Surgery', *BMJ*, 2 Aug. 1890, p. 271.
50. The relations between hospital use and changing urban social structures in the nineteenth century has been less studied for Britain than for America. Helpful are Pickstone, *Medicine and Industrial Society*, and Marland, *Medicine in Wakefield and Huddersfield*. For America, see Charles Rosenberg, *The Care of Strangers: the rise of America's hospital system* (New York, 1987). For neither country do we have detailed quantitative studies of, for example, the marital status of hospital patients. For the eighteenth century, however, see Mary Fissell, *Patients, Power, and the Poor in Eighteenth-Century Bristol* (New York, 1991).
51. Irvine Loudon, 'Historical Importance of Outpatients', *BMJ*, 15 Apr. 1978, pp. 974–77.
52. Although the problem was more obvious in the metropolis, it also existed in provincial voluntary hospitals. While at the London Hospital between 1860 and 1900 new outpatient attendances rose nine-fold, from 25 000 to over 220 000, at the Gloucester Infirmary over this same period outpatient attendances rose from just over 1000 to over 9000. Loudon, 'Outpatients', and Abel-Smith, *Hospitals*, pp. 153ff.
53. See, for example, Robert Bridges, 'An Account of the Casualty Department', *St Barts Hosp. Repts.*, 14 (1878), pp. 167–82; and Bernard Cashman, *Private Charity and the Public Purse: the development of Bedford General Hospital, 1794–1988* (Bedford, 1988), p. 145.
54. Quoted in Nuffield Provincial Hospitals' Trust, *Casualty Services and Their Setting: a study in medical care* (Oxford, 1960), p. 26.
55. Loudon, 'Outpatients', p. 975.
56. *Lancet* commenting on Barts-in 1865, quoted in Nuffield Trust, *Casualty Services*, p. 25. Cf. J. G. Wood, 'Inner Life of a Hospital', *Cornhill Mag.*, 5 (1862), pp. 462–77 at p. 463, according to whom, 'casualties' were those cases with the slightest problems requiring the least assistance, whereas the graver cases were classified as 'outpatients'.
57. The first figure is from the 'Summary of Statistics of Out-Patients at 65 Hospitals for the Year 1910', contained in the King Edward's Hospital Fund, *Report of Committee on Out-patients, July 1912*, vol. 2, appendix section A, pp. 480–1. The latter figure is calculated from the 'Account of Accidents Brought to the London Hospital, 1842–1908' contained in the Hospital's *Annual Report for 1909*. (For 1906 to 1908 the figures had been 17 276, 18 702 and 18 501.)
58. COS, *Report of the Sub-Committee Appointed to Consider the Means by which the abuses of the Out-Patient Department of General Hospitals May best be Remedied* (1870), p. 3.
59. *Lancet*, 21 May 1887, p. 1058.

60. The COS's 1870 *Report on Out-Patient Departments* led directly to the most famous commission on 'out-patient abuse', that of 1871 headed by Sir William Fergusson. Its major recommendations were: (i) improving the administration of Poor Law medical relief; (ii) placing free dispensaries under the control of the Poor Law authorities; (iii) establishing an adequate system of provident dispensaries; (iv) curtailing unrestricted gratuitous relief; and (v) the payment of the medical staff engaged in both inpatient and outpatient work, and the payment of fees by patients in the pay-wards and in the consultation departments of the voluntary hospitals. See H. Burdett, 'Hospital', *Ency. Brit.* (11th edn 1911), vol. 13/14, p. 796.

61. 'The Out-Patient Department', *Guy's Hosp. Gaz.*, 29 Mar. 1890, pp. 79–80. See also *Lancet*, 5 Feb. 1887, p. 288; *Out-Patient Reform: letter to the 'Times' and speech by Sir William Gull, Bart., at the annual meeting of the Charity Organization Society* (1878); and Ernest H. Jacob, 'The Abuse of Hospital Out-Patient and Casualty Departments', *BMJ*, 8 Apr. 1893, pp. 770–1.

62. See M. Jeanne Peterson, *The Medical Profession in Mid-Victorian London* (Berkeley, 1978), pp. 175–6, and Adrian Forty, 'The Modern Hospital in England and France: the social and medical uses of architecture', in Anthony D. King (ed.), *Buildings and Society* (1980), pp. 61–93 at pp. 76–7. On the staffing of outpatient departments, see also the editorial in *BMJ*, 25 March 1893, pp. 649–50.

63. Edward H. Sieveking, 'Medical Education: out-patient hospital practice', *BMJ*, 13 June 1868, p. 592. See also W. Spencer Watson, 'Out-Patient Hospital Practice', *BMJ*, 11 July 1868, pp. 44–5; 'The Reform of the Hospital Out-Patient Department', *BMJ*, 22 Feb. 1913, pp. 403–4; and David Rosner, *A Once Charitable Enterprise: hospitals and health care in Brooklyn and New York, 1885–1915* (New York, 1982), pp. 10ff.

64. See Colcott Fox, 'On Medical Education and Specialism', *BMJ*, 4 Oct. 1890, pp. 784–6, esp. p. 785; Tait, 'Address in Surgery', *BMJ*, 2 Aug. 1890, pp. 267–73, esp. pp. 270–1; and *Third Report from the Select Committee of the House of Lords on Metropolitan Hospitals, together with . . . Minutes of Evidence*, 1891, Cmd 457, p. 34 *et passim*.

65. Lawson Tait, 'Address in Surgery', p. 271. He attributed these tendencies to the ovariotomist, Spencer Wells. The keeping of record cards was developed largely at the Massachusetts General Hospital in the 1900s. It became all the more necessary in contexts where there was no place for intimate doctor/patient relations and where routine work in specialist care might be delegated to several assistants. See S. J. Reiser, 'Creating Form Out of Mass: the development of the medical record', in E. Mendelsohn (ed.), *Transformation and Tradition in the Sciences: essays in honor of I. Bernard Cohen* (New York, 1984), pp. 301–16; Rosner, *Hospitals and Health Care*, pp. 55 *et passim*; and Joel Howell, 'Machines and Medicine: technology transforms the American hospital', in Diana Long and Janet Golden (eds), *The American General Hospital: communities and social contexts* (Ithaca, New York, 1989), pp. 109–34, esp. pp. 121ff. See also David Armstrong, 'Space and Time in British

General Practice', in Margaret Lock and D. Gordon (eds), *Biomedicine Examined* (1975), pp. 207–25.

66. See, for example, 'Hospital Reform', editorial, *BMJ*, 16 July 1881, p. 87.
67. Quoted in S. E. Finer, *The Life and Times of Sir Edwin Chadwick* (1952), p. 93.
68. See Ruth Richardson and Brian Hurwitz, 'Joseph Rogers and the Reform of Workhouse Medicine', *BMJ*, 16 Dec. 1989, pp. 1507–10.
69. See Leslie Hannah, *The Rise of the Corporate Economy* (2nd edn., 1983), Alfred D. Chandler, Jr, *The Visible Hand: the managerial revolution in American business*, Cambridge, Mass., 1977; and idem., '*Fin de siècle*: industrial transformation', in Mikulás Teich and Roy Porter (eds), *Fin De Siècle and its Legacy* (Cambridge, 1990), pp. 28–41.
70. See Stephen Kern, *The Culture of Time and Space, 1880–1918* (Cambridge, Mass., 1983); Ian R. Bartky, 'The Adoption of Standard Time', *Technology and Culture*, 30 (1989), pp. 25–56; and Schivelbusch, *Railway Journey*. See also Robert H. Wiebe, *The Search For Order, 1877–1920* (New York, 1967).
71. Eric Hobsbawm, 'The Making of the English Working Class' in his *Worlds of Labour: further studies in the history of labour* (1984), pp. 194–213 at p. 204.
72. See Joel A. Tarr and Gabriel Dupuy (eds), *Technology and the Rise of the Networked City in Europe and America* (Philadelphia, 1988); Joel A. Tarr *et al.*, 'The City and the Telegraph: urban telecommunications in the pre-telephone era', *J. Urban Hist.*, 14 (1987), pp. 38–80; Thomas P. Hughes, *Networks of Power: electrification in western society, 1880–1930* (Baltimore, 1983); Bill Luckin, *Pollution and Control: a social history of the Thames in the nineteenth century* (Bristol/Boston, 1986), and idem, *Questions of Power: electricity and environment in inter-war Britain* (Manchester, 1990).
73. Geoffrey Rivett, *The Development of the London Hospital System, 1823–1982* (1986), p. 133.
74. F. Taylor, 'A piece-work system', *Trans. Amer. Society of Mechanical Engineers*, no. 16 (1895), pp. 856–903. See also Samuel Haber, *Efficiency and Uplift: scientific management in the progressive era* (Chicago, 1964).
75. Lydall Urwick, *The Meaning of Rationalisation* (1929).
76. For America, see Charles Rosenberg, 'Inward Vision and Outward Glance: the shaping of the American Hospital, 1880–1914', *Bull. Hist. Med.*, 53 (1979), pp. 346–91; Morris Vogel, 'Managing Medicine: creating a profession of hospital administration in the United States, 1895–1915', in Lindsay Granshaw and Roy Porter (eds), *The Hospital in History* (1989), pp. 243–60, esp. p. 246; Stephen J. Kunitz, 'Efficiency and Reform in the Financing and Organization of American Medicine in the Progressive Era', *Bull. Hist. Med.*, 55 (1981), pp. 497–515; Susan Reverby, 'Stealing the Golden Eggs: Ernest Armory Codman and the science and management of medicine', *Bull. Hist. Med.*, 55 (1981), pp. 156–71; idem, 'A Legitimate Relationship: nursing, hospitals, and science in the twentieth century', in Long and Golden (eds), *American General Hospital*, pp. 135–56; and idem, *Ordered to Care: the dilemma of American nursing, 1850–1945* (New York, 1987), pp. 150ff. For a

contemporary comment, see, J. M. T. Finney, 'The standardization of the surgeon', *JAMA*, 63 (1914), pp. 1433–7.

77. Knutsford, *Black and White*, pp. 132–9, 142; and A. E. Clark-Kennedy, *London Pride: the story of a voluntary hospital*, 2 vols (1963).

78. See his evidence to the Home Office, Departmental Committee on the London Ambulance Service, 1909, *Minutes of Evidence*, pp. 111–14. Cmd 4564.

79. Christopher Maggs, *A Century of Change: the story of the Royal National Pension Fund for Nurses* (1988), of which Burdett was the founder. See also, Rivett, 'Biographical notes on Sir Henry Burdett, 1847–1920', in his *London Hospital System*, pp. 373–4. A full biography of Burdett is long overdue.

80. Burdett, *The Uniform System of Accounts, Audit, and Tenders for Hospitals and Institutions* (1893; 4th edn, 1916); and his 'Hospital', *Ency. Brit.*, p. 799.

81. Burdett, *Hospital Sunday*, pp. 13–16.

82. Ibid., p. 796; see also his *Hospitals and Asylums of the World*, 4 vols, 1891–3; idem., *Cottage Hospitals . . . their progress, management, and work in Great Britain and Ireland, and the United States of America* (3rd edn, 1896); his journal, *The Hospital*, esp. vol. 29 (1900), p. 124; vol. 38 (1905), pp. 455–6; and vol. 50 (1911), pp. 143–4; and his *The Medical Attendance of Londoners. An economical system of medical relief freed from existing abuses and adequate to protect all classes and interests* (1903).

83. Frank Prochaska, *Philanthropy and the Hospitals of London: the King's Fund, 1897–1990* (Oxford, 1992), p. 70.

84. See the Annual Reports of the King Edward's Hospital Fund for London for 1903–5; Rivett, *London Hospital System*, p. 163; and Prochaska, *King's Fund*, p. 68.

85. Whether street accidents were actually increasing more rapidly than increases in population is unclear. Of 22 000 casualties received at five London hospitals in 1880, 3000 were said to be due to horse and vehicle accidents. Benjamin Howard, 'A Hospital and Accident Ambulance Service for London', *BMJ*, 4 Feb. 1882, pp. 152–3.

86. Benjamin Vincent, *Haydn's Dictionary of Dates* (16th edn, 1881), pp. 5–6.

87. Benjamin Howard, 'The New York Ambulance System', *BMJ*, 16 July 1881, pp. 72–3.

88. *BMJ* 16 July 1881, p. 88.

89. See G. M. Ayers, 'Ambulance Services', in her *England's First State Hospitals, 1867–1930* (1971), p. 188; and H. W. Hart, 'The Conveyance of Patients to and from Hospitals, 1720–1850', *Med. Hist.*, 22 (1978), pp. 397–407.

90. See Ayers, 'Ambulance Services', pp. 188–92.

91. Howard, 'Hospital and Accident Ambulance Service for London', *BMJ*, 4 Feb. 1882, p. 155.

92. *BMJ*, 3 Dec. 1881, p. 910 and 24 Dec. 1881, p. 1027.

93. G. J. H. Evatt, *Ambulance Organisation, Equipment, and Transport* (1884), pp. 34–5. The service, which was funded through private

subscription, operated from a few police stations, charged a minimum of 5s per call, and was largely unknown.

94. See James Whitson, 'The Ambulance Movement in Scotland', *Edin. Med. J.*, 30 (1884/5), pp. 193–99, and the memoranda on the Liverpool and Manchester ambulance services in *Report of the Ambulance Committee*, 1909, vol. I, pp. 99–107.

95. See Dr P. Murray Braidwood, *Hospital Ships* (Hospitals' Association pamphlet, *c.* 1885); and John Burne, 'River Ambulance and Tramway' in his *Dartford's Capital River: paddle steamers, personalities and smallpox boats* (Buckingham, 1989), pp. 23–7.

96. See *BMJ*, 7 Feb. 1880, p. 216 and 14 Feb. 1880, p. 263. Ambulance transport facilities had been one of the displays at the 1867 Exhibition in Paris.

97. Mayo's article appeared in the *BMJ*, 3 May 1884, p. 854. A later example is Richard Davy, 'Ambulance Work in Civil Practice', *BMJ*, 15 Feb. 1888, pp. 345–47.

98. See John Ellis, *The Social History of the Machine Gun* (1975); Anne Summers, *Angels and Citizens: British women as military nurses, 1854–1914* (1988); and Schivelbusch, *The Railway Journey*.

99. See Cole-Mackintosh, *St. John Ambulance Brigade*, pp. 17ff.; Furley, *Autobiographical Sketches*; Furley's evidence before the *Departmental Committee on the London Ambulance Service*, vol. II: *Minutes of Evidence*, pp. 53–9; and Clifford, *Furley, Lechmere and Duncan*. Among Furley's contemporary contributions to the ambulance movement, see his *On the Use of Ambulance Litters and Horse Carriages for the Removal of Sick and Injured Persons, especially in reference to the metropolis* (1882).

100. See Sir Neil Cantlie, *A History of the Army Medical Department*, vol. 2 (1974), pp. 406–10. Knutsford was the director of the ambulance department of the Order of St John. See Furley, *Autobiography*, p. 284. As such he was probably responsible for the training in first aid that took place on the docks around the Poplar Hospital (see *BMJ*, 11 June 1881, p. 929). On Sieveking and the Association, see *BMJ*, 19 Nov. 1881.

101. This was among the several works commissioned for the International Health Exhibition held in London in 1884. Another was *Accidental Injuries: their relief and immediate treatment* by James Cantlie, a surgeon of Charing Cross Hospital who was also the originator of the volunteer medical staff corps which became the Royal Army Medical Corps Territorial Force in 1907 (see Cantlie, *Army Medical Department*, vol. 2, pp. 360–4). See *Catalogue of the International Health Exhibition Library* (1884), p. 63, and *Burdett's Hospitals Annual for 1890*, p. xxx, which advertised Evatt's *The Ambulance and Hospital Arrangements of an English Army in the Field*, Capt. William Joynson's *Horse Ambulances in Connection with Hospitals*; and Mr Thomas Ryan's *The Conveyance of Injured Persons to the Metropolitan Hospitals. Defects in the existing system and proposals for their removal*. Joynson was the Chairman of the Northern Hospital, Liverpool, and Ryan was the Secretary of St Mary's Hospital, London, and one-time president of the Hospital Officers'

Association (for information on whom I am indebted to Kevin Brown, Archivist at St Mary's Hospital).

102. See, for example, the idea for London's hospitals in Henry S. Souttar, *A Surgeon in Belgium* (1915), pp. 213–14. See also below on the 'Dawson Plan', Chs 8 and 9.
103. Edited by W. A. Morris with assistance from Furley and published by Her Majesty's Stationery Office.
104. 'Chief Events and Progress in 1893', in the preface to his *Hospitals and Charities Annual for 1894*, pp. lx–lxi.
105. Paul O'Brien, 'Orthopaedic Service', in J. J. Byrne (ed.), *A History of the Boston City Hospital, 1905–1964* (Boston [1964]), pp. 227–31. See also K. E. Digby, 'Note on Ambulances at Boston, Montreal and New York' in *Report of the Ambulance Committee*, 1909, vol. I, Appendix viii, pp. 107–9.
106. 'Chief Events and Progress in 1893', p. lxi.
107. Ayers, *State Hospitals*, pp. 189–91.
108. 'Hospital', *Ency. Brit.*, pp. 793, 796–7.
109. *Manchester City News*, 16 June 1894, cited in D. A. Farnie, *The Manchester Ship Canal and the Rise of the Port of Manchester, 1894–1975* (Manchester, 1980), p. 7.
110. F. Watson, *The Life of Sir Robert Jones* (1934), ch. 4.
111. In this connection it is worth noting that the establishment of the Poplar Hospital was justified on the grounds that the two-mile distance from the London Hospital was too far to transport an accident victim; the Albert Dock Hospital, a mile and a half downstream from the Poplar, was similarly justified. On the former, see Knutsford, *Black and White*, p. 132.
112. Terry Coleman, *Railway Navvies: a history of the men who made the railways* (Harmondsworth, 1968), pp. 73–4.
113. See Raymond Munts, *Bargaining for Health: labour unions, health insurance, and medical care* (Madison, Wisconsin, 1967); and Paul Starr, *The Social Transformation of American Medicine* (New York, 1982), pp. 201ff. Similar circumstances at the railway works at Crewe led to the erection there in 1899 of a small company hospital (see below Ch. 10). Other prewar examples of company hospitals in Britain are rare.
114. The application is contained in Jones's papers in the Liverpool Medical Institute. Immediately before securing the Ship Canal appointment, Jones had applied for the post of surgeon to the police; in his application he stressed his all-round surgical ability and the fact that he had previously filled the post of Port Sanitary Officer for the Government of the United States.
115. According to one account, given in Watson, *Life of Jones*, p. 62, Jones's name was put forward by the woman who headed The Navvy Mission, who had seen Jones attend an accident case while she was staying at a hotel in Norway. Other sources indicate that it was while he was at the Stanley Hospital that there came under his care an employee of one of the main contractors for the Ship Canal, and that it was through this connection that he was asked to take charge of the provisional hospital

erected near the Liverpool end. Full control of the project was handed to Jones after he had proven his competence in this limited sphere.

116. Such 'contracting out' – obtaining from employees an undertaking not to avail themselves of the terms of the Act for which in return employers often subsidized workers' contributions to Friendly Society insurance schemes – was particularly common in the north-west of England, especially among miners. See Bartrip and Burnam, *Soldiers of Industry*, pp. 170ff.

117. Cited in B. T. Leech, *History of the Manchester Ship Canal* (Manchester, 1907), vol. 2, pp. 142–3. Considerably lower figures were stated by the secretary of the Ship Canal Company in 1892. Cited in David Owen, *The Manchester Ship Canal* (Manchester, 1983), pp. 91–2.

118. Leech, *Ship Canal*, vol. 2, pp. 69–70.

119. Bartrip and Burnam, *Soldiers of Industry*, p. 69.

120. E. J. Hobsbawm, 'Custom, Wages and Work-load in Nineteenth-century Industry', in his *Labouring Men: studies in the history of labour* (1968), pp. 344–70, at p. 345. Anson Rabinbach has observed that 'By 1870, especially in England, the wasteful expenditure of energy, particularly in the form of labor power, became a source of grave national concern. The increasing costs of labor power resulting from the impact of trade union agitation and a general rise in the working-class standard of living contributed to this concern. Even more important was the recognition that the costs of reproducing labor power could be turned into profit, and the working class constituted a potentially lucrative domestic market, particularly during the years of prosperity after 1871.' 'The Body without Fatigue: a nineteenth-century utopia', in Seymour Drescher *et al.* (eds), *Political Symbolism in Modern Europe* (New Brunswick, NJ, 1982), pp. 42–62 at p. 50.

121. Farnie, *Manchester Ship Canal*, pp. 2–3.

6 The Great War

1. See especially Modris Eksteins, *Rites of Spring: the Great War and the birth of the modern age* (New York, 1989); Paul Fussell, *The Great War and Modern Memory* (Oxford, 1975); and G. D. Josipovici, 'The Birth of the Moderns: 1885–1914' in Clive Emsley *et al.* (eds), *War, Peace and Social Change in Twentieth-Century Europe* (Milton Keynes, 1989), pp. 72–88.

2. For a review and critique of this literature, see Roger Cooter, 'Medicine and the Goodness of War', *Canadian Bull. Hist. Med.*, 7 (1990), pp. 147–59, and idem, 'War and Modern Medicine' in W. F. Bynum and Roy Porter (eds), *Encyclopedia of the History of Medicine* (London/New York, forthcoming).

3. The best overviews of specialization during the war appear in the works of Rosemary Stevens. See, in particular, *Medical Practice in Modern England* (New Haven, 1966), ch. 3, and *American Medicine and the Public Interest* (New Haven, 1971), pp. 127ff.

4. See, for example, Stevens, *Medical Practice in Modern England*, p. 38.

5. Sir Harry Platt, 'British Orthopaedic Association: First Founders' Lecture', *JBJS*, 41B (1959), reprinted in Platt, *Selected Papers* (Edinburgh, 1963), p. 118. See also L. O. Betts, 'Orthopaedics and the Great War', *Med. J. Australia*, 13 July 1940, pp. 35–8; and R. Jones, 'The Cripple (a retrospect and a forecast)', *Cripples' J.*, no.5 (1928), pp. 5–18.

6. Joel Goldthwait, *The Division of Orthopaedic Surgery in the A.E.F.* (Norwood, Mass., 1941), p. 13; Sir W. G. Macpherson, *History of the Great War, Medical Services* [hereafter *Off. Med. Hist.*], *General History*, vol. 1 (1923), p. 390.

7. See G. Murray Levick, D. Macrae Aitken, and J. P. Mennell, 'Organisation for Orthopaedic Treatment of War Injuries', in *Off. Med. Hist: Surgery of the War*, vol. 2 (1922), pp. 381–408; Robert Osgood, 'The Orthopedic Centers of Great Britain and their American Medical Officers', *AJOS*, 16 (1918), pp. 132–40; 'Orthopaedic Centres in Scotland for the Treatment of Disabled Soldiers and Sailors', *Lancet*, 27 July 1918, pp. 118–19; and Robert Jones, 'Military and Orthopaedic Surgery: its scope and aims', *Inter-Allied Conference on the After-care of Disabled Men – 2nd annual meeting, reports presented to the conference* (HMSO, 1918), pp. 519–28.

8. Sir Harry Platt, interview 7 Nov. 1984. The 'halo', he suggested, was comparable to that around heart transplant in the 1970s.

9. See T. H. Openshaw's presidential address to the orthopaedic sub-section of the RSM, 3 Oct. 1922. *Proc. Roy. Soc. Med.*, 16, pt 3 (1922–3), pp. 1–10.

10. Army Council Instructions (AC I 72, 1916) devised by Jones. Cited in the 'Report of the Orthopaedic Committee of the Royal College of Surgeons', 4 July 1918, in MS *Committee for Temporary Purposes*, vol. 6, 1907–22, pp. 312–15, RSC. See also Jones, 'Military and Orthopaedic Surgery', p. 519.

11. Field Marshall French, as quoted in Frederick Watson, *The Life of Sir Robert Jones* (1934), p. 171. French soon became an enthusiast of orthopaedics; in 1917 he was among those who inspected the RNOH. See John A. Cholmely, *A Brief History of the Royal National Orthopaedic Hospital from its Origin in 1905 to 1982* (1982), p. 5.

12. Recounted in T. B. Layton, *Sir William Arbuthnot Lane* (1956). On the use of Balkan frames in the war see Fred M. Albee, *A Surgeon's Fight to Rebuild Men: an autobiography* (1950), pp. 91ff. On the origins and popularization of the frame, see Daniel de Moulin, *A History of Surgery, with emphasis on the Netherlands* (Dordrecht, 1988), p. 322.

13. Intelligence Summary, DMS Lines of Communication, 11 Jan. 1916: PRO:WO/95/3979. Jones, 'Memo. on the Treatment of Injuries of War', published by AMS in July 1915. On 8 Oct. 1915 an exhibition of fracture apparatus was held at the RSM, the dominant feature of which was the Thomas knee splint. See 'Exhibition of fracture apparatus', *BMJ*, 16 Oct. 1915, pp. 573–4; *Proc. Roy. Soc. Med.*, 9 (1915–16), p. 1; and *Off. Med. Hist., Gen. Hist.*, vol. 1, pp. 176, 411.

14. See Robert Osgood, 'Orthopedic Service in the British General Hospitals', *Military Surg.*, 45 (1919), pp. 262–7 at p. 265; Sir Anthony

Bowlby, 'Growth of the Surgery at the Front in France', *BMJ*, 2 Aug. 1919, pp. 127–31 at p. 129; Watson, *Life of Jones*, pp. 152–9; and S. Maynard Smith, 'Fractures of the Lower Extremity', in *Off. Med. Hist.: Surgery of the War*, vol. 2, pp. 339–80 at pp. 343–5.

15. See below Chs 8 and 10.
16. Goldthwait, *Orthopaedics in AEF*, p. 9.
17. Ralph Fitch, 'Presidential Address', *JBJS*, 5 (1923), pp. 397–9 at p. 398.
18. Biesalski's orthopaedic centres were organized throughout Germany before the Shepherd's Bush centre was opened. His *The Care of War Cripples* (1915) is said to have rapidly sold some 150 000 copies. See Leo Mayer in *JBJS*, 37A (1955), p. 377. (Mayer worked with Biesalski from 1913 until America's entry into the war in 1917.)
19. See Platt, 'Orthopaedics in Continental Europe, 1900–1950', *JBJS*, 32B (1950), pp. 570–86, reprinted in *Selected Papers*, pp. 65–93; Leo Mayer, 'Reflections', *JBJS*, 37A (1955), p. 378; W. Rogers, 'Dr Leo Mayer: an appreciation', *Bull. Hosp. for Joint Dis.*, 21 (1960), pp. 81–5; and Albee, *Autobiography*, p. 80.
20. Levick *et al.*, 'Organisation for Orthopaedics', pp. 382–3.
21. H. A. T. Fairbank and T. H. Openshaw each did a year's service in South Africa in 1900. Ernest Muirhead Little, of the RNOH, had served with the National Aid Society (the precursor to the British Red Cross) in the Turko-Serbian War of 1876, but this was four years before he took his medical qualification. Among the orthopaedists with non-military experience of trauma were George Newbolt (who had served under Jones on the Manchester Ship Canal project), John Poland (who headed the orthopaedic department at the Miller General Hospital after 1908) and, again, Openshaw (who had formerly worked at the Poplar Accident Hospital).
22. Platt, 'BOA – founders' lecture', p. 120.
23. See W. Watson Cheyne, 'Hunterian Oration on the Treatment of Wounds in War', *Lancet*, 27 Feb. 1915, pp. 419–430, and D'Arcy Power, *Wounds in War* (Oxford, 1917), p. 7. According to Albee, *Autobiography*, pp. 80–1, the Germans were much better prepared for the type of injuries to be experienced.
24. As, for example, in the battle of the Somme. See Bowlby, 'Surgery at the Front', p. 128 (where the figures for injuries in other battles are also given). See also John Laffin, *Surgeons in the Field* (1970), pp. 211ff.
25. William Heneage Ogilvie, 'The Surgery of War Wounds', reprinted in his *Surgery: orthodox and heterodox* (1948), pp. 204–15 at p. 207. (Ogilvie was a BOA member who began his career working at the Military Orthopaedic Hospital in Birmingham. He subsequently became a surgeon at Guy's Hosptial.) Sir Alfred Keogh made the same point in his introduction to the English translation of August Broca, *The After-Effects of Wounds of the Bones and Joints* (1918), p. ix.
26. C. S. H. Frankau. 'Gunshot Wounds of Joints', in *Off. Med. Hist., Surgery*, vol. 2, pp. 297–325 at p. 298, and Albee, *Autobiography*, p. 100.
27. Cited in E. Muirhead Little, *Artificial Limbs and Amputation Stumps* (1922), p. 22. See also R. Elmslie, 'Amputations and Artificial Limbs' in *Off. Med. Hist., Surgery*, vol. 2, pp. 460–98.

28. S. Dumas and K. O. Vedel-Petersen, *Losses of Life Caused by War*, ed. Harald Westergaard (Oxford, 1923), pp. 94–5; Cooter, 'War and Modern Medicine'; and Thomas Longmore, *Amputations: an historical sketch* (Glasgow, 1876).
29. See Little, *Artificial Limbs*. For a first-hand account by a patient, see Albee, *Autobiography*, pp. 136–7. The principal limb-fitting hospitals were Queen Mary's, at Roehampton (initially organized by Openshaw, then headed by Little), and the Erskine Hospital, near Glasgow, of which William Macewen was in charge. See John Calder, *The Vanishing Willows, the story of Erskine Hospital* (Renfrewshire. 1982). On concern among consultant advisers to the War Office over the so-called 'cucumber stump', see the Intelligence Summary of 18 Feb. 1916 in PRO:WO/95/45.
30. Frankau, 'Gunshot Wounds of Joints', p. 298; Bowlby, quoted in S. Maynard Smith, 'Fractures of the Lower Extremity', pp. 354–6. In all, some 41 300 war amputees were registered by the Director-General of Medical Service of the Ministry of Pensions, 27.5 per cent of the upper extremity, 72.5 per cent of the lower extremity: quoted in E. Muirhead Little, *Artificial Limbs*, p. 23. Twenty-five thousand of these amputations were carried out at Queen Mary's Hospital, Roehampton, between 1916 and 1922 (ibid., p. v).
31. Quoted in Watson, *Life of Jones*, p. 158. Maurice G. Pearson, at a demonstration on fractures at the Imperial College of Science, London, 10 Apr. 1919, claimed that 'Estimates of the mortality from fractured femurs up to 1916 varied from 50 percent to 80 percent'; *BMJ*, 19 Aug. 1919, p. 672. For comparative statistics, see 'Surgery of War', in O. H. Wangensteen and S. D. Wangensteen, *The Rise of Surgery from Empiric Craft to Scientific Discipline* (Folkestone, Kent, 1978), ch. 24.
32. See A. Bowlby, 'The Mortality of Cases of Fractured Femur', *BMJ*, 25 Jan. 1919, p. 112, in which he reported that before 1916 his statistics on 1008 cases of fractured femurs indicated a mortality rate of only 16 per cent.
33. Alexis Carrel and G. Dehelly, *The Treatment of Infected Wounds*, trans. Herbert Child, with an introduction by Anthony A. Bowlby (1917). On Henry Dakin and Carrel, see Albee, *Autobiography*, pp. 56, 100ff. The Carrel–Dakin method had become general in medical units of the BEF in France by the summer of 1917. See Intelligence Summary for 28 Aug. 1917 in PRO:WO/95/46.
34. See Intelligence Summary 4 Jan. 1916 in PRO:WO/95/45.
35. On the wartime organization of hospitals, see Brian Abel-Smith, *The Hospitals, 1800–1948* (1964), chs 16 and 17; and R. McNair Wilson, 'Medical War Organization' in *Ency. Brit.*, supplement to the 11th edn covering the period 1910–21 (1922), pp. 902–4.
36. See William Thorburn, 'The 2nd Western General Hospital', *Brit. J. Surg.*, 2 (1914–15), pp. 491–505 at p. 492.
37. Redmond McLaughlin, *The Royal Army Medical Corps* (1972), pp. 38–9. Beyond this organization, but linked to it in various financial and other ways, were the auxiliary voluntary hospitals in France and in Britain, coordinated by the Joint Committee of the Red Cross and Order of St John.

38. Sydney Holland, *In Black and White* (1926), p. 272.
39. See Sir Cuthbert S. Wallace and Sir John Fraser, *Surgery at a Casualty Clearing Station* (1918); Bowlby, 'Surgery at the Front', p. 128.
40. Holland, *Black and White*, pp. 268–9.
41. Jones, 'Orthopaedic Surgery in its Relation to War', *Recalled to Life*, no. 1 (1917), pp. 50–9, at p. 50.
42. Goldthwait, 'The Backgrounds and Foregrounds of Orthopaedics (the Robert Jones Lecture, October 27, 1932)', *JBJS*, 15 (1933), pp. 279–301 at p. 282.
43. Watson, *Life of Jones*, p. 146.
44. Ibid., pp. 147ff. Obstructing Jones's assault on the problem of disabled soldiers was the Army's practice of quickly discharging its disabled so as not to be medically and financially responsible for them. The conflict between military and civilian medical interests that this raised was not sorted out in Britain until 1919, when a six-month limit was fixed by the Ministry of Pensions. In practice the wartime policy eventually became: 'A man who is no longer fit for military service is discharged only when he no longer requires in-patient treatment.' PRO:WO/32/4746.
45. Quoted in Guy Hartcup, *The War of Invention: scientific developments, 1914–18* (1983), p. 166. On Keogh (1857–1936) see also *DNB*; *Med.Press and Circular*, 3 July 1957, p. 21; and R. K. May, 'The Forgotten General: Lieutenant General Sir Alfred Keogh', *Br. Army Rev.*, no. 76 (1984), pp. 51–59.
46. Allan McLane Hamilton, 'London in War Time', *Med. Record*, 19 Sept. 1914, p. 517.
47. Watson, *Life of Jones*, p. 148.
48. See 'A Military Orthopaedic Hospital: life at Alder Hey', *Recalled to Life*, no. 3 (1918), pp. 287–97. See also T. P. McMurray on Alder Hey in *Liverpool Medico-Chirurg. J.*, 41 (1933), pp. 38–40; and Albee, *Autobiography*, p. 83.
49. From his speech to the American College of Surgeons, Nov. 1919, quoted in 'Orthopedic Surgery', in *The Medical Department of the United States Army in the World War* (hereafter *US Off. Med. Hist.*), vol. 9: *Surgery* (Washington, DC, 1927), pp. 549–748 at p. 552n.
50. Keogh was with Jones in Liverpool in October 1917 to discuss the extension of orthopaedic provision. See Goldthwait, *Orthopaedics in AEF*, p. 32.
51. See *Off. Med. Hist.*, vol. 1, *Gen. Hist.*, pp. 3–4.
52. See H. Platt, 'Moynihan: the education and training of the surgeon', in his *Selected Papers*, p. 163.
53. Ibid., p. 162.
54. Quoted in Ronald Rompkey, *Grenfell of Labrador: a biography* (Toronto, 1991), p. 188.
55. *House of Commons: Parliamentary Debates*, 5th ser., vol. 80, 1916, cols 2119–38; May, 'Keogh', p. 57. The complaints were raised by the Tory MP, Ronald McNeill, during the debate on Army Estimates, 1916–17.
56. Unfortunately the minutes of the Advisory Board for the war years have not been traced; some idea of the subjects discussed can be gleaned from the Intelligence Summaries (or 'Diaries') of the various Director-

Generals and their deputies. See, for example, Director-General Sir Arthur Sloggett's entry for 13 Jan. 1917 on the meeting called in order to respond to 'the attacks made on the medical administration by Sir Almroth Wright', in PRO:WO/95/46.

57. Abel-Smith, *Hospitals*, pp. 254–5.

58. Among other such friends and patrons were the COS activist and littérateur, Lord Charnwood, and Sir Walter Lawrence, an elder statesman involved with the Ministry of Pensions. In 1917 Charnwood eagerly took up the editorship of *Recalled to Life*, a largely Jones-inspired journal intended to rouse public support for the 'proper' civilian care and physical 're-education' of the disabled. John Galsworthy, whose wife was one of Jones's patients, took over the editorship in August 1918, changing the name to *Reveille*. On Lord Charnwood (i.e., Godfrey Rathbone Benson) (1864–1945), Lawrence (1857–1940), and Stanley (1869–1947), see *DNB*. On Lawrence's support for Jones's orthopaedics, see also Goldthwait, *Orthopaedics in AEF*, pp. 22–3, and Watson, *Life of Jones*, pp. 194, 224, 243, 258.

59. See *JBJS*, 30B (1948), p. 201, and 'The Red Cross and Crippled Children', *Cripples' J.*, 3 (1927), pp. 1–3.

60. One such tour, in 1917, led to the purchase by the East Lancashire branch of the Red Cross of Grangethorpe House, Manchester, where Harry Platt, Geoffrey Jefferson and John Stopford carried out much of their work on peripheral nerve injury. See Patricia Gray, 'Grangethorpe Hospital Rusholme 1917–1929', *Lancs & Cheshire Antiquarian Society*, 78 (1975), pp. 51–64.

61. The following paragraphs draw on 'King Manoel's Campaign', *The Hospital*, 11 Nov. 1916, p. 113, and 2 Dec. 1916, p. 176; 'Work in Orthopaedic Centres', and 'King Manoel's Report' in *Reports by the Joint War Committee and the Joint War Finance Committee of the British Red Cross Society and the Order of St. John of Jerusalem in England on Voluntary Aid rendered to the Sick and Wounded at Home and Abroad and to British Prisoners of War, 1914–1919* (HMSO, 1921), pp. 248–53 and 732–44; and 'Final Report on Work of Military Orthopaedic Hospital, Shepherd's Bush 1916–1919', PRO:WO/32/5334.

62. The mission consisted of Reginald Elmslie, then the head of the orthopaedic outpatients department at St Bartholomew's Hospital; Grainger Stewart, a recently qualified physician with a special interest in nervous diseases, who was also the honorary neurologist to one of the cripples' homes affiliated with Barts; and James Mennell, medical officer to the Physical Exercise Department at St Thomas's Hospital, who had recently published on the treatment of fractures by mobilization and massage (see above Ch. 5, note 11).

63. In total, Manoel disbursed some £38 000 of the Joint Committee's funds for the purchase and equipping of orthopaedic hospitals (£17 000 of which went to Shepherd's Bush) and through his provincial campaigns he raised another £167 000 for the purpose. See 'King Manoel's Report', p. 742.

64. Exactly when Jones was appointed Military Director of Orthopaedics is not specified in any of the sources I have consulted.

65. Osgood, 'Orthopedic Centers in Britain', p. 140.
66. Jones, 'Military Orthopaedic Surgery', p. 528. On the percentages, see ibid., p. 522, Lord Charnwood in *Recalled to Life*, no. 2 (1917), p. 225, and W. Colin Mackenzie, 'Military Orthopaedic Hospitals' in *British Medicine in the War, 1914–1917 (papers collected from the BMJ)* (1917), pp. 78–87 at p. 85.
67. Levick *et al.*, 'Organisation for Orthopaedics', p. 390. 'Category E' were those thought unfit for any kind of military or manual labour at home or abroad, and who were regarded as 'not likely to be fit within six months'. *Off. Med. Hist, Gen. Hist.*, vol. 1, p. 124.
68. 'King Manoel's Report', p. 734.
69. Ibid., pp. 251–2, 734. The wartime demand for standard splints alone was enormous, nearly a million and a half of them being supplied under contract in England. There were also splint-making shops with experienced workmen in Boulogne, Calais, Alexandria and Salonika. See *Off. Med. Hist., Gen.Hist.*, vol. 1, p. 177.
70. 'King Manoel's Report', p. 253.
71. For Germany, see Robert W. Whalen, *Bitter Wounds: German victims of the Great War, 1914–1939* (Ithaca, 1984); for France, see Antoine Prost, *Les Anciens Combattants et la Société Francaise, 1914–1939* (Paris, 1977), 3 vols.
72. Goldthwait, *Orthopaedics in AEF*, p. 7.
73. Ibid., p. 7.
74. Jones, 'The Problem of the Disabled', *Recalled to Life*, no.3 (1918), pp. 266–86 at p. 280. See also Keogh's Introduction to Jones's *Notes on Military Orthopaedics* (1917), p. xiii, and *The Times*, 2 Mar. 1922, p. 8.
75. Watson, *Life of Jones*, pp. 147, 166.
76. Sir Herbert Barker, *Leaves from My Life* (1927), pp. 178–86. The agitation was carried on in 1917 by the journalist and Irish Nationalist MP, Jeremiah MacVeagh. See letter from Lord Knutsford to *Daily Telegraph*, 6 Mar. 1917, in which Jones's work is praisingly contrasted to Barker's. On the attempt by American osteopaths to secure a place in military medicine (and hence acquire official sanction), see Franklin Martin, *The Joy of Living: an autobiography* (New York, 1933), vol. 2, pp. 212–13.
77. See *House of Commons: Parliamentary Debates*, 5th ser., vol. 81, 28 March 1916, cols 554–5. Hodge, a trades union leader, was also notable for opposing the 'Rothbaud Scheme' (or King's Roll) for employing the disabled. See Arnold Bennett, 'A National Responsibility: future employment of the disabled', reprinted from *Daily News and Leader*, 6 June 1917. In the 1920s Hodge was Chairman of the Labour Party's Advisory Committee on Public Health which called for the merger of public and voluntary hospitals. See A. Marwick, 'The Labour Party and the Welfare State in Britain, 1900–1948', *Amer. Hist. Rev.*, 73 (1967–8), pp. 380–403 at p. 389.
78. See Joel Howell, '"Soldier's Heart": the redefinition of heart disease and speciality formation in early twentieth-century Great Britain', in W. F. Bynum *et al.* (eds), *The Emergence of Modern Cardiology, Med. Hist.* Suppl. no. 5 (1985), pp. 34–52, and C. Lawrence, 'Moderns and

Ancients: the "new cardiology" in Britain 1880–1930', in ibid., pp. 1–33.

79. R. McNair Wilson, 'Medical War Organization', p. 904. On asphyxia, see Steve Sturdy, 'From the Trenches to the Hospitals at Home: physiologists, clinicians and oxygen therapy, 1914–1930', in J. V. Pickstone (ed.), *Medical Innovations in Historical Perspective* (1992), pp. 104–23, 234–45. On shellshock, see Martin Stone, 'Shellshock and the psychologists', in W. F. Bynum *et al.* (eds), *The Anatomy of Madness*, vol. 2 (1985), pp. 242–71; Harold Merskey, 'Shell-shock', in German E. Berrios and H. Freeman (eds), *150 Years of British Psychiatry, 1841–1991* (1991), pp. 245–67; and Peter English, *Shock, Physiological Surgery, and George Washington Crile: medical innovation in the Progressive Era* (Westport. Conn., 1980), ch. 9.

80. Osgood, *AJOS*, 16 (1918), p. 140.

81. The lectures, which consisted of biographical sketches of the 'pioneers', were subsequently published (with the lecture title as subtitle) as *Menders of the Maimed* (1919). Keith nowhere defined what he meant by 'physiological principles', the implication being that modern orthopaedics was 'applied physiology' as derived from Hunter, Thomas and others. He became an honorary member of the BOA and was an enthusiast of James Mackenzie's 'new cardiology'. See Keith, *An Autobiography* (1950), pp. 253ff., and Lawrence, ' "New Cardiology" ', p. 16.

82. 'The Future of the Medical Profession', the Cavendish Lectures delivered to the West London Medico-Chirurgical Society, *BMJ*, 13 July 1918, pp. 23–6, 56–60, at p. 23.

83. See Jones on 'team work' in 'Military Orthopaedic Surgery', *Lancet*, 27 July 1918, p. 117; and Stiles, 'Surgical Training', *BMJ*, 18 Oct. 1919, p. 503.

84. Intelligence Summary 21 Jan. 1915 in PRO:WO/95/44. According to R. McNair Wilson ('Medical War Organization', p. 902), team work was apparently first implemented by Keogh for wartime research on war diseases. The idea of surgical 'teams' was borrowed from America (especially the Mayo Clinic) and developed among the new surgical elite before the war. Moynihan claimed that he himself 'coined the phrase "team work" for my own practice in Leeds . . . [that he] carried this into the war and was officially responsible for the team principle . . . The seed fell on fertile soil'. Quoted in Donald Bateman, *Berkeley Moynihan, surgeon* (1940), p. 285. The idea of 'surgical teams', as relatively self-reliant mobile units, appears in an Intelligence Summary of 12 June 1917, where 'It is proposed to form "Surgical Teams" from the personnel of Hospitals on the Lines of Communication, in order to meet sudden emergencies at Casualty Clearing Stations. The team will consist of: 1 Operating Surgeon, 1 Anaesthetist, 1 Sister, 1 Operating room attendant. They will train and work together, will be sent by road when their services are urgently required at Casualty Clearing Stations, and will return to their Units, it is hoped, before the rush of patients arrives from the front. The team will be drawn from Imperial, Colonial, or American Units.' PRO:WO/95/3980.

85. Goldthwait, *Orthopaedics in AEF*, p. 42. Chutro (1880–1937) is among the wartime worthies in medicine mentioned by Fielding Garrison in his *Introduction to the History of Medicine* (4th edn, Philadelphia, 1929), p. 792. For an obituary, see *Bull. Acad. Méd* (Paris), 119 (1938), p. 26.

86. Camus, *Physical and Occupational Re-Education of the Maimed* (New York, 1919).

87. On Amar, see George Humphreys, *Taylorism in France: the impact of scientific management on factory relations and society* (New York, 1986), pp. 217–18, and Anson Rabinbach, *The Human Motor* (Berkeley, 1992), pp. 185–8, 246–9, and 337n. *et passim*.

88. Whalen, *Bitter Wounds*, p. 61.

89. These were the consultants John Lynn-Thomas, Henry Gray, and T. H. Openshaw. See Intelligence Summary 6 July 1916: PRO:WO/95/3980. Prior to Jones's installation at AMS headquarters, such orders had carried little weight. For instance, it was decided by the DMS Lines of Communication that Jones's proposal to segregate cases of gunshot fractures of the femur was simply impracticable. See Intelligence Summary, DMS Lines of Communication, 16 Jan. 1916: PRO:WO/95/3979.

90. See Macpherson, Introduction, *Off. Med. Hist., Gen.Hist.*, vol.1 (1923), pp. 49, 70. Earlier in the war 'specialist' was often used to denote a surgeon who was not a 'generalist' in the old sense, but may not have been of 'consultant' status. For example, in 1915 there was a call at the front for 'specialists in operative surgery'. See Intelligence Summaries, 6 Mar. and 4 Apr. 1915 in PRO:WO/95/44.

91. Garrison, *History of Medicine*, 4th edn, p. 790.

92. Intelligence Summaries on 'Transfer to England of Fracture Femur Cases', 30 Jan. 1917 – 8 Sept. 1917: PRO:WO/95/3980. It was not until the end of 1917, however, that Bowlby advised that all fracture cases should be concentrated in one hospital in each base and that carefully selected surgeons and nurses be chosen for duty. See Intelligence Summaries 8 Feb. 1916 and 20 Dec. 1917 in PRO:WO/95/45 and 46 respectively; S. Maynard Smith, 'Fractures of Lower Extremity', pp. 353–4 and 349; and George Crile, 'Standardization of the Practice of Military Surgery', *JAMA*, 28 July 1917, pp. 291–2. Makins regarded policy decisions on fracture treatment and amputations as very much his special territory. See Intelligence Summaries for 25 June 1915 and 25 Mar. 1917 in PRO:WO/95/44 and 45.

93. This was especially so with regard to the London orthopaedists attached to the RNOH (see below, Ch. 8). Among those that Jones could rely upon: Girdlestone was put in charge of military orthopaedics in Oxford; Naughton Dunn was appointed for Birmingham, Hey Groves for Bristol, and S. Alwyn Smith for Cardiff. D. McCrae Aitken went to Shepherd's Bush to help out E. Laming Evans, Elmslie, B. Whitechurch Howell and Jones himself; Harry Platt and Robert Ollerenshaw were retained in Manchester, and McMurray and T. Armour in Liverpool; Lynn-Thomas was appointed Assistant Orthopaedic Inspector for Wales and Sir Arthur Chance for Ireland, while Harold Stiles doubled as Inspector for Scotland and director of the centre in Edinburgh.

94. Jones, 'Military Orthopaedic Surgery', *Lancet*, 27 July 1918, pp. 115–17 at p. 117.
95. Jones, 'Address to American College of Surgeons', op cit., note 49 above, p. 553n.
96. Information on these men, many of whom are referred to in Osgood, 'Orthopedic Centers in Britain', is taken from the *Medical Directory*. On Paterson, see also Watson, *Life of Jones*, p. 230, and Thomas Kelly, *For Advancement of Learning: the University of Liverpool 1881–1981* (Liverpool, 1981), p. 481 *et passim*.
97. In the 1920s he became medical director of the London Clinic of Physiotherapy. See Gerald Larkin, 'Physiotherapy', in his *Occupational Monopoly and Modern Medicine* (1983), pp. 104–5.
98. Platt, interview 7 Nov. 1984. For Bristow's life (1883–1947) and career, see his obituary in *JBJS*, 30B (1948), pp. 200–2.
99. See Watson, *Life of Jones*, ch. 13: 'The American Contingent', pp. 183–91.
100. Jones, 'Address to American College of Surgeons', p. 553n.
101. Quoted in Watson, *Life of Jones*, p. 183.
102. Ibid., p. 184. Presumably this arrangement would not have been more diplomatically difficult than the arrangements operating since 1915 for the American medical units to work under the British in France. However, it was not always clear to whom the Americans were responsible, since it was actually under the aegis of the American Red Cross in France that their Units were formed.
103. See *US Off. Med. Hist.*, vol. 9, p. 554; Osgood, 'Orthopedic Service in the British General Hospitals', p. 262; and idem, 'A Survey of the Orthopedic Services in the U.S. Army Hospitals. General, Base and Debarkation', *JOS*, 1 (1919), pp. 359–82.
104. *US Off. Med. Hist.*, vol. 9, p. 553. Osgood, 'Orthopedic Centers in Britain', p. 136. See also Goldthwait, *Orthopaedics in AEF*, pp. 4–6.
105. See his obituary in *JBJS*, 25 (1943), pp. 702–3.
106. Robert W. Johnson, Jr, subsequently Professor of Clinical Medicine at Johns Hopkins and the Director of its Division of Orthopedic Surgery from 1947 to 1953, may have been representative: a recent medical graduate commissioned as a Lieutenant in the US Medical Reserve, he spent two and a half months gaining experience in orthopaedics at Shepherd's Bush before being dispatched to Oxford in September 1917 to work for the next 18 months under G. R. Girdlestone at the Wingfield Military Orthopaedic Hospital. Cited in J. Trueta, *Gathorne Robert Girdlestone* (Oxford 1971), p. 22.
107. 'Civilian Practitioners and Military Hospitals', *The Hospital*, 29 Sept. 1917, p. 520. Cited in Abel-Smith, *Hospitals*, p. 270.
108. See Osgood, 'Orthopedic Centers in Britain', pp. 136, 139.
109. Ibid., p. 133.
110. Ibid., p. 137. The latter was Sydney M. Cone (1869–1939), a graduate of Johns Hopkins (1890), Pennsylvania (1893), Heidelberg (1895) and Strasbourg (1896). He became an Associate in Surgical Pathology at Johns Hopkins, 1896–99, studied under Lorenz in Vienna in 1900, and became Professor of Orthopedic Surgery at the Baltimore Medical College. See obituary in *JBJS*, 22 (1940), p. 759.

111. Orr discusses his war work in *On the Contributions of Thomas, Jones and Ridlon to Modern Orthopedic Surgery* (Springfield, Illinois, 1949), p. 153 *et passim*.
112. Osgood, 'Orthopedic Centers in Britain', p. 140.
113. See *US Off. Med. Hist.*, vol. 9, p. 554.
114. Jones, 'Address to American College of Surgeons', p. 553n.; see also Jones 'Military Orthopaedic Surgery', p. 521.
115. Goldthwait, *Orthopaedics in AEF*, p. 10 and pp. 25ff. Harvey Cushing, head of the Harvard unit base hospital in France and a frequent visitor to AMS headquarters, noted in his diary, 29 May 1917, that the first 20 orthopaedists had entered the country 'free from any regular army tangle, though commissioned'. *From a Surgeon's Journal, 1915–1918* (1936), p. 110. See also H. P. H. Galloway, 'Readjustment to Changing Conditions: AOA President's address', *JBJS*, 1 (1919), pp. 395–400 at p. 396.
116. In May 1916 the AOA met in Washington to appoint a 'preparedness committee, whose duty it would be to consider the needs and equipment of orthopedic hospitals should such be required in any future emergency'. E. G. Brackett, 'Division of Military Orthopedic Surgery', in *US Off. Med. Hist.*, vol. 1: *The Surgeon General's Office* (Washington, DC, 1923), pp. 424–36, at p. 424; see also p. 549.
117. The letter from the Director-General of 19 Feb. 1918 transferring these responsibilities to the orthopaedists is printed in Goldthwait, *Orthopaedics in AEF*, pp. 74–5.
118. Ibid., p. 426.
119. Quoted in Martin, *Autobiography*, vol. 2, p. 226, and in Martin, *Council of National Defense*, p. 261.
120. Cf. Glenn Gritzer and Arnold Arluke, *The Making of Rehabilitation* (Berkeley, 1985), pp. 43–4.
121. Martin, *Autobiography*, vol. 2, p. 226.
122. *US Off. Med. Hist.*, vol. 9: *Surgery*, p. 550.
123. Ibid., p. 552; and Gritzer and Arluke, *Rehabilitation*, pp. 44–5. Goldthwait put the figure at 400 (cited in Watson, *Life of Jones*, p. 189). On education in orthopaedics in America, see Charles V. Heck, *50th Anniversary of the AAO* (Chicago, 1983), pp. 80ff.
124. Quoted in Watson, *Life of Jones*, pp. 188–9, and see Jones, 'Military Orthopaedic Surgery', *Lancet*, 27 July 1918, pp. 115–17. On Goodwin, see *Lives of the Fellows of the Royal College of Surgeons* (1960), pp. 156–7; on the cordial relations between him and Jones, see Watson, *Life of Jones*, pp. 227, 219, 221, 226.
125. Galloway, 'Readjustments', p. 399; and Stevens, *American Medicine and the Public Interest*, pp. 127–8.
126. See Alfred Cox, *Among the Doctors* (1950), pp. 115–16; Lindsay Granshaw, *St. Mark's Hospital London* (1985), pp. 185–6; cf. Abel-Smith, *Hospitals*, p. 281.
127. H. Burdett, 'War and Orthopaedics', *The Hospital*, 26 Jan. 1918, p. 353. Burdett was drawing on Jones's lament about the 'number of young men who have failed to resist the lure of the abdomen'. Jones, 'The Orthopaedic Outlook in Military Surgery', *BMJ*, 12 Jan. 1918, pp. 41–45.

128. See Jones, 'Remarks on Orthopaedic Surgery in Relation to Hospital Training', *BMJ*, 20 Nov. 1920, pp. 773–5; idem, 'The Necessity of Orthopaedic Training', *BMJ*, 5 Feb. 1921, pp. 181–6; and 'King Manoel's Report', p. 740.

129. Jones's plans for Shepherd's Bush as an orthopaedic postgraduate teaching centre may have triggered the idea of turning the Infirmary into a new postgraduate medical school in London. The Post Graduate Medical Education Committee was appointed by the Ministry of Health in 1925 and recommended the use of the Hammersmith site in 1930 (GLRO:PH/HOSP/3/26). See also James Calnan, *The Hammersmith: the first fifty years [1935–1985] of the Royal Postgraduate Medical School of Hammersmith Hospital* (Lancaster/Boston, 1985).

130. Grangethorpe reverted to a girls' private school in 1929, while Alder Hey, Edmonton and Tooting were returned to the Poor Law authorities. The centre at Fishponds, near Bristol, and that at Bangour, outside Edinburgh, reverted to mental asylums. The fate of the other centres is not clear. For a partial list of War Hospitals established in asylums, fever hospitals and Poor Law premises, see *Off. Med. Hist.*, vol.1: *Gen. Hist.*, pp. 80–1.

131. Stiles, 'Practice and Teaching of Orthopaedic Surgery', *BMJ*, 19 Aug. 1919, p. 671; on the GMC recommendation, see *BMJ*, 4 Sept. 1926, p. 147.

132. The degree was offered through the Faculty's Board of Orthopaedic Studies (of which Jones was the Chairman) within the Department of Surgery. See *Minutes of the Faculty Board*, 5 Dec. 1921 to 24 Feb. 1949, University of Liverpool Archives. See also Kelly, *University of Liverpool*, p. 154. On orthopaedics in medical education see below, Ch. 12.

133. See *Liverpool Daily Post*, cutting in Bickerton MS 942, Liverpool Public Library: Bio 7f. 157, and, for Newcastle, *BMJ*. 13 Apr. 1918, p. 440.

134. H. Jackson Burrows, 'The Orthopaedic Department', *St Barts Hosp. J.*, 69 (1965), pp. 355–6. The appointment in 1920 of A. S. Blundell Bankart as the first orthopaedic surgeon to the Middlesex Hospital, while notable, is hardly an exception, since he was granted only one outpatient clinic a week and was given no hospital beds. See the interview of Jones and G. R. Girdlestone with Sir George Newman, 5 Mar. 1920: PRO:ED/50/153. Jones's appointment in 1919 as Director of Orthopaedic Surgery at St Thomas's seems to have been little more than token.

135. Cited in H. A. H. Harris, 'Sketch of William Henry Trethowan', MS, BOA Archives, p. 4. Trethowan's position was doubtless weakened by the fact that, prior to his 1913 appointment, he had been the surgeon-in-charge of the Physical Exercise Department at Guy's Hospital.

136. Bristow, 'The Influence of War Surgery on Treatment of Fractures in Great Britain', *JAMA*, 3 Dec. 1927, p. 1920. See also Bristow in the discussion on fractures in *BMJ*, 22 Aug. 1925, p. 331.

137. G. L. Cheatle, *BMJ*, 1920, quoted in Stevens, *Medical Practice in Modern England*, p. 38n. Jones wrote in 1917 that in pursuit of the object of the total restoration of the locomotor system 'the orthopaedic surgeon has already invaded the domain of the surgery of the peripheral nerves; he has not yet in this generation encroached on the central nervous

system, but he will do so unless the neurological surgeon pays more attention to the orthopaedic aspect of some of the disorders with which he deals'. Jones, 'Orthopaedic Surgery in its Relation to the War', p. 51. See also the letter from P. B. Roth, *BMJ*, 11 Dec. 1920, p. 921 complaining that Jones's BOA address did not go far enough in claiming territory for orthopaedics; and the discussion on orthopaedic inroads into gynaecology (via the treatment of backache in women) in *Lancet*, 13 Dec. 1924, p. 1220.

138. Royal College of Surgeons, *Minutes of Council*, 16 July 1918, pp. 159–60.
139. Jones, 'Hospital Training', p. 32.
140. Royal College of Surgeons, *Minutes of Council*, 16 July 1918, pp. 159–60. In an article in the *Lancet* in 1918 Jones stated: 'We have the teaching staffs represented in all these centres, but the more so-called general surgeons we can rope in the more effective will we become.' 'Military Orthopaedic Surgery', 27 July 1918, pp. 115–17 at p. 117.
141. 'Hospital Training', pp. 2–3. Careful language was also required when the orthopaedic surgeons applied to become a full section of the RSM in 1919; as H. Osmond-Clarke pointed out, in their claim ' "to the surgery of congenital and acquired deformities of the extremities and spine" there was nothing . . . to arouse suspicion or increase hostility from general surgeon colleagues'. 'Half A Century', *JBJS*, 32B (1950) p. 659.
142. 'Hospital Training', p. 2.
143. *The Times*, 14 Feb. 1922, p. 7; 2 Mar. 1922, p. 8; 3 Mar. 1922, pp. 8 and 13; 9 Mar. 1922, p. 8; 13 Mar. 1922, p. 8; 15 Mar. 1922, p. 14; 24 Mar. 1922, pp. 6, 7, and (whence the quotation) p. 13. On the subject of Army expenditure on Poor Law premises, see the circular from the Local Government Board (31 Mar. 1915) in PRO:MH/57/176. Shepherd's Bush cost the Ministry of Pensions £8000 rent in 1919: Watson, *Life of Jones*, p. 215.
144. See Paul B. Johnson, *Land Fit for Heroes: the planning of British reconstruction 1916–1919* (Chicago, 1968).
145. Report of an interview with Mr Quine of the Ministry of Health, 3 June 1936: GLRO:PH/HOSP/1/72.
146. GLRO:PH/HOSP/1/72.
147. A. B. Beresford-Jones (1881–1974), for example, after being badly gassed in France in 1916, acted as an honorary surgeon to the Kent and Canterbury Hospital. A general surgeon, he had a special interest in orthopaedics and in 1922 established one of the first hospital fracture clinics in Britain, but he was never a member of the BOA. F. Marcus Hall *et al.*, *Kent and Canterbury Hospital* (Canterbury, 1987), p. 202.
148. '[T]he importance of massage as a remedial measure' was noted in the Intelligence Summary of Sloggett as early as 23 Apr. 1915. See PRO:WO/95/44.
149. On the Depôts, see W. Colin Mackenzie, 'Military Orthopaedic Hospitals', p. 84. Robert Tait McKenzie (1867–1938), a Canadian sculptor and surgeon, was Professor of Physical Education at the University of Pennsylvania. He had been a student of Osler's at McGill and it was Osler who brought his book on *Exercise* to the attention of the military. His war work was conducted at Heaton Park, Manchester,

where there was accommodation for 5000 soldiers. See his 'The Treatment of Convalescent Soldiers by Physical Means', *BMJ*, 12 June 1916, p. 215; the *Proc. Roy. Soc. Med.*, 9, pt 3 (1915–16), pp. 31–70; and in his chapter in R. Fortesque Fox, *Physical Remedies for Disabled Soldiers* (1917). See also the article by Frank Radcliffe (Fox's chief assistant at Heaton Park), 'Hydrotherapy as an Agent in the Treatment of Convalescents', *BMJ*, 21 Oct. 1916, p. 554; McKenzie, *Reclaiming the Maimed* (1918); and Watson, *Life of Jones*, p. 175. For information on McKenzie, I am grateful to Stewart A. Davidson and A. H. T. Robb Smith. Dr Robert Fortesque Fox (1858–1940) was the Chairman of the Committee on the Section of Balneology of the War Measures Committee, and honorary medical director of the Red Cross Clinic for the Physical Treatment of Disabled Officers in London (formerly the private enterprise Alexandra Therapeutic Institute, established some years before the war). See *Reports by the Joint War Committee*, p. 256.

150. Shepherd's Bush and Alder Hey each employed more than 50 masseurs in their physiotherapy departments. See Osgood, 'Orthopedic Centers in Britain', p. 137; Levick *et al.*, 'Organisation for Orthopaedics', pp. 393–8. Almost all were recruited from the Chartered Society of Massage and Medical Gymnastics. See Jane H. Wickstead, *The Growth of a Profession, being the history of the Chartered Society of Physiotherapy 1894–1945* (1948).

151. See David Cantor, 'The MRC's Support for Experimental Radiology During the Inter-War Years', in L. Bryder and J. Austoker (eds), *Historical Perspectives on the Role of the MRC* (Oxford, 1989), pp. 181–204. By the 1930s leading advocates in this field were identifying orthopaedics as one of 'the stronger boys' who had stolen the 'toys' of the radiologists. See A. E. Barclay, 'The Dangers of Specialisation in Medicine', *Br. J. Radiology*, 4 (1931), pp. 60–82 at p. 78.

152. From an appeal for Jones's orthopaedics in *The Times*, 2 Mar. 1917, quoted in Watson, *Life of Jones*, p. 208.

153. *The Times*, 8 June 1918, quoted in A. Keith, *Menders of the Maimed* (1919), p. 313n.; and see Watson, *Life of Jones*, pp. 208–9.

154. Keogh, Introduction to Jones, *Notes on Military Orthopaedics* (1917), p. xiv.

155. Gritzer and Arluke, *Rehabilitation,* pp. 43–5; see also Goldthwait, *Orthopaedics in AEF*, p. 41.

156. A. H. Freiberg, 'Orthopaedic Surgery in the Light of its Evolution' (Robert Jones Lecture, 18 Feb. 1937), *JBJS*, 19 (1937), pp. 279–96 at p. 292.

7 Industry and Labour, Part I

1. Evidence of C. F. A. Hore to the (Holman Gregory) *Departmental Committee on Workmen's Compensation, Minutes of Evidence*, 9 Dec. 1919, p. 504. For the orthopaedists' belief that they would move into this area, see Ernest Hey Groves, 'A Surgical Adventure: an autobiographical sketch', reprinted from the *Bristol Medico-Chirurg. J.*, 50 (1933), p. 22; and E. M. Little's presidential address to the BOA in 1919: *JOS*, 2

(1920), p. 212. See also, Anthony Bowlby, 'On the Application of War Methods to Civil Practice', *Proc. Roy. Soc. Med.*, 13, pts 1–2 (1919–20), pp. 35–48.

2. 'Salvage of Disabled Workmen', *Engineering*, 23 May 1919, p. 675.

3. James R. Kerr, 'The Treatment of Industrial Accidents', *BMJ*, 26 Aug. 1922, pp. 377–8.

4. 'The Working Capacity of the Disabled Man', *Lancet*, 11 Oct.1919, pp. 650–1.

5. Largely as a result, it seems, of William Colin MacKenzie's *The Action of Muscles, including muscle rest and reeducation* (1918; 3rd edn, 1940). MacKenzie, an Australian, visited Jones before the war and worked with him during the war at Shepherd's Bush where he set up a department of muscle retraining.

   On the non-legal history of the word 'rehabilitation', see Joseph Trueta, 'Rehabilitation – Past and Future', *Physiotherapy: the journal of the Chartered Society of Physiotherapy*, 49 (1963), pp. 346–51. Trueta claims that the earliest usage was in a paper by E. M. Law on 'Problems of Rehabilitation of the War Victims', published in the *J. Florida Med. Assn.* in 1922, but Lloyd George had appointed E. C. Geddes to 'the rehabilitation of industry' in December 1918, and an article by the Italian orthopaedist, Vittorio Putti on the 'National Organization of Rehabilitation of the Disabled in Italy' appeared in the *Med. Record* in Sept. 1919, shortly before the Inter-Allied Conference on Rehabilitation was held in Rome. In America there was much discussion in 1920 on the Federal Vocational Rehabilitation Act. See also R. E. Matkin, 'Rehabilitation: an ambiguous term and an unfulfilled ideal', *Rehab. Lit.*, 46 (1985), pp. 314–20; and P. A. Nelson, 'History of the Archives – A Journal of Ideas and Ideals', *Arch. Physical Med. and Rehab.*, 50 (1970), pp. 3–42, at p. 14.

6. Keogh, Preface to Robert Jones, *Military Orthopaedics* (1918), p. xiv. See also A. Gwynne James (a county court judge), 'Orthopaedics in Industrial Life', *Recalled to Life*, no. 2 (Sept. 1917), pp. 299–301, and 'King Manoel's Report' (of July 1919) in *Reports by the Joint War Committee . . . 1914–1919* (1921), p. 740.

7. The Lancashire and Cheshire Miners' Permanent Relief Fund in their *Annual Report* for 1918 stated that 'the problem of the injured miner is exactly similar to that of the wounded soldier, and it would be as well if the various organisations would co-operate in arranging for the more difficult cases of injured members to be taught new trades and occupations, especially the younger men, and one might venture to suggest that there should be no insuperable difficulty in the admission of injured miners into the various institutions in the country which at the moment are reserved exclusively for wounded soldiers' (p. 8). I am grateful to Eirau Eynon for this reference and for access to background material for her forthcoming doctoral thesis on medical provision in the Lancashire coalfield from the 1860s to 1945.

8. Namely, for a Ministry of Mines with a medical department. In his evidence before the (Sankey) Royal Commission on Mining. *Report of the Minutes of Evidence on the Second Stage of the Enquiry*, 1919. Cmd

360, pp. 745–56. (The Coal Mines Act, 1911, in fact, gave full power to the Secretary of State to provide for the safety and health and convenience of persons employed in the mines, but the powers were not taken up.)

Shufflebotham received his MD from Cambridge in 1900; during the war he was physician to the Stoke-on-Trent War Hospital. He was subsequently assistant physician to the North Staffordshire Infirmary and was the medical adviser to the National Society of Cotton Operatives, the North Staffordshire Coal Owners' Association, and the National Union of Pottery Workers. *Medical Directory*, 1930.

9. 'Salvage of Disabled', p. 677.
10. *Lancet*, 19 Apr. 1919, p. 673.
11. 'Industrial Medicine in America and Here', *Lancet*, 27 Dec. 1919, pp. 1199–1200.
12. See 'Review of Reconstructive Surgery', *Lancet*, 19 April 1919, pp. 670–3.
13. Albert H. Freiberg, 'The Casualties of War and Industry and Their Relation to Orthopedic Surgery', address of the chairman of the AMA Section on Orthopedic Surgery read before the joint meeting of the Section with the Section on Preventive Medicine and Public Health at the 69th Annual Session of the AMA, June 1918, *JAMA*, 10 Aug.1918, pp. 417–19 at p. 418.
14. Roland Hammond, 'Relation of Orthopedic Surgery to Industrial Surgery', chairman's address read before the AMA Section on Orthopaedic Surgery at the 71st Annual Session of the AMA, *JAMA*, 24 July 1920, pp. 213–4.
15. *JOS*, 1 (1919), p. 752.
16. Osgood, 'The Orthopedic Outlook', *JOS*, 1 (1919), pp. 1–6. See also D. Hinton, 'Application of War Surgery to Industrial Practice', *Penn. Med. J.*, 23 (Jan. 1920), pp. 188–93, and J. J. Moorhead, 'Is War Time Surgery Applicable to Industrial Surgery?', *J. Indust. Hygiene*, 1 (1919), pp. 158–62.
17. *JAMA*, 15 Nov.1919, pp. 1518–22.
18. Freiberg, 'Casualties of War and Industry', p. 418.
19. See *JAMA*, 3 Apr.1909, pp. 1134–6, and *Med. Record*, 1 May 1909, pp. 780–2.
20. See David Rosner and Gerald Markowitz, 'The Early Movement for Occupational Safety and Health, 1900–1917' in Judith W. Leavitt and Ronald L. Numbers (eds), *Sickness and Health in America* (Madison, 1985), pp. 507–21.
21. See Albert G. Love and Charles B. Davenport, *Defects Found in Drafted Men* (US War Department, Washington, 1920), and Roy N. Anderson, *The Disabled Man and His Vocational Adjustment: a study of the types of jobs held by 4,404 orthopedic cases in relation to the specific disability* (New York, 1932), pp. 12–13 which revealed from case histories at the Employment Center for the Handicapped in New York City, between 1917–30, that the most frequent cause of crippling was infantile paralysis (polio); 30 per cent were the result of public accidents, and only 20 per cent the result of industrial accidents. According to another source 'at

least 80 per cent of all disabled persons applying for rehabilitation were handicapped as a result of disease and congenital deformities': D. M. Blankenship, 'Industrial Clinics as a Factor in Rehabilitation', *Rehab. Rev.*, 1 (Jan. 1927), pp. 26–7. See also J. Whitley, 'What America is doing for her Civil and Industrial Cripples', *Cripples' J.*, 4, nos 14–16 (Oct. 1927 – Apr. 1928), pp. 115–22, 193–204 and 311–16, at p. 121 where it is claimed that in 1926 49 per cent of the 24 034 disabled persons rehabilitated in America had been injured in employment; 19 per cent in public accidents; 27 per cent by disease; and only 5 per cent from congenital deformities.

22. *Med.Record*, 1 May 1909, p. 782.

23. Cited in A. Gottlieb, 'Industrial Injuries: prevention of loss and restoration of function', *North West Med.* (Seattle), 18 (Dec. 1919), pp. 267–70 at p. 268.

24. In 1927 it became the Employment Center for the Handicapped. See Anderson, *Disabled Man*, p. 4.

25. '. . . for the object of furthering co-operative methods in introducing into industrial establishments the most effective methods for the treatment of injuries, for caring for the ailments of employees, for the promotion of improved sanitary conditions in industry, and to aid in the prevention of industrial diseases': *JBJS*, 6 (1924), p. 719, at which time the Conference was celebrating its 10th anniversary. The Conference proceedings were regularly reported in *JBJS* and there were close relations with the AOA.

26. A. Nugent, 'Fit for Work: the introduction of physical examinations in industry', *Bull. Hist. Med.*, 57 (1983), pp. 578–95.

27. For these and other instances of the growth of interest in industrial medicine at this time, see H. E. Mock, 'Industrial Medicine and Surgery: a resumé of its development and scope', *J. Indust. Hygiene*, 1 (May 1919), pp. 1–8, 251–4.

28. Glen Gritzer and Arnold Arluke, *The Making of Rehabilitation* (Berkeley, 1985), p. 40. In 1926 the industrial medicine specialism became the Board on Industrial Medicine and Traumatic Surgery, to which many orthopaedic surgeons were attached. See Mock, 'Industrial Medicine and Surgery', pp. 5ff., and F. Martin, *Fifty Years of Medicine and Surgery: an autobiographical sketch* (Chicago, 1934), pp. 350–2.

29. See G. Grey Turner, 'A Visit to America in War Time', *Univ. of Durham College of Med. Gaz.*, 19 (15 Mar. 1919), p. 22; and T. Lyle Hazlett and William W. Hummel, *Industrial Medicine in Western Pennsylvania, 1850–1950* (Pittsburgh, 1957), p. 74 *et passim*. Sherman visited Britain during the war and delivered a paper on the sterilization of war wounds at the RSM on 31 Oct. 1916. He was a member of the fracture committee of the American College of Surgeons and was on the certifying Board on Industrial Medical and Traumatic Surgery after its establishment in 1926.

30. The Birmingham survey found that 133 (16.1 per cent) of its sample of 828 cripples over the age of 16 were crippled as a result of accidents, 51 (6.1 per cent) being accidents at work. City of Birmingham Education

Committee, *Report of a Special Sub-committee of Enquiry Concerning Physically-Defective Adults and Children* (Birmingham, 1911), pp. 10, 18.

31. See TUC, *Safety First* (1928); Helen Jones, 'Employers' Welfare Schemes and Industrial Relations in Inter-War Britain', *Business Hist.*, 25 (1983), pp. 61–75; Arthur McIvor, 'Industrial Health in Britain: theory and workshop practice, 1918–39', unpublished paper; and Political and Economic Planning, *Report on the British Health Services* (1937), pp. 74, 86.

32. H. E. Mock, 'Penalty the American Nation Pays for Pursuing its "Speed-Mad" Way', *Rehab.Rev.*, 1 (Apr. 1927), p. 157. For comparative statistics, see Henry H. Kessler, *The Crippled and the Disabled: rehabilitation of the physically handicapped in the United States* (Columbia, NY, 1935), pp. 77–8n.

33. *Annual Report, IFRB*, no. 2, 1921, p. 17. I am grateful to Steve Sturdy for this reference.

34. *JBJS*, 6 (1924), p. 721; Mock, 'Speed-mad', p. 251. For a sample of the range of interests surrounding injured workers in America in the mid-1920s, see *Relation of Medicine to Industry: special report number thirty, of the National Industrial Conference Board* as reported in *JBJS*, 7 (1925), p. 231. An idea of differences in the scale of industrial operations between America and Britain is reflected in the fact that the Tennessee Coal, Iron, and Railroad Company in the early 1930s required the full-time employment of 15 dentists and 25 doctors: *JBJS*, 15 (1933), p. 270.

35. Rosemary Stevens, *American Medicine and the Public Interest* (New Haven, 1971), p. 162. The exact figure was 326. By 1929 numbers had risen to 504, by 1934 to 722 and by 1940 to 1078.

36. Albee visited both Alder Hey and Shepherd's Bush in 1916: *Autobiography*, pp. 83–5, 127, 158. In April 1918 he opened a four-bed rehabilitation clinic for industrial injury downstairs from a New Jersey State workers' compensation office. See Edward D. Berkowitz, 'The Federal Government and the Emergence of Rehabilitation Medicine', *The Historian*, 43 (1981), pp. 530–45 at pp. 531–2.

37. On the struggles for the control of physiotherapy and vocational rehabilitation, see Gritzer and Arluke, *Rehabilitation*, pp. 38–60. For an example of orthopaedic control over physiotherapy at a base hospital, see E. B. Mumford, 'Application of Curative Therapy in the Workshop', *JOS*, 1 (1919), pp. 676–81. See also J. L. Porter, 'The Reconstruction Problem for the Disabled Soldier', *J. Iowa Med. Soc.*, 9 (1919), pp. 365–70; G. Harris, *The Redemption of the Disabled* (New York, 1919); and Douglas McMurtrie, *The Disabled Soldier* (New York, 1919).

38. Joel E. Goldthwait, 'Organization of the Division of Orthopedic Surgery in the U.S. Army with the Expeditionary Force', *AJOS*, 16 (1918), p. 288, quoted in Gritzer and Arluke, *Rehabilitation*, p. 43. See also E. G. Brackett, 'Productive Occupational Therapy in the Treatment of the Disabilities of the Extremities', *JOS*, 1 (1919), pp. 40–5.

39. The Federal Vocational Rehabilitation Act, which distributed $1 000 000 annually to the states according to population, and for which the states

had to match an equal sum, was for the vocational, not the physical, restoration of beneficiaries. However, under the terms of the Act and those of most states, public money could 'be applied to every phase of training from pre-vocational experiences in hospitals to placement and follow-up work after training': see C. A. Prosser, 'The Rehabilitation of Disabled Persons', *Rehab. Rev.*, 1 (June 1927), pp. 197–207 at p. 201. Rehabilitation agents were conscious that they would bring the system into disrepute if they placed a person in employment who could have had his or her disability improved by medical means. Thus, according to one witness, 'various devices are adopted in the various states for securing the co-operation of the surgeon in the orthopaedic hospital. One state visited had organized five diagnostic clinics, to which applicants were called for examination by an itinerant consultant. In another state clinics had been established originally to cater for workmen's compensation cases in which complete therapeutic treatment for the disabled was offered, while yet another state had established a central hospital for the examination and complete treatment of rehabilitation cases. In all states, however, an effort is made to secure the opinion and, if necessary, the treatment services of the specialist in orthopaedic surgery.' J. Whitley, 'What America is Doing', p. 200.

40. Freiberg, 'Casualties of War and Industry', p. 418; Osgood, 'Orthopedic Outlook', p. 2.
41. Albee, *Autobiography*, p. 180. Or as Gottlieb put it, 'The restoration of function should not be left to the process of after-treatment'. 'Industrial Injuries', p. 268.
42. R. W. Johnson, Jr, 'The Time Element in Reconstructive Surgery', *JOS*, 2 (1920), pp. 33–42 at p. 41. See also H. P. H. Galloway's AOA presidential address 'Readjustment to Changing Conditions', *JOS*, 1 (1919), pp. 395–401 at p. 397.
43. Lever Stewart, 'The Industrial Surgeon's Treatment of Fractures', *J. Indust. Hygiene*, 8 (1926), pp. 283–7, at p. 285. Cf. Dr Henry Kessler, *The Knife is Not Enough* (New York, 1968). Kessler began his career as assistant to Albee at the latter's rehabilitation clinic. By the Second World War he was one of the leading Americans in the field of rehabilitation.
44. 'Industrial Medicine in America and Here', *Lancet*, 27 Dec. 1919, pp. 1199–1200.
45. One of the few places where industrial health was emphasized in the 1930s was in the DPH course at the London School of Tropical Medicine and Hygiene, of which nearly one-half of the class were women. Neville Goodman, *Wilson Jameson: architect of national health* (1970), p. 78. It is also worth noting the absence of British orthopaedists from the 5th International Medical Congress for Industrial Accidents and Occupational Diseases of 1928, in spite of the fact that the meeting was addressed by the fracture specialist, Lorenz Böhler. The only British speaker was Sir Thomas Oliver, the author of *Dangerous Trades* and *The Health of Workers*. See the report in *JBJS*, 10 (1928), p. 147.
      In Britain as in America, however, industrial surgery was better organized and more prestigious than industrial medicine: see Stevens,

American Medicine and the Public Interest, pp. 330–1, and Morris Fishbein (ed.), *Doctors at War* (New York, 1945), p. 26.

46. See Jane Lewis, *What Price Community Medicine?* (Brighton, 1986). For a specific example of the economic rationale for running down an industrial medical service, see Ch. 10 below on the LMS Railway Hospital at Crewe.

47. See comments by T. P. McMurray, in the discussion on H. E. Moore's paper, 'Avoidable Wastage in Connection with Industrial Injuries', *Liverpool Medico-Chirurg. J.*, 41 (1933), p. 39; and the letter of Robert Jones to Neville Chamberlain, Jan.1927, quoted in Watson, *Life of Jones*, p. 257. Jones also urged Alan Malkin in 1922 not to confine himself merely to crippled children whilst setting up the orthopaedic service in Nottingham. See Malkin, 'The Care of Crippled Children in England', *JBJS*, 36A (1954), p. 1282.

48. James Rutherford Kerr, 'The Treatment of Industrial Accidents', *BMJ*, 26 Aug. 1922, pp. 377–80 at p. 377.

49. USA 200 000; UK 1 170 000, according to the figures gathered by the International Labour Office. Cited in Kessler, *Crippled and Disabled*, pp. 146–7.

50. See Rodney Lowe, 'The Erosion of State Intervention in Britain, 1917–1924', *Econ. Hist. Rev.*, 31 (1978), pp. 270–86.

51. Jones's friend John Galsworthy resigned as editor of *Reveille* in 1919, bitter and discouraged at the Ministry of Pensions' regard of the journal (which was committed to diffusing information on ways and means of rehabilitating the disabled) as 'too unofficial'. Watson, *Life of Jones*, p. 207.

52. J. Whitley, 'What America is Doing', p. 203.

53. Titmuss, 'War and Social Policy', in his *Essays on the Welfare State* (2nd edn, 1963), pp. 75–87. Idem, *Problems of Social Policy* (1950), esp. p. 474. For elaboration of the view that the Great War spawned little social reform, see Arthur Marwick, 'The Impact of the First World War on British Society', *J. Contemp. Hist.*, 3 (1968), pp. 51–63, and the discussion in P. W. J. Bartrip, *Workmen's Compensation in Twentieth Century Britain: law, history and social policy* (Aldershot, 1987), pp. 92ff. and p. 238.

54. In Britain, 'nothing has been done since [1917] to widen the basis of eligibility or otherwise interfere with the year-by-year reduction in the cost of the system': Kessler, *Crippled and Disabled*, p. 161. There were 41,000 permanently disabled war veterans in Britain in 1924: John Calder, *The Vanishing Willows, the story of Erskine Hospital* (Bishopton, Renfrewshire, 1982), p. 28. For a synopsis of the involvement of the Ministry of Pensions in 'Convalescent Centres for Concurrent Treatment and Training', see PRO:MH 76/157. On Ministry of Labour planning, see PRO:CAB/24/125. For international comparisons, see Douglas C. McMurtrie, *The Relation of Earning Power to Award of Compensation for Disability Incurred in Military or Naval Service: a memorandum on the pensions practice of other nations* (New York, 1919).

55. The 'appeal' was succeeded by the National Scheme for the Employment of Disabled Ex-Servicemen, launched by Royal Proclamation. Admin-

istered by local employment committees, the 'King's Roll' began to be effective when the Ministry of Labour decided that government contracts should only be given to participants in the scheme. By 1928 some 27000 firms were on the roll employing 380,000 disabled ex-servicemen (figures scarcely different a decade later). For an outline of the government's response to the employment of the disabled ex-servicemen and for the relevant files in the PRO, see Public Record Office, *Documents of Interest to Social Scientists: employment and unemployment, 1919–1939* (1977), pp. 35–7. See also Arnold Bennett, 'A National responsibility: future employment of the disabled', reprinted from the *Daily News and Leader*, 6 June 1917; and Hermann Levy, *Back to Work? The case of the partially disabled worker*, Fabian Society Research Series, no. 56 (1941), p. 20.

In Germany, Biesalski played a leading role in the law for the care of the war crippled (*Kriegs-Kruppelfursorge Gesetz*) passed in 1923, which made it obligatory for every industrial organization of 50 or more employees to employ at least one crippled war veteran. See *JBJS*, 12 (1930), pp. 448–9.

For Italy, see V. Putti, 'National Organization of Rehabilitation of the Disabled in Italy', *Med. Record*, 27 Sept.1919, abstracted in *JOS* 1 (1919), p. 701. For other international comparisons, see Kessler, *Crippled and Disabled*, and the *Cripples' J.*

56. On the Workshops, see PRO:T/161/867/S10064/05.
57. Arnold Wilson and Hermann Levy, *Workmen's Compensation*, vol. 2 (Oxford, 1941), p. 222. Similarly defeated was the proposal in 1923 to keep open Sir Bernard Oppenheim's diamond-cutting factory in Brighton, which, from its commencement in 1918, had employed mainly disabled men.
58. See Kessler, *Crippled and Disabled*, p. 258; A. F. McBride, 'The Workmen's Compensation Law and Its Relation to the Doctor and the Hospital', *Rehab. Rev.*, 1 (Mar. 1927), pp. 100–3; and R. L. Numbers, 'The Third Party: health insurance in America', in Morris J. Vogel and Charles Rosenberg (eds), *The Therapeutic Revolution: essays in the social history of American medicine* (Philadelphia, 1979), pp. 177–200.
59. Wilson and Levy, *Workmen's Compensation*, vol. 1, pp. 199ff. See also Bartrip, *Workmen's Compensation in Twentieth Century Britain*.
60. Sickness benefit, fixed at 15s per week (12s for women) in 1920, along with basic medical treatment, was available for up to 26 weeks a year to all workers earning less than £250 per annum, in return for a joint contribution from employer and employed. After the initial six months, persistent cases could claim disability benefit at half the previous rate.' Noel Whiteside, 'Counting the Cost: sickness and disability among working people in an era of industrial recession, 1920–39', *Econ. Hist. Rev.*, 40 (1987), pp. 228–46 at p. 230.
61. J. Middleton papers, 5 July 1939, JSM/WCOM/93 (Labour Party Headquarters, London), for which reference I am grateful to Helen Jones.

62. *Social Insurance and Allied Services: report by Sir William Beveridge* (New York, 1942), p. 38. The *Departmental Committee on Workmen's Compensation* (Holman Gregory Report), 1920, wanted the additional cost of treating industrial accidents and occupational diseases within NHI to be born by the employers. See also Hermann Levy, *National Health Insurance: a critical study* (1944), pp. 138–42, and *Royal Commission on National Health Insurance*, 1928, Cmd 2596, p. 256.

63. Evidence of Drs George C. Anderson and John Wardle Bone, speaking for the BMA, to the Workmen's Compensation Committee, *Minutes of Evidence*, 22 Jan. 1920, p. 303. On the latter point, see also A. G. James, 'Orthopaedics in Industrial Life', p. 301.

64. *The Labour Movement and the Hospital Crisis* (1922), p. 15.

65. V. Warren Low (senior surgeon, St Mary's Hospital, London, and consulting surgeon to the LMS Railways) 'Workmen's Compensation and the Surgeon' (presidential address to the section of surgery of the RSM), *Lancet*, 23 Oct. 1926, pp. 844–50, at p. 845.

66. See R. N. Gray (medical referee, Aetna Life Insurance Company), 'Disability and Cost of Industrial Fractures: a comparison based upon an impersonal study of statistics of fractures treated by the specially trained surgeon and the general practitioner', paper delivered to the AOA meeting, June 1927, *JBJS*, 10 (1928), pp. 27–39; idem., 'Rehabilitation after Industrial Fractures: a plea for care, caution and further scientific study in the use of physical therapy', *Rehab. Rev.*, 1 (Sept.1927), pp. 285–8.

67. See Wilson and Levy, *Workmen's Compensation*, vol. 2, pp. 270ff.

68. *J. Brit. Dermatology*, 1937, pp. 426–7, quoted in Levy, *Back to Work*, p. 13.

69. 'Organization of the Treatment of Fractures', *BMJ*, 20 April, 1935, p. 817.

70. See 'Memorandum of Evidence by the British Hospitals' Association', in the *Minutes of Evidence taken before the Royal Commission on Workmen's Compensation* (1939/40), pp. 1078ff.

71. R. F. Herndon, 'Back Injuries in Industrial Employees', *JBJS*, 9 (1927), p. 234. In America, the insurance companies began to act as a third party in medicine in 1911 after Massachusetts introduced workmen's compensation. See Morris Vogel, *The Invention of the Modern Hospital: Boston, 1870–1930* (Chicago, 1985), pp. 121ff.; David Rosner, *A Once Charitable Enterprise* (New York, 1982), p. 94; and (Cave) *Voluntary Hospitals' Committee*, 1921, p. 28.

72. Meetings of 22 Dec. 1930 at Dr T. Carnwath's office at the Ministry of Health (a copy of the minutes of which were circulated in County Hall): GLRO:PH/HOSP/1/66.

73. *The Times*, quoted in 'Memorandum of Evidence from the British Hospitals' Association', *Minutes of Evidence Taken Before the Royal Commission on Workmen's Compensation*, p. 1078. See also 'Memorandum of Evidence from Accident Officers' Association', ibid., pp. 828ff.; and Wilson and Levy, *Workmen's Compensation*, vol. 2, ch. 21.

74. See letter from Robert R. Hyde, Industrial Welfare Society, 'Treatment of Fractures', *The Times*, 18 March 1935, p. 8.

75. 'Rehabilitation of the Disabled: an American experiment', *BMJ*, 6 Apr. 1935, p. 726.
76. Quoted in Wilson and Levy, *Workmen's Compensation*, vol. 2, p. 219.
77. Mock, 'Industrial Medicine and Surgery', 1919, p. 254. 'There is a tendency to stress the humanitarian aspect of such activity', wrote the president of the National Safety Council in 1927; 'this may affect some employers, but the majority can only be reached by demonstrating clearly to them that rehabilitation is a sound investment': W. G. King, 'Industry's Obligation in the Rehabilitation of Workmen', *Rehab. Rev.*, 1 (Feb. 1927), pp. 45–46.
78. E. W. Ryerson, 'Treatment of Fractures from an Industrial Standpoint', *JBJS*, 6 (1924), pp. 188–91, and the AOA discussion on this paper, pp. 191–3.
79. According to Whitley, ('What America is Doing', 1927, pp. 200–1), the Americans too had concentrated on the child cripple to the relative neglect of the adult. However, he argued that this defect was 'rapidly being remedied' as a result of workmen's compensation regulations providing funds for the complete treatment of industrial cripples; the anxiety of insurance companies to restore as far as possible the disabled covered by their policies; and the demands of the rehabilitation service for specialist medical treatment for large numbers of disabled persons previously left untreated, and the extension of hospital facilities to meet this demand. See also W. C. Campbell, 'The Care of the Crippled Adult', *Rehab. Rev.*, 1 (Mar. 1927), pp. 113–15. Many American orthopaedic surgeons appear to have early specialized in combined orthopaedic and industrial surgery. As early as 1921, R. B. Osgood wrote on 'The Standardization of Methods of Treatment in Orthopedic Surgery and Industrial Surgery of the Extremities and Spinal Column', *Illinois Med. J.*, Apr. 1921, pp. 342–52.
80. For example, the orthopaedic clinic at the Royal Albert Edward Hospital in Wigan provided by the Ministry of Pensions in 1919. The acting head of the clinic, however, was recruited from the general surgical staff and had no particular training or experience in orthopaedics. Although Robert Jones accepted an invitation to demonstrate the use of splints there in 1919, little else is heard of the clinic until the 1930s when the Liverpool-trained orthopaedic surgeon and BOA member W. J. Eastwood took it over. The Lancashire and Cheshire Miners' Permanent Relief Fund may have made use of the clinic for its members, since from 1912 it had given money to the Wigan Infirmary as part payment for specialist services. Again, my thanks to Eirau Eynon for this information.
81. In his evidence to the Sankey Commission, 1919, as reported in 'Salvage of Disabled Workmen', *Engineering*, 23 May 1919, p. 675.
82. On Kerr (who should not be confused with his namesake, the pioneer school medical officer and member of the 1920s Labour Party Advisory Committee on Public Health referred to in Ch. 4 above), see the obituary notice in the *BMJ*, 3 Oct. 1942, pp. 412–13; and 'Medical Centre at St.Helens has a face-lift operation', *Cullet News*, Feb. 1967, pp. 12–13. On the hospital at St Helens, Lancs., see also 'Salvage of Disabled', p.

675; and (with illustrations), Kerr, 'Orthopaedics in Relation to Man-Power', *Engineering*, 3 Jan 1919, pp. 6–9; and Kerr, 'The Pilkington Special Hospital, St. Helens, Lancashire', *Reveille*, no. 3 (Feb. 1919), pp. 451–6. The hospital, founded and equipped by Messrs Pilkington Brothers Ltd, was approved by Keogh and Jones during the war as a special orthopaedic centre. Jones 'and other eminent authorities on orthopaedic science' advised on the Hospital's construction, and Jones remained a consultant to it. The first patients were admitted in 1917, all of whom were 'soldiers who had been injured in service, treated in hospital and discharged from the Army [from two months to three years before], but whose disablements interfered with bread-earning'.

Both Pilkington brothers were themselves injured in France during the war. In 1919 Lionel Pilkington was also the president of the Lancashire and Cheshire Miners' Permanent Relief Fund, a major function of which was medical care.

83. See Workmen's Compensation Committee, *Minutes of Evidence*, 10 Mar. 1920, pp. 381–91; Samuel James Woodall, *The Manor House Hospital: a personal record* (1966); and below, Ch. 10.

84. See 'Glasgow Orthopaedic Clinic', *Glasgow Med. J.*, Oct. 1926, pp. 258–9. See also, ibid., July 1925, p. 50, Apr., June and Dec. 1926, pp. 303, 452, 379, Mar. 1930, p. 154, Mar.1931, pp. 146–7, and May 1932, p. 252. The Clinic, which also ran a domiciliary service, was formally opened by Walter Elliot, Parliamentary Under-Secretary of State for Scotland, in November 1926. In the 1930s it became the Glasgow Orthopaedic and Rheumatic Clinic, and subsequently the Glasgow Physiotherapy and Rehabilitation Clinic. I am grateful to Derek Dow for drawing my attention to this clinic.

8 Colonization Among Cripples

1. Board of Education, *Annual Report of the CMO for 1919*, p. 101.

2. At the 'Conference on Education in Hospital Schools and Schools for Physically Defective Children', held in the ICAA's Carnegie House, London, 12 Dec. 1925, reported in *Cripples' J.*, 1 (1925), pp. 279–311 at p. 282.

3. Fifty-four of the 94 senior members had such appointments in 1928, 16 of them holding more than one such post.

4. 'The Cure of Crippled Children. Proposed National Scheme', *BMJ*, 11 Oct. 1919, pp. 457–60.

5. Josep Trueta, *Gathorne Robert Girdlestone* (Oxford, 1971), p. 31.

6. Although Bertrand Dawson in 1918, in his 'revolutionary' plan for primary health-care centres, had called for staffing 'where possible by the medical men of the district on the basis of a part-time service paid by salary', the question of whether hospital consultants should be thus remunerated remained 'a vexed question' overridden by fears within the profession of a state salaried medical service. See Dawson, 'The Future of the Medical Profession', *BMJ*, 13 July 1918, p. 25, and Dawson's Consultative Committee on Medical and Allied Services for the Ministry

of Health: PRO:MH/73/48. The latter pointed out that 'The BMA laid down many years ago the principle that where the cost of maintenance of a patient in an institution was borne by a Local Authority, the medical staff should be paid for their professional services, the general principle being that they should only give their services as a matter of charity where the remainder of the cost of treatment was defrayed by the benevolent public as a charity. . . . It may perhaps be agreed that experts should be paid by the authority, and general practitioners paid, if at all, by the patient . . . and full time medical officers of health at a fixed annual salary.'

Before the mid-1930s, when orthopaedic surgeons again pressed this point, only the Socialist Medical Association and certain members of the Labour Party argued for a salaried medical service that included hospital consultants. The employment of full-time salaried medical staff had been implied in the document produced by the Webbs and the Labour Party's Advisory Committee on Public Health in 1919, *The Organization of the Preventive & Curative Medical Services & Hospital & Laboratory Systems under a Ministry of Health*, but Sidney Webb subsequently went out of his way to reassure the profession that the Labour Party had not committed itself to a full-time 'State Army of Salaried Clinicians'. Quoted in Brian Abel-Smith, *The Hospitals, 1800–1948* (1964), pp. 286–8.

7. Girdlestone, 'The Robert Jones Tradition', *JBJS*, 30B (1948), pp. 187–95 at p. 192. See also below Ch. 11.
8. Daniel Fox, *Health Policies, Health Politics: the British and American Experience, 1911–1965* (Princeton, 1986). As Charles Webster has shown, however, the tendency to 'hierarchical regionalism' was by no means universal nor inevitable: 'Conflict and Consensus: explaining the British health service', *Twentieth Century British History*, 1 (1990), pp. 121–33.
9. Trueta, *Girdlestone*, p. 34.
10. See Girdlestone to Jones, 29 Aug. 1929. Girdlestone Papers, BOA.
11. Direct references to Biesalski at this time are not apparent, however. Knowledge of his work seems to have come to Britain via America (see, for example, 'Modern Methods of Cripple Care in Germany', *J. State Med*, 20 (1912), reviewed in *Br. J. Childr. Dis.*, 10 (1913), p. 90) and through the New York orthopaedist, Leo Mayer, who worked with Biesalski in Berlin from 1913 to 1917.
12. See Robert Jones to H. A. L. Fisher, 16 Dec. 1919, in PRO:ED/50/153; and Trueta, *Girdlestone*, p. 56. For the first published membership list (1921), see Joan Anderson, *A Record of Fifty Years' Service to the Disabled 1919–1969 by the Central Council for the Disabled* (1970), p. 79.
13. Jones to Fisher, op. cit.; Ursula Townsend (Secretary of the CCCC) to Girdlestone, 19 June 1920, Girdlestone papers, BOA.
14. Anderson, *Record*, p. 59; Trueta, 'The National Scheme and the CCCC', in *Girdlestone*, pp. 30–44; and (for Elmslie), '[Proceedings of] Joint Conference, Invalid Children's Aid Association and Central Committee for the Care of Cripples', *Cripples' J.*, 3 (1927), pp. 163–239, at p. 176.
15. Jones to Fisher, 16 Dec. 1919.

16. See Board of Education, *Annual Report of the CMO for 1914*, p. 179n., *1917*, p. 89, and *1918*, pp. 112–13. See also Girdlestone, 'Methods by Which Assistance Can be Obtained from Public Bodies' in his *The Care and Cure of Crippled Children: the scheme of the Central Committee for the Care of Cripples* (Bristol, 1925), pp. 52–5. Newman in his *Annual Report for 1923* (pp. 76ff.) outlined five means by which financial assistance from public bodies could be obtained for orthopaedic hospitals.

17. Arthur Marwick, *The Deluge: British society and the First World War* (1965), pp. 240–1.

18. See Frank Honigsbaum, *The Struggle for the Ministry of Health, 1914–1919* (1970).

19. David Lloyd George was a close colleague of Fisher, and was regarded by Jones and Girdlestone as an ally, especially in relation to schemes for cripples in Wales. See, for example, the letter from Girdlestone to Fisher, 13 May 1937, Fisher Papers, Bodelian Library, Oxford.

20. Board of Education, *Annual Report of the CMO for 1919*, p. 107.

21. Letter in BOA Archives, quoted in Trueta, *Girdlestone*, p. 37.

22. In PRO:ED/50/153. The other six points were: the division of England and Wales into administrative units, each with a COH and hospital school; each hospital school to have attached to it scattered outpatient clinics, day invalid schools and a residential school; treatment to be conducted by those experienced in orthopaedic work; provision to be made for paying patients; wards to be available for adolescent and adult cripples; and all existing organizations for cripples to be welcomed into the scheme.

23. Memorandum of Interview with Newman, 5 March 1920, PRO:ED/50/153. Girdlestone's account of the interview (as well as that of 27 February 1920) is contained in the Girdlestone papers.

24. 'Orthopaedics and the Child', pp. 89–102.

25. In Sir James Marchant (ed.), *The Claims of the Coming Generation* (1923), pp. 15–36 at p. 35. See also *Cripples' J.*, 2 (Apr. 1926), p. 242.

26. The drop in subscribers to voluntary hospitals upon the formation of the Ministry was noted by Dawson's Consultative Council. See minutes in PRO:MH/73/49.

27. Trueta, *Girdlestone*, p. 31.

28. From the Red Cross the Oswestry Orthopaedic Hospital alone received £25 000 in 1918 for its after-care clinics. See Dorothy Wrenn, 'The Lyrical Adventure, the story of Agnes Hunt', MS, *c.* 1970, copy in Shropshire Local History Library, p. 170. On the involvement of the Red Cross in the CCCC, see S. Tower, 'The Red Cross in the After-Care Clinics', *Cripples' J.*, 4 (1928), pp. 9–16. Newman, while generally encouraging voluntary activity, resented the way in which the Red Cross after the War was 'trying to absorb everything, which is bad for it & everything,' and noted that there was 'a good deal of criticism of this'. Note of 29 Sept. 1920 in PRO:ED/50/153.

29. COS, *Charities Annual and Register for 1930*, pp. 67, 337; Burdett, *Hospitals and Charities Annual for 1930*, p. 803. Incomplete statistics in the *Charities Annual and Register* for contributions to about a dozen

agencies and homes for cripples, show receipts rising from £56,000 in 1914 to £176 000 in 1922. William de Courcy Wheeler referred to 'encouraging correspondence with the Carnegie Trustees' in his pre-war discussions on a scheme for cripples in Ireland (*BMJ*, 18 Oct. 1919, p. 508); the ICAA moved into Carnegie House, London, around 1922.

30. On the People's League of Health, see PRO:MH/58/153. Inaugurated by Miss Olga Nethersole in January 1920, the League was dedicated to preventive health propaganda. Among those whose names appear on its letterhead are Moynihan, John Lynn-Thomas, Lord Charnwood, Leonard Hill and Benjamin Moore. For Jones's letter to Chamberlain, see Frederick Watson, *The Life of Sir Robert Jones* (1934), pp. 255–7.

31. Fifty had done so by 1924, 21 of them during the previous year; another 35 followed in 1925. See Board of Education, *Annual Report of the CMO for 1924* (p.82) and for *1925* (p. 51).

32. Newman's assistant at the Ministry of Health, A. S. MacNalty, was acutely aware that 'We are so often unjustly accused of stultifying voluntary work': MacNalty to Girdlestone, 17 Apr. 1924, Girdlestone Papers, BOA archives. For details of the kinds of problems existing between voluntary and statutory bodies in this period, see Madeline Rooff, *Voluntary Societies and Social Policy* (1957), ch. 21: 'Influences Affecting the Relationship'.

33. Sir Ryland Adkins to Lord Dawson's Consultative Council in 1921, confidential memo on the administration of health services, in PRO:MH/73/38. Appreciating the voluntary societies' social and financial influence and power, Adkins hoped that the Ministry of Health would recognize and encourage them 'provided that they do not in any way dominate the responsible Local Authorities, and provided that the Local Authorities are empowered'. See also Sir Arthur Newsholme, *The Last Thirty Years in Public Health* (1936), p. 238.

34. Girdlestone, 'Voluntary Effort in an Orthopaedic Scheme' (paper to the joint ICAA and CCCC conference, Nov. 1926), *Cripples' J.*, 3 (1927), pp. 240–9, at pp. 248–9.

35. Ibid., p. 241. Girdlestone obtained his list of the pros and cons from E. Elkington of the Lancashire County Council School Medical and Child Welfare Department. See letter of 15 Sept. 1926 in Girdlestone Papers, BOA archives. See also Girdlestone, 'The Value of Voluntary Organisation', *Cripples' J.*, 4 (1928), pp. 269–71; Trueta, *Girdlestone*, p. 82; and Dawson, 'Future of the Medical Profession', p. 57.

36. Anderson, '1935 Annus Mirabilis and the Beginning of a New Era' in *A Record*, pp. 24ff.

37. Adkins, PRO: MH/73/38.

38. Mr S. W. Dow, at the opening of the Children's Orthopaedic Hospital at Kirkbymoorside, in Yorkshire, quoted in *Cripples' J.*, 3 (1926), p. 79.

39. Robert Veitch Clark, the MOH for Manchester, in discussion at 'Joint Conference, 1926', p. 195. See also, for the high proportion of rickety children in 'PD' schools in Leeds and Manchester relative to other areas in 1920, PRO:ED/50/153.

40. See, for example, Kenneth Fraser, Deputy School Medical Officer, Cumberland County Council, *The Cripple Scheme* (1922) and the highly enthusiastic review in *Med. Officer*, 21 July 1923, p. 29.
41. *Cripples' J.*, 4 (1928), p. 445.
42. At Biddulph Grange, near Stoke-on-Trent. This country mansion-turned-open-air hospital was offered to the North Staffordshire Cripples' Aid Society in 1921, but within six months of their opening it in 1924, it ran into serious financial difficulties. While the Staffordshire County Council procrastinated about taking it over, the Lancashire County Council purchased it. See A.J. Carter, 'A History of Hartshill', *Midlands Med.*, 18 (1991), pp. 136–9. See also, *Cripples' J.*, 4 (1928), p. 83.
43. See Central Council for the Care of Cripples, *Directory of Orthopaedic Institutions, Voluntary Organisations and Official Schemes for the Welfare of Cripples* (1935), pp. 63–76.
44. Compare this view with that in the 'Memorandum on Provision of Specialist Services by the CMO, 1939–42', PRO:MH/80/24.
45. CCCC, *Directory for 1935*. See also, 'Joint Conference, Invalid Children's Aid Association and Central Committee For the Care of Cripples', *Cripples' J.*, 3 (1927), p. 178.
46. Glasgow Children's Hospital, *48th Annual Report for the Year Ending 1930*. For similar, see George Newman, *The Building of a Nation's Health* (1939), p. 226 and Alfred Tubby, *The Advance of Orthopaedic Surgery* (1924), p. 96.
47. Henry Gauvain lamented in 1919, for instance, that 'the pressure of work resulting from the war' had held back research into the chemotherapeutic treatment of tuberculosis; and Jones concluded in 1920 that the procedure for dealing with joint abscesses by opening, draining and treating them with germicidal solution (comparable to the procedure for dealing with wound infection during the war) was 'indefensible'. Gauvain, 'Chemotherapy in Cutaneous Tuberculosis', *Lancet*, 15 Mar. 1919, p. 412; and F.B. Smith, *The Retreat of Tuberculosis, 1850–1950* (1988), pp. 136ff.
48. Guildhall Conference, London, on the Care of Crippled Children, 16 Nov. 1920, as quoted in *Liverpool Courier*, 17 Nov. 1920, in PRO:ED/50/153.
49. Jones, 'The Cripple (A Retrospect and a Forecast)', *Cripples' J.*, 5 (1928), p. 8. See also F. Watson, 'The Influence of the War on the Cripple', *Cripples' J.*, 6 (1930), pp. 476–84, p. 476, and H. Gauvain and E.M. Holmes, 'The Evolution of Hospital Schools', *Lancet*, 13 Apr. 1929, pp. 789–92, 838–40, at p. 791.
50. *48th Annual Report*, 1930.
51. Ministry of Health, *An Outline of the Practice of Preventive Medicine: a memorandum addressed to the Minister of Health by Sir George Newman* (HMSO, 1919).
52. Elizabeth Fee and Dorothy Porter, 'Public Health, Preventive Medicine and Professionalization: England and America in the nineteenth century', in Andrew Wear (ed.), *Medicine in Society: historical essays*

(Cambridge, 1991), pp. 249–75. See also the essays in Elizabeth Fee and Roy M. Acheson (eds), *A History of Education in Public Health* (Oxford/New York, 1991).

53. Newman, *Practice of Preventive Medicine*, p. 6.

54. A full account of the political construction and ideological implications of Preventive Medicine has yet to be written. Brief but important insights are to be gained from Jane Lewis, *What Price Community Medicine? The philosophy, practice and politics of public health since 1919* (Brighton, 1986). I owe many debts to Steve Sturdy for his insights on this subject.

55. See Paul Weindling, 'From Isolation to Therapy: children's hospitals and diphtheria in *fin de siècle* Paris, London and Berlin' in Roger Cooter (ed.), *In the Name of the Child* (1992), pp. 124–45. On Salvarsan, see Jonathan Liebenau, *Medical Science and Medical Industry* (1987).

56. Newman, *Practice of Preventive Medicine*, p. 5.

57. Ministry of Health, Consultative Council on Medical and Allied Services, *Interim Report on the Future Provision of Medical and Allied Services* (HMSO, 1920), para. 92. Dawson was appointed the Chairman of the Ministry's Consultative Council when it was established in 1919, and the *Interim Report* was very largely his own product.

58. Webster, 'Conflict and Consensus', p. 123. See also Francis Watson, *Dawson of Penn* (1950), p. 155, and Vicente Navarro, *Class Struggle, The State and Medicine* (Oxford, 1978), ch. 2.

59. Webster, 'Conflict and Consensus', p. 122.

60. In their National Scheme Jones and Girdlestone drew attention to tuberculous milk supply, insufficient food, inadequate sunlight and lack of ventilation as among the avoidable causes of crippling. They estimated that 70 per cent of bone and joint tuberculosis could be attributed to infected milk: Watson, *Life of Jones*, pp. 250ff. See also 'The Campaign for Pure Milk', *Cripples' J.*, 1 (1925), pp. 211–12, and the various articles on 'The Problem of Slums' in *Cripples' J.*, 5 (1929), pp. 239–73. In 'Propaganda Lectures for Voluntary Associations. I – What Makes Cripples', *Cripples' J.*, 4 (1928), p. 327, prospective lecturers were told, 'It will be your duty to try and explain that sunshine and fresh air are the greatest attributes to health, and to the proper and healthy growth of the human body.' Agnes Hunt declared in 1926 that the object of what Jones' called the 'holy war' against crippledom was to take the bread from the mouths of orthopaedists. Quoted in Robert Osgood, 'Progress in Treatment of the Crippled Child', *Cripples' J.*, 4 (1927–8), p. 231. Dr Rollier of Leysin similarly remarked to Margaret Beavan in 1925: 'It is for you workers in the great cities, to put me out of a job.' Beaven, 'Education of Public Opinion', paper to the Joint Conference of ICAA and CCCC, 1926, *Cripples' J.*, 3 (1927), p. 200.

61. Newman, *Practice of Preventive Medicine*, pp. 87–8. By extending 'the spirit of Preventive Medicine', he said, Jones's orthopaedics 'mark[ed] an epoch in English medicine'. For similar statements, see Robert Veitch Clark (MOH for Manchester) in the discussion on Jones' paper at 'Joint Conference, 1926', p. 194; and James Rognväld Learmonth, 'The Contribution of Surgery to Preventive Medicine' (Heath Clark Lec-

ture, 1949), reprinted in *The Thoughtful Surgeon*, ed. Donald M. Douglas (Glasgow, 1969), p. 99.

62. Dawson, *Interim Report*, para. 14.

63. Watson, *Dawson*, p. 146; Lewis, *Community Medicine*, pp. 18–19. See also Dorothy Porter, 'How Soon is Now? Public Health and the *BMJ*', *BMJ* 3 Oct. 1990, pp. 738–40; idem. (née Watkins), 'The English Revolution in Social Medicine, 1880–1911', PhD thesis, University of London, 1984; and J.M. Winter, *The Great War and the British People* (1985), pp. 173ff.

64. The ultimate threat came from Somerville Hastings, one of the leaders of the Socialist Medical Association, who proposed that GPs become salaried servants of the state like MOsH. But the idea was rejected by the Labour Party's Advisory Committee on Public Health in the early 1930s. See Marwick, 'Labour Party and the Welfare State', p. 389; and Charles Webster, *The Health Service Since the War*: vol. 1: *Problems of Health Care, The National Health Service before 1957* (1988), p. 82.

65. Newman to Jones, 27 May 1919, in PRO:ED/50/152.

66. Newman, *Practice of Preventive Medicine*, p. 88.

67. Second interview with Jones and Girdlestone, 5 Mar. 1920 in PRO:ED/50/153; and Board of Education, *Annual Report of the CMO for 1922*, p. 97. Girdlestone, when he took over the Wingfield Hospital from the Ministry of Pensions in 1922 guessed, on the basis of the Oswestry Hospital, that it would cost about £15000 per annum to operate 'of which £13,000 should be provided by patients' fees paid either by themselves, their friends, or local authorities.' *Annual Report for 1922*, contained in PRO:ED/62/67. The Waldorf Astor Committee in its *Interim Report of the Departmental Committee on Tuberculosis*, 1912 (Cmd. 6164), p. 16, estimated that the capital cost per bed for children with bone and joint tuberculosis would be about £150 'on average, and the maintenance charge per bed per week at from 25*s* to 30*s*'.

68. See Gauvain, 'Conservative Treatment of Tuberculous Cripples', *BMJ*, 15 Oct. 1910, pp. 1124–26; and 'Contributions of the late Sir Henry Gauvain to Orthopaedic Surgery', *JBJS*, 30B (1948), p. 385.

69. Gauvain to Girdlestone, 12 May 1924. Girdlestone Papers.

70. Cited in *JBJS*, 30B (1948), p. 202. For comparisons in the field of rheumatology, see David Cantor, 'The Contradictions of Specialization: rheumatism and the decline of the spa in inter-war Britain', in Roy Porter (ed.), *The Medical History of Waters and Spas. Med. Hist.*, Suppl. No. 10 (1990), pp. 127–44.

71. For example, James B. Mennell of St Thomas's Hospital, the civilian officer in charge of the massage department at Shepherd's Bush during the war, the author of various works on massage and remedial therapy, and one of the advisory chairmen of the British Association for the Advancement of Radiology and Physiotherapy from 1929 to 1935. In 1938 Mennell applied for membership in the BMA Orthopaedic Group Committee stating 'If eligible, I should like to belong to the Group as an orthopaedic physician': BMA Archives. A BOA member in the same mould was Sir Morton Smart, 'Manipulative Surgeon to the King' and a

consultant to the London Clinic for Injuries. He wrote various papers on physical medicine in wartime, and was described by the Ministry of Health in a letter of 27 Nov. 1939 as 'a pukka doctor, not a bonesetter': PRO:MH/76/321. Another was Alfred George Timbrell Fisher, the honorary orthopaedic surgeon, Order of St John Clinic for Rheumatism, a member of the Executive Committee of the Empire Rheumatism Council, and author of *Treatment by Manipulation* (5th edn, 1948). Yet another was James Cyriax, whose *Textbook of Orthopaedic Medicine* (1947) is still regularly reprinted (7th edn 1978). Cyriax claims in his preface that orthopaedic medicine was 'born in 1929' when he was an 'orthopaedic physician' at St Thomas's Hospital. See also, Norman Capener, 'Forty Years on', *JBJS*, 40B (1958), pp. 615–17.

72. Platt, 'The Early Mechanical Treatment of Acute Anterior Poliomyelitis', *BMJ*, 16 Feb. 1924, p. 266. D. McCrae Aitken, one of Robert Jones's pupils in Liverpool, expressed the conventional view of British orthopaedic surgeons when he wrote in 1936: 'No surgeon can become really great unless he be a sound physician although many a mechanically-minded person can by practice become an efficient operator': 'Hugh Owen Thomas' in D'Arcy Power (ed.), *British Masters of Medicine* (1936), p. 161.

73. Osler, 'The Problem of the Crippled', *Recalled to Life*, 2 (1917), pp. 265–6. Osler had been appointed to the Philadelphia Orthopaedic Hospital in 1887, although its (and his) concerns at that time were mainly with neurology.

74. Robert Osgood, presidential address to the AOA, 1921, *JBJS*, 3 (1921), p. 265.

75. For example, Kellogg Speed (1879–1955), one of the most prolific authors on orthopaedics in the USA and a principal figure in the establishment of fracture services in the 1930s. See *JBJS*, 38A (1956), p. 245.

76. D. P. Willard, *JBJS*, 17 (1935), p. 533.

77. Jones, cited in Watson, *Life of Jones*, p. 274.

78. See, for example, AOA presidential addresses by Fred Albee, *JBJS*, 11 (1929), p. 698; Willis Campbell, *JBJS*, 13 (1931), p. 421; and Arthur Steindler, *JBJS*, 15 (1933), p. 568. And see C. L. Lowman, 'Preventive and Prophylactic Orthopaedic Practice', *JOS*, 3 (1921), pp. 576–83.

79. Girdlestone once remarked to Jones, 'I pinned my faith to my emphasis on the conservative treatment as the one and only safeguard for *life* and the foundation of the treatment of limbs': Girdlestone to Jones, 31 July 1929. Girdlestone Papers. See also Jones, 'Rest for Acute Poliomyelitis', *BMJ*, 20 Nov. 1926, pp. 947–8, 1018–19, and 11 Dec. 1926, pp. 1142–3. As late as 1950 it was said against Girdlestone's anti-operative stance: '[he] remains contented to condemn an innocent child to five years' imprisonment with immobility for developing skeletal tuberculosis': G. Perkins review of Platt (ed.), *Modern Trends in Orthopaedics*, in *JBJS*, 32B (1950), p. 444. Platt's statement was delivered at the annual dinner of the BOA in 1948. See *JBJS*, 31B (1949), p. 132.

80. Cited in the obituary of W. A. Cochrane, *JBJS*, 27 (1945), pp. 524–5. A student of Stiles, Cochrane also worked in Boston with Goldthwait and Osgood.

81. Beveridge, *Voluntary Action: a report on methods of social advance* (1948), p. 243.

82. Cited in ibid., p. 254. Osgood quoted 80,000 as the number of crippled children in England and Wales in 1927: 'Progress', p. 221. Kessler, *Crippled and Disabled* (New York, 1935), p. 39, cites estimates of 100 000 adult and child cripples in England in 1935, which was also the figure provided by Jones in his address to the Essex County Council in 1926. Quoted in *JAMA*, 7 Aug. 1926, p. 424.

83. Elmslie, 'A Survey of Physically Defective Children', Report of the School Medical Officer to the LCC, No. 2201 (1923), quoted in Board of Education, *Annual Report of the CMO for 1922*, p. 99.

84. Jones, 'Future Developments', paper at 'Joint Conference, 1926', p. 192.

85. Osgood, 'Progress', pp. 227–8. See also Clement Smith, *Children's Hospital of Boston* (Boston, 1983), pp. 189ff.

86. In 1947 there were some 7207 confirmed cases and 8752 suspected cases of poliomyelitis in Britain, over 3000 cases being in the north-west of England: Platt, 'The Care of the Physically Handicapped Child', *Manchester Univ. Med. Gazette*, 45 (1964), pp. 4–11 at p. 6. See also *Reports to the Local Government Board on Public Health and Medical Subjects: further reports and papers on epidemic poliomyelitis* (HMSO, 1918); and Trueta, *Girdlestone*, p. 84. Severe epidemics in America in 1910–13 and 1916 permitted orthopaedists such as Robert Lovett (who headed the Harvard Infantile Paralysis Commission Clinic) to become deeply involved with the disease and its treatment. See John R. Paul, *A History of Poliomyelitis* (New Haven, 1971).

87. Gauvain, 'Anterior Poliomyelitis in England', *Intern. Bull. for Economics, Medical Research and Public Hygiene*, A40 (1939), pp. 82–94 at pp. 85–6. See also, 'Report of the B.O.A. on Poliomyelitis, an orthopaedic problem', *JBJS*, 36B (1954), pp. 666–7.

88. See S. J. Cowell, 'The Prevention of Rickets', paper to 'Joint Conference, 1926', pp. 211–15. See also Edward Mellanby, 'The Cause of Rickets as Revealed by Experiment', *Cripples' J.*, 3 (1926), pp. 29–35. For background on the debate over the causation of rickets, see A.J. Ihde and S.L. Becker, 'Conflict of Concepts in Early Vitamin Studies', *J. Hist. Biology*, 4 (1971), pp. 1–33.

89. Celia Petty, 'Primary Research and Public Health: the prioritization of nutrition research in inter-war Britain', in J. Austoker and L. Bryder (eds), *Historical Perspectives on the Role of the MRC* (Oxford, 1989), pp. 59–82. See also David Smith and Malcolm Nicolson, 'The "Glasgow School" of Paton, Findlay and Cathcart: conservative thought in chemical physiology, nutrition and public health', *Social Stud. Sci.*, 19 (1989), pp. 195–238, and Naomi Aronson, 'Nutrition as a Social Problem: a case study of entrepreneurial strategy in science', *Social Policy*, 29 (1982), pp. 474–87.

90. In fact, these concerns had already been marginalised in the debate over the causation of rickets in the 1910s and 1920s. See Smith and Nicolson, 'The "Glasgow School"', pp. 200ff.

91. See, for example, the reports contained in 'The Orthopaedic Service', in 'Health of the School Child', *Annual Report of the CMO for 1935*, pp. 90–101.

92. *Cripples' J.*, 4 (1928), p. 8.

93. Cited in Elmslie, 'The British Empire at the Present Time', paper at 'Joint Conference, 1926', p. 176, and in the speech by the Duchess of Atholl at the Joint Conference, p. 208.

94. Political and Economic Planning, *Report on the British Health Services* (1937), p. 300; Newman, *Nation's Health*, pp. 227–8; and 'Health of the School Child', *Annual Report for 1935*, p. 90.

95. Dr Lititia Fairfield, Assistant MO, LCC, at the 'Conference on Education in Hospital Schools for Physically Defective Children', *Cripples' J.*, 1 (1925), p. 289. See also Arthur Black, 'Provision for Crippled Children in England and Wales', *Cripples' J.*, 4 (1927–28), p. 32.

96. Anon., *British Red Cross Scheme for the Provision of Central and Auxiliary Clinics for the Treatment of Rheumatism (Adults)* (c. 1930), n.p. I am grateful to David Cantor for this reference.

97. E. Hey Groves, 'Should Medicine be a Mendicant?', *Lancet*, 17 May 1930, p. 1106. For this same problem in the tuberculosis movement, see Linda Bryder, *Below the Magic Mountain: a social history of tuberculosis in twentieth century Britain* (Oxford, 1988), pp. 81ff.

98. Elmslie, 'British Empire', p. 177.

99. Girdlestone to Jones, 9 July 1926, Girdlestone Papers.

100. Dr Greenwood (MOH, Kent) at 'Joint Conference, 1926, pp. 167–7. For an American comparison, see Thomas N. Bonner, *The Kansas Doctor: a century of pioneering* (Lawrence, 1959), p. 86.

101. Gauvain, 'Anterior Poliomyelitis', p. 86.

102. Gauvain and Holmes, 'Hospital Schools', p. 839.

103. The proposal had come from Sir Arthur Stanley. See the interview with Newman of 5 March 1920: PRO:ED/50/153; and, in the same file, the letter from Ursula Townsend to Dr. Alfred Eichholz, 29 June 1920.

104. See Pugh, 'Orthopaedics at a Country Children's Hospital, President's Address', *Proc. Roy. Soc. Med.*, 20, pt 1 (1926–7), pp. 131–6; and G. Ayers, *England's First State Hospitals, 1867–1930* (1971), p. 210. On Pugh see, P. D. G. Pugh, *Pugh of Carshalton . . . Medical Superintendent of Queen Mary's Hospital for Children, Carshalton, Surrey, England, from 1909 until 1937* (privately printed, 1973).

105. Elmslie, 'British Empire', p. 180, complained in particular of the situation in Scotland where, with no central organization for the welfare of cripples, 'the large general hospitals have not . . . devoted many beds or much out-patient accommodation to orthopaedic work'.

106. See letters between Jones and Arthur Morely, Secretary of the RNOH, 10–24 October 1927, in the Girdlestone Papers, and John Cholmeley, *History of the Royal National Orthopaedic Hospital* (1985), pp. 118–20.

107. W. S. Baer, *JBJS*, 6 (1924), pp. 503–9 at p. 508.

9 The Fracture Movement

1. Adolf Lorenz, *My Life and Work* (New York, 1936), pp. 335–6.
2. William H. Ogilvie, 'Orthodoxy and Heterodoxy in Surgery', *Lancet*, 25 Apr. 1931, pp. 897–902.
3. 'British Orthopaedic Association: first founders' lecture', *JBJS*, 41B (1959), pp. 231–6, reprinted in *Selected Papers* (Edinburgh, 1963), pp. 116–25 at p. 124.
4. In 1935 the orthopaedic unit at St James's received no fewer than 735 inpatients for fracture treatment. Sinclair was the consulting orthopaedic specialist and William Gissane (on whom see below) was the senior assistant MO. Gissane later claimed the credit for introducing 'modern methods of fracture treatment to this hospital' in 1934, but this would seem to have consisted mostly of streamlining what Sinclair had already established – essentially 'a miniature of No.8 Stationary Hospital [Wimereux]', consisting of 92 beds in two inpatient wards, an outpatient ward, X-ray services and a small splint room. See 'BMA Report of Committee on Fractures', *BMJ*, suppl., 16 Feb. 1935, pp. 53–62 at p. 57, and the report on 'Orthopaedic After Care, 1932–47', GLRO:PH/HOSP/1/66. On Sinclair and his methods, see Robert Jones, 'Introduction' to Sinclair, *Fractures* (1931), pp. xxxiii–iv; Harvey Cushing, *From a Surgeon's Journal, 1915–1918* (1936), entry for 4 June 1917, p. 113; H. Winnett Orr, *On the Contributions of Thomas, Jones and Ridlon* (Springfield, Ohio, 1949), pp. 149–51; and Arbuthnot Lane, 'The Treatment of Fractures in Warfare', *Lancet*, 5 Jan. 1918, p. 4. Rutherford Morison said of Sinclair: 'He is a Regular Army surgeon and a mechanical genius. In addition to this, he is a unique character and entirely self-centred. He never reads surgical literature, does not care what anyone says or thinks, and knows that nobody can be right except himself.' Quoted in G. Gordon-Taylor, 'The Rutherford Morison Tradition', *Newcastle Med. J.*, 24 (1954), pp. 248–59 at pp. 254–5. Although he regarded himself as professionally indebted to Jones, Sinclair did not become a member of the BOA until 1935.
5. Biographical information on Platt comes from the appreciations by H. Osmond-Clarke and others in Platt's birthday volume of the *JBJS*, 48B (1966), pp. 613–22; various incidental lectures and addresses by Platt (mostly unpublished); and several interviews conducted by myself and others between 1974 and 1985. See also the introduction to Stella Butler, 'A Handlist for the Papers of Sir Harry Platt', typescript, 1984. On the RNOH see J.A. Cholmeley, *The History of the Royal National Orthopaedic Hospital* (1985); on St Peter's, see A. Clifford Morson (ed.), *St. Peter's Hospital for Stone, 1860–1960* (Edinburgh, 1960); and on St Mark's, see Lindsay Granshaw, *St. Mark's Hospital, London: the social history of a specialist hospital* (1985).
6. Platt aspired and eventually succeeded in following Robert Jones in building up a lucrative private practice in orthopaedics (a possibility for no more than one or two orthopaedists in any major urban centre outside London). His family (manufacturers) were well enough off to

allow him to consider becoming a hospital consultant, but in his early years as a consultant he derived his income from performing the occasional appendectomy, conducting coroners' post-mortems, assisting private surgeons, medical coaching, and by writing medico-legal reports. The financial disincentive to specializing in hospital fracture work was the reason why Platt and his orthopaedic colleagues came to endorse a salaried service for hospital consultants (see below). For a comparative portrait and comments on the shaping of the new concept of the 'pure consultant' around 1918, see the obituary on Ernest Finch in *Br. J. Surg.*, 52 (1965), p. 83.

7. 'Orthopaedic Surgery in Boston', *Med. Chron.*, 58 (March 1914), pp. 473–9.

8. On Barclay and the others, and on the Ancoats Hospital, see John V. Pickstone, *Medicine and Industrial Society* (Manchester, 1985), pp. 145–6, 204–7.

9. The average was 30 new fracture cases a week: Platt, 'On the Organisation of a Fracture Service', *Lancet*, 17 Sept. 1921, pp. 620–1, reprinted in Platt, *Selected Papers*, pp. 1–5.

10. For an example of the traditional routine (at the Liverpool Royal Southern), see William Mayo, 'Present-Day Surgery in England and Scotland, from notes made on a recent short visit', reprinted from *J. Minn: State Med. Assn*, 1 Dec. 1907, p. 4. There, the surgical service was divided into three divisions of from 40 to 50 beds each, and the chief surgeon of each division took all patients admitted on two days each week.

11. See anon., *History of Ancoats Hospital 1873–1900* (Manchester, n.d.), p. 34.

12. *Biographical Memoirs of the Fellows of the Royal Society*, 7 (1961), pp. 127–31; obituary in *Br. J. Surg.*, 48 (1961), pp. 586–8; and Pickstone, *Medicine and Industrial Society*, p. 145. Jefferson, who was a member of the BOA from the early 1920s, opened the discussion on 'Fractures of the Spine' at the orthopaedic meeting at the RSM in 1927: *BMJ*, 17 Dec. 1927, pp. 1152–3. Like Platt, Morley and Barclay, he too succeeded to a consultancy at the Manchester Royal Infirmary.

13. J. L. Thornton, 'Orthopaedic Surgeons at St Bartholomew's Hospital, London', *St Barts Hosp. J.*, 59 (1955), pp. 195–204, at p. 200.

14. See P. Gray, 'Grangethorpe Hospital Rusholme, 1917–1929', *Trans. Lancs. & Ches. Antiquarian Soc.*, 78, (1975), pp. 51–64.

15. Platt, 'On the Organisation of a Fracture Service', *Lancet*, 17 Sept. 1921, pp. 620–1, reprinted in Platt, *Selected Papers*, pp. 1–5.

16. 'BMA Report of Committee on Fractures', *BMJ*, suppl., 16 Feb. 1935: pp. 53–62. Of the 17 members of the BMA Fracture Committee, 10 were members of the BOA; of the others, only Henry S. Souttar (the Chairman), Henry Brackenbury and John Bishop Harman (a Harley Street surgeon) did not have obvious vested interests in the campaign for fracture clinics.

17. *Interim Report of the Inter-Departmental Committee on the Rehabilitation of Persons Injured by Accidents* (HMSO, 1937); *Final Report* (HMSO, 1939). The 21 members of the Delevingne Committee were drawn from

the Ministries of Health, Labour and Pensions, the Department of Health for Scotland, the BMA, the British Hospitals' Association, the Accident Officers' Association, LMS Railways, Midland Colliery Owners' Mutual Indemnity Co. Ltd., Durham Miners' Union, National Union of Railwaymen, and the General Council of the Scottish TUC. BOA interests were represented by W. A. Cochrane and Ernest Hey Groves. For a personal view of Delevingne, see Hilda Martindale, 'Power of Social Reform: Sir Malcolm Delevingne', in her *Some Victorian Portraits and Others* (1948), pp. 31–9.

18. 'Lady Jones Lecture on Crippling Due to Fractures: its prevention and remedy', *BMJ*, 16 May 1925, pp. 909–13.

19. Platt, *Selected Papers*, p. 124; and see Bristow's attack on the London teaching hospitals in his 'The Influence of War Surgery on Treatment of Fractures in Great Britain', *JAMA*, 3 Dec. 1927, pp. 1920–4.

20. George Gask in the 'Discussion on the Treatment of Fractures: with special reference to its organization and teaching', *BMJ*, 22 Aug. 1925, pp. 317–31, at p. 317.

21. 'Discussion on the Treatment of Fractures: with special reference to its organization and teaching', *BMJ*, 22 Aug. 1925, pp. 317–31.

22. Among the Americans present was Robert Osgood, who in 1921 organized a two-day conference on the treatment of fractures at the Massachusetts General Hospital, which brought together over 50 general surgeons and orthopaedists (see obituary of Osgood, *JBJS*, 39A (1957), pp. 726–33). From the Orthopaedic Service of the Mayo Clinic (est. 1912) came Melvin Henderson; from the Montreal General Hospital (where a fracture service had been established in 1919) came A. T. Bazin.

The discussion at Bath appears to have prompted the questionnaire of the American orthopaedist, John Prentiss Lord, in which evidence was sought for the merits of hospital fracture services in the hands of orthopaedists. Lord's highly favourable findings constituted the basis of his address to the section on orthopedic surgery of the AMA in 1927 which, in turn, was the basis for the editorial on fracture treatment in the *BMJ*. See Lord, 'Factors in the Advancement of Orthopaedic Surgery', *JAMA*, 27 Aug. 1927, pp. 651–4; and 'The Treatment of Fractures', *BMJ*, 15 Oct. 1927, p. 695.

23. For the notion made explicit, see Melvin Henderson, 'Leadership in Orthopaedic Surgery', *JBJS*, 16 (1934), pp. 495–98.

24. Gask, 'Discussion on Fractures' at Bath, p. 318. Gask's point of reference was to (Frederick Marsh's close colleague) Sir T. Clifford Allbutt's *The Historical Relations of Medicine and Surgery* (1905). Although he was probably partly responsible for St Bartholomew's Hospital not having a fully segregated fracture service until after the Second World War, in the 1930s he took his staff to visit Reginald Watson-Jones's orthopaedic department at the Liverpool Royal Infirmary, which was the home of the country's foremost fracture service. See Geoffrey Keynes, *The Gates of Memory* (Oxford, 1983), p. 264 (Keynes was Gask's chief assistant). See also the entry on Gask in the *DNB*.

25. W. McAdam Eccles, in the 'Discussion on Fractures' at Bath, p. 329.
H. H. Sampson made the same claim, ibid., p. 326.
26. Platt in the 'Discussion on Fractures' at Bath, p. 325.
27. Notably by H. E. Moore in his influential 'Avoidable Wastage in
Connexion with Industrial Injuries', *Liverpool Medico-Chirurg. J.*, 41
(1933), pp. 19–50. For discussion on Moore's work, see below Ch. 10.
28. Cf. Lever Stewart, 'The Industrial Surgeon's Treatment of Fractures', *J.
Indust. Hygiene*, 8 (1926), pp. 283–7; R. N. Gray, 'Disability and Cost of
Industrial Fractures: a comparison based upon an impersonal study of
statistics of fractures treated by the specially trained surgeon and the
general practitioner', paper delivered to the AOA meeting, June 1927,
JBJS, 10 (1928), pp. 27–39; idem, 'Rehabilitation after Industrial
Fractures: a plea for care, caution and further scientific study in the
use of physical therapy', *Rehab. Rev.*, 1 (1927), pp. 285–8; W. A. Rogers,
'End Results and the Follow-up in Orthopaedic Surgery', *JBJS*, 10
(1928), pp. 104–7. See also the comments in Thomas N. Bonner, *The
Kansas Doctor* (Lawrence, 1959), pp. 226ff.
29. Most orthopaedists followed Jones in believing that 'we must reach the
practitioner because we have to trust him for an early diagnosis';
moreover, 'the prevention of surgical tragedies is largely dependent
upon [him]': Jones, 'The Necessity of Orthopaedic Training', *BMJ*, 5
Feb. 1921, p. 181. Jones always maintained 'a close and confidential co-
allegiance with the family doctor', according to his biographer. '"He
never let the general practitioner down" appeared above the names of a
body of North Wales doctors when [Jones] died.' Frederick Watson, *Life
of Sir Robert Jones* (1934), p. 263.
30. See, for example, C. Max Page and W. Rowley Bristow, *The Treatment
of Fractures in General Practice* (1923); and William Heneage Ogilvie,
Treatment of Fractures in General Practice, 2 vols. (1932). John Hosford
(a general surgeon at St Bartholomew's Hospital) began the preface to
his *Fractures and Dislocations in General Practice* (1939) with the words:
'I can answer the question why has yet another book on "Fractures"
been written by saying that it is in response to the request of a number of
students and recently qualified men.'
31. See editorials in *Practitioner*, 137 (1936), pp. 402–3, and *Med. Officer*, 26
June 1937, p. 255. The *Lancet* in its editorial on 'The Fracture Problem'
rightly criticized the BMA's Report on Fractures for completely
avoiding the question of 'the function of the GP in his duty to his
middle-class patients [sustaining fractures]': 16 Feb. 1935, pp. 383–4. The
criticism was never dealt with.

In America, as late as 1940, as many as 95 per cent of compound
fractures continued to be treated by GPs: *JAMA*, 30 Nov. 1940, p. 1855.
However, the establishment of hospital fracture clinics under orthopaedic control appears to have been less problematic than in Britain because
orthopaedic surgeons had already secured places in hospitals and
university medical schools. Many of the American recruits who worked
with Jones during the war engaged in hospital fracture work upon their
return to the United States. Among these were Fraser B. Gurd of
Montreal, and Murray Danforth (who became Chief of the Fracture

Service at the Rhode Island Hospital upon its inception in 1931). See *JBJS*, 15 (1933), p. 327 and 25 (1943), pp. 702–3.

32. Anne Digby and Nick Bosanquet, 'Doctors and Patients in an Era of National Health Insurance and Private Practice, 1913–1938', *Econ. Hist. Rev.*, 41 (1988), pp. 74–94 at p. 91. Attention was paid to the legal risks in the primers cited above.

33. See Frank Honigsbaum, *The Division in British Medicine* (1979), pp. 146–8 *et passim*. Since 1923 there had been a BMA Orthopaedic Section (of which Openshaw was the first president) which was intended to interest GPs *and* specialists in orthopaedic matters.

34. Wilson and Parker, 'Cardiff Corporation Accident Service': 'The private practitioner is notified of the diagnosis and is acquainted with the treatment that is being carried out. It is found that the practitioners make increasing use of the fracture clinic and readily refer their patients for treatment and after-care' (pp. 35–6).

35. Delevingne, *Final Report*, p. 50.

36. 'The Treatment of Fractures', *BMJ*, 15 Oct. 1927, p. 695.

37. Jones's 'Discussion at Bath', pp. 319–22; Hey Groves, 'The Treatment of Fractures: a problem of organization', *BMJ*, 1 Dec. 1928, pp. 993–5.

38. See BMA, 'Report of the Committee on Treatment of Simple Fractures', *BMJ*, 30 Nov. 1912, pp. 1505–41.

39. The more or less modern use of X-rays in fracture treatment came about after members of the BOA visited H. Waldenström's orthopaedic clinic in Stockholm in 1936; see Rowley Bristow, 'Orthopaedic Surgery – Retrospect and Forecast', *Lancet*, 6 Nov. 1937, pp. 1061–4 at p. 1063. In Cardiff, in 1937, X-rays were still only used after reduction of fractures, and not for primary diagnosis; see J. Greenwood Wilson (MOH, Cardiff) and A. O. Parker (orthopaedic surgeon), 'The Cardiff Corporation Accident Service', *Med. Officer*, 16 and 23 Jan. 1937, pp. 25–27 and 35–7. Cf. Joel Howell, 'Early Use of X-ray machines and electrocardiographs at the Pennsylvania Hospital', *JAMA*, 255 (1986), pp. 2320–3.

40. *Voluntary Hospitals Committee, Final Report*, 1928. Cmd 1335. See also John Pater, *The Making of the National Health Service* (1981), pp. 4–13.

41. See A. Marwick, 'The Labour Party and the Welfare State in Britain, 1900–1948', *Amer. Hist. Rev.*, 73 (1967–8), pp. 380–403, esp. at pp. 386–90.

42. *Interim Report of the Future Provision of Medical and Allied Services* (HMSO, 1920), para. 43. This was not unlike the use of ambulances envisioned in late nineteenth-century plans (such as Burdett's) for linking suburbanites to central hospital facilities. See above, Ch. 5.

43. *The Labour Movement and the Hospital Crisis* (TUC and Labour Party, 1922), p. 7. See also 'The Labour Party and the Hospital Problem, Conference at Caxton Hall', *BMJ*, suppl., 3 May 1924, pp. 213–22; and *The Labour Movement and Preventive & Curative Medical Services: a statement of policy with regard to health* (TUC and Labour Party, (1922)), esp. p. 6: 'Public hospitals when established should become the health centre or institution of each local health authority, and should provide accommodation within their walls for all medical activities.'

44. *The Labour Movement and the Hospital Crisis*, p. 11.
45. King Edward's Hospital Fund for London, *Ambulance Case Disposal Committee: Report of a Special Committee*, 1924. For discussion of this and other literature relating to accident and emergency services in the interwar period, see K. S. Cliff, 'The Development and Organisation of Accident and Emergency Services', DM thesis, Southampton University, 1981, pp. 38–76.
46. Directly as a result of the King's Fund *Report*, and in order to secure further monies from the Fund, King's College Hospital, London, immediately added 14 accident beds: *BMJ*, 13 Sept. 1924, p. 483.
47. The latter system became widely known in 1925 through a joint British and American publication: P. D. Wilson (Harvard) and W. A. Cochrane (Edinburgh), *Fractures and Dislocations* (Philadelphia/London, 1925), which was reviewed in the *BMJ*, 16 June 1925, pp. 928–9. For background to the fracture service in Boston, see 'Discussion on Fracture Symposium', *JOS*, 3 (1921), pp. 556–9 (Jones was present at this meeting). See also *Experience in the Management of Fractures and Dislocations*, by the staff of the fracture service, Massachusetts General Hospital, Boston, under the general editorship of P. D. Wilson (Philadelphia, 1938).
48. King's Fund *Report*, p. 35.
49. 'Accidents and Hospitals', *Lancet*, 23 Oct. 1926, p. 864. See also *Lancet*, 26 Feb. 1927, p. 463; 'Medical Practitioners and Road Accidents', *BMJ*, suppl., 25 July 1931, pp. 62–5; and 'Emergency Treatment for Road Accidents', *BMJ*, 4 Aug. 1934, pp. 213–14. For the implications of motor accidents for cottage hospitals, see *Lancet*, 27 June 1931, p. 1410.
50. Quoted in W. Plowden, *The Motor Car and Politics in Britain 1896–1970* (1973), p. 271. See also his Appendix D, 'Road Casualties, 1928–69', p. 483.
51. Of 19 286 fractures cases in 1937, 14.9 per cent were the result of road traffic accidents: Delevingne, *Interim Report*, p. 8. An LCC inquiry into 1068 fracture cases treated at four of their hospitals during summer and winter sample periods in 1936 and 1937 revealed that industry was responsible for 13.5 per cent (145), road traffic for 22.9 per cent (245), while 63.4 per cent (678) were the result of other causes. GLRO:PH/HOSP/1/72. In Manchester, where there were about 5000 fracture cases annually in the early 1930s, 70 per cent were domestic, 18 per cent street accidents, and 12 per cent industrial: Pickstone, *Medicine and Industrial Society*, p. 287.
52. Plowden, *Motor Car*, p. 276; and 'Accidents and Hospitals', *Lancet*, p. 864. For statistics on the victims treated and the costs recovered in Manchester and Salford in 1932, see Pickstone, *Medicine and Industrial Society*, p. 273.
53. The Road Traffic Act provided for payment up to a maximum of £5 for outpatient cases and £50 for inpatient care. A fee of 12s 6d per patient was payable by the user of the car to the doctor or hospital who first attended the case. Delevingne, *Final Report*, pp. 102–3. See also 'Emergency Treatment for Road Accidents', *BMJ*, 4 Aug. 1934, pp. 213–14.

54. See C. V. Mackay, 'Dr. Böhler's Fracture Clinic in Vienna', *BMJ*, 19 Apr. 1935, p. 522; Platt, 'Orthopaedics in Continental Europe, 1900–1950', in his *Selected Papers*, pp. 65–88 at pp. 84–5; Friedrich Lorenz, *Lorenz Böhler: der Vater der Unfallchirurgie* (Vienna, 1955); and the BMA 'Report on Fractures', 1935, pp. 57–8.

55. *Lancet*, 16 Feb. 1935: p. 383. In 1931 Böhler was invited to address the annual meeting of the BOA (*JBJS*, 13 (1931), pp. 382–3), and in 1934 he was one of the three persons considered for the delivery of the Hugh Owen Thomas Lecture in Orthopaedic Surgery (Minutes of the Faculty's Board of Orthopaedic Studies, 15 June 1933, University of Liverpool Archives, 53389).

56. 'The Log of Vienna, 26–29 September 1929', MS typescript, BOA archives, Royal College of Surgeons. According to Platt, T. Porter McMurray, one of Jones's successors in Liverpool, continued to use splints rather than plaster in the 1930s, because he had 'never . . . visited Böhler's clinic': Platt, interview with the author, 7 Nov. 1984. However, there were also important commercial interests behind the promotion of plaster of Paris; see Richard Bennett and J. A. Leavey, 'Gypsona and the Fracture Clinics', in their *A History of Smith & Nephew, 1856–1981* (1981), pp. 19–22. Indeed, T. J. Smith of Smith & Nephew funded the fracture clinic at the Hull Royal Infirmary: Platt correspondence, 14 June 1941, EMS files, University of Manchester Library. Then, of course, there was the movie, *Plastered in Paris* (USA, 1928), directed by Ben Stoloff for Twentieth-Century Fox.

57. See E. M. Bick, *Source Book of Orthopaedic Surgery* (Baltimore, 1948), pp. 293–4; W. H. Ogilvie, 'Physiology and the Surgeon', *Edin. Med. J.*, 43 (1936), reprinted in his *Surgery: orthodox and heterodox* (Oxford, 1948), pp. 147–67 at p. 152; W.J. Eastwood, 'Orthopaedics: old and new', *Liverpool Medico-Chirurg.J.*, 45 (1937), pp. 186–94 at pp. 190–1; and Platt, 'The Evolution of the Treatment of Fractures', *Manchester Univ. Med. School Gaz.*, 17 (1938), pp. 56–62 at p. 59. Böhler's work on fractures was still widely consulted in the 1950s.

58. Böhler, *The Treatment of Fractures*, 4th English edn, translated by Ernest Hey Groves from the 4th enlarged and revised German edn of 1933 (Bristol, 1935), pp. 14ff, and Jim Fyrth, *The Signal was Spain: the aid Spain movement in Britain, 1936–39* (1986), pp. 147–9, which discusses the take up of Böhler's methods by A. Tudor Hart and Douglas Jolly for use in the Spanish Civil War.

59. 'Accidents and Hospitals', *Lancet*, 23 Oct. 1926, p. 864.

60. 'The Organization of the Treatment of Fractures', *BMJ*, 20 Apr. 1935, p. 817.

61. Böhler, The Treatment of Fractures, p. i.

62. Delevingne, *Final Report*, 1939, pp. 14ff. See also, for 1935, statistics on the incidence of fractures at 33 LCC hospitals: GLRO:PH/HOSP/1/72. H. A. T. Fairbank in 1944, on evidence from H. E. Griffiths of the Albert Dock Hospital, London, reported that fractures constituted only 4 per cent of all industrial accidents and that, 'in 1938, there were 18 269 fractures due to industrial accidents – about 10% of the total fractures in the country': 'Rehabilitation of the Injured in this War and the Last',

Lancet, 29 July 1944, pp. 131–4 at p. 132. William Gissane reported in 1952 that in Birmingham 'the treatment of fractures represents no more than 30 per cent of the surgical problems involved in accidents': *JBJS*, 34B (1952), p. 336.

63. T. P. McMurray in the discussion on the paper by H. E. Moore, 'Avoidable Wastage', p. 38. See also in this discussion Bristow's reaction to separate accident hospitals as called for by Moore, p. 36.

64. As orthopaedic professor J. Trueta reflected on the situation in Oxford around 1940, when an accident service was being proposed, the creation of a hospital department for fractures only was 'inadequate – too many cases would escape the orthopaedic surgeon's care'. He therefore favoured the creation of a complete accident service, under the care of an orthopaedic surgeon. Trueta, *Trueta: surgeon in war and peace* (1980), p. 142.

65. See GLRO:PHD/HOSP/1/72. Groves reported to the LCC on the fracture problem after studying 7 LCC hospitals in 1936 for material for his lectures on fracture organization to the Royal Postgraduate Medical School.

66. As reported in *The Times* under the heading 'Cooperation in First Aid', 1 June 1935, p. 11.

67. Gissane to Dr William Brander of the LCC, 31 July 1936: GLRO:PH/HOSP/4/27, and Gissane to H. W. Bruce (Principal MO to the LCC), 30 Oct. 1938. 'I do not say that Gissane is wrong', Bruce stated in a note attached to the latter, 'but I feel that he is trying to rush us for personal reasons; ... he is probably a good surgeon, but there is also some evidence that he is ... an uncertain teacher ... [his] proposal could get him out of his present position into an anomalous one and give him, possibly, a [?] claim to one of the big jobs'. GLRO:PH/HOSP/4/33. On the character and career of Gissane (1898–1981) see his obituary in *JBJS*, 63B (1981), pp. 623–4.

68. Although statistics supplied by the Ministry of Labour to the Ministry of Health in 1940 (PRO:MH 76/322) showed that Birmingham had the highest accident rate in the country, the creation of the Birmingham Accident Hospital and Rehabilitation Centre in the premises of the former Queen's General Hospital was largely the fortuitous consequence of the latter's vacancy and availability for conversion under the Emergency Medical Service. Local industrialists interested in wartime manpower shortages played a part in its foundation. See Ministry of Health, *Hospital Survey: West Midlands* (HMSO, 1945), p. 15; P. Clarkson, 'Out-Patient Arrangements and Accident Services', *Guy's Hosp. Gaz.*, 62 (1948), pp. 202–11 at pp. 208–9; *JBJS*, 63B (1981), p. 624; and Alan Ruscoe Clarke *et al.*, 'Organisation of Accident Services', in Clarke *et al.* (eds), *Modern Trends in Accident Surgery and Medicine* (1959), pp. 1–8.

   The Cardiff Corporation Accident Service, established in 1935 – the first such service under municipal control – was also something of an historical accident in that its creation was due to the fact that the Poor Law Hospital at City Lodge 'had been thrown on the scrap heap and was no longer being usefully employed'. However, this service was mainly for fractures: 'the aim is to attract as many accident cases as possible to City Lodge fracture unit [and orthopaedic clinic] in the first instance': Wilson and Parker, 'Cardiff Corporation Accident Service', p. 25.

69. One of the first to thus endorse separate accident centres was W. McAdam Eccles, colleague of Gask and surgeon to the Orthopaedic Department at St Bartholomew's Hospital (1903–12) before becoming a consulting surgeon specializing in fractures and a member of BMA Fracture Committee. He advocated establishing a Böhler-like industrial accident clinic at the Royal Postgraduate Medical School at Hammersmith. See discussion in Donald C. Norris (Principal MOH to the Bank of England), Presidential Address to the Hunterian Society, 'Some Medical Problems in Accident Insurance', *Trans. Hunter. Soc.*, 2 (1937–8), pp. 10–36, at p. 32. Hey Groves, shortly before he died, also came to endorse the idea of a great accident hospital for London. See *JBJS*, 30B (1948), p. 3.

70. See BOA, *Memorandum on Fracture and Accident Services Committee* (1943); and below, Ch. 11.

71. Platt, 'Orthopaedics in Europe', p. 85, my emphasis. See also, *idem.*, 'The Organisation of Orthopaedic Services in a Large City', *Med. Officer*, 14 Jan. 1939, pp. 15–16; and the discussion on the organization of orthopaedic provision in the north east of England in *Br. J. Indust. Med.*, 1 (1944), pp. 143–4.

72. See 'Appendix: Non-Medical Factors of Prolonged Disability', pp. 60–2.

73. Ibid., p. 60.

74. Hey Groves, 'A Surgical Adventure: an autobiographical sketch', reprinted from *Bristol Med.-Chirurg. J.*, 50, (1933), p. 22.

75. See J. C. Nicholson, 'Fracture of the Neck of the Femur: a personal experience', *BMJ*, 27 Aug. 1938, pp. 464–6, and Hey Groves comments on this article, *BMJ*, 17 Sept. 1938, pp. 633–4.

76. Hey Groves, 'Should Medicine be a Mendicant? A review of our hospital service', *Lancet*, 10 May 1930, pp. 1107. See also the argument for state-funded orthopaedic institutions in Ireland managed by private societies and under the control of orthopaedists: W. C. Somerville-Large, 'Study of a National Orthopaedic System', *Irish J.Med.Sci.*, 6th ser. (1937), pp. 161–72, and idem., 'The Orthopaedic Problem in Ireland', ibid., 6th ser. (1935), pp. 82–8.

77. Hey Groves, 'Should Medicine be a Mendicant?', pp. 1106, 1051. Among other consultants who called for state aid without state control at this time was Henry S. Souttar, surgeon to the London Hospital, and subsequently Chairman of the BMA Fracture Committee. See Arthur Newsholme, *Medicine and the State* (1932), pp. 48–9.

78. James Russell to Professor R. C. Alexander of Dundee, letter of 26 Dec. 1944, Dundee University Library Archives, MS 16/17 (4). I am grateful to David Cantor for drawing this correspondence to my attention.

79. Meeting held at County Hall, 16 June 1938: GLRO:PH/HOSP/4/25.

80. By 1939 there was, as Sir Frederic Menzies, the head of the Medical Department of the LCC, stated, 'a growing demand for remuneration of the 'Honorary' medical staffs': PRO:MH/80/24. In general, the more pay-patients there were in public hospitals, the greater was the demand

by doctors and consultants for payment, and on these grounds, in 1935, both the BMA and the Liverpool Hospitals' Commission approved the payment of hospital consultants. See Political and Economic Planning, *Report on The British Health Services* (1937), pp. 240–1.

Since 1933 the LCC had approved the appointment to its hospitals of certain medical and surgical specialists. However, it was reported in 1939 that 'very little further progress in this direction has . . . been possible, owing to the enormous amount of work involved in the reorganisation of the hospitals. Indeed, the only case in which action has been taken to implement the policy was the appointment of a surgeon specialist at Lambeth hospital': 'Extract of the Report of the Hospital and Medical Service Committee on LCC Hospitals Division. Agenda for 14 March 1939', GLRO:PH/HOSP/4/28.

81. Voluntary Hospitals' Committee for London, *Organised Fracture Services for London: Report by Fracture Sub-Committee* (June 1939), p. 6.
82. Honigsbaum, *Division*, p. 240.
83. Delevingne, *Interim Report*, 1937, p. 7; and *Final Report*, 1939, p. 26. The Minister of Health, Walter Elliot, in reply to a parliamentary question, announced on 26 Apr. 1939 that 77 fracture clinics had been established 'substantially on the lines advocated in the Interim Report of the Committee on the Rehabilitation of Injured Persons'. He provided a regional breakdown subdivided according to provision by local authorities and voluntary hospitals. Only seven fracture clinics had been provided by local authorities (three of which were in Lancashire). Lancashire also topped the voluntary hospital list with 14 fracture clinics in 70 hospitals: Hansard, *House of Commons* vol. 346, 26 Apr. 1939, pp. 1172–3.

St Bartholomew's Hospital had a fracture service from 1927, but the inpatient treatment of fractures was shared by general surgeons, with only special cases being referred to the Orthopaedic Department. At St Thomas's and the Westminster hospitals a similar division of labour prevailed in the 1930s.

84. Delevingne, *Final Report*, p. 76.
85. 'Report of the Hospitals and Medical Services Committee on LCC General Hospitals Division, Agenda for 14 March 1939': GLRO:PH/HOSP/4/28.
86. See discussion on J. L. Smyth's memorandum on 'Fracture Treatment and Rehabilitation': PRO:LAB/14/425.
87. See Arnold Wilson and Hermann Levy, *Workmen's Compensation*, vol. 2 (Oxford, 1941), p. 203.
88. J. Cohen, 'Orthopaedics,' in J. Walton, P. B. Brown and R. B. Scott (eds), *Oxford Companion to Medicine* (Oxford, 1986), p. 959.
89. Delevingne, *Final Report*, 1939, pp. 4, 23–5, 121. The Committee took it as given that the principles of fracture organization as laid down in the BMA Report 'were accepted by the Government Departments concerned'.
90. Delevingne, careful always to avoid controversy and dissent in his committee, was well aware of 'the difficulty which existed as to whether an orthopaedic surgeon or a general surgeon should undertake the

treatment of fractures'. At a meeting with Sir Frederic Menzies, 6 May 1938, he let it be known 'that it was the intention of his committee to use the term "fracture surgeon" only': GLRO:PH/HOSP/4/28.
91. Delevingne, *Interim Report*, 1937, p. 11. See also the editorial in the *BMJ*, 12 Aug.1939, pp. 402–3. It was widely recognized that the word 'honoraria' in this context was a euphemism for 'salaries'. See, for example, *Med. Officer*, 26 June 1937, p. 255.
92. For references and for discussion on both planning bodies, see Pickstone, *Medicine and Industrial Society*, pp. 287ff., 301ff., and Neville M. Goodman, *Wilson Jameson: architect of national health* (1970), p. 133 *et passim*.
93. Examples would include Geoffrey Jefferson (neurosurgery), John Morley (general and abdominal surgery and paediatrics), E.D. Telford (general surgery and orthopaedics); James Spence (paediatrics) and (although without the provincial background) Ernest Rock Carling (a London consultant with a special interest in radiology).

10 Industry and Labour, Part II

1. Quoted by Harry Platt at the opening of the Rehabilitation Department at the Preston Royal Infirmary, March 1946. Typescript in Platt Papers dated 8 March 1946.
2. *Medical Press and Circular*, editorial, 17 Mar. 1943, p. 161.
3. Richard Titmuss, *Problems of Social Policy* (1950), p. 478: 'The year 1941 was a year during which . . . the manpower shortage began to make itself felt and rehabilitation – in a wide sense – became a watchword, the most fashionable word in medicine, covering many ideas and purposes.' Or, as Reginald Watson-Jones stated in 1944 with regard to rehabilitation: 'within a few months of the outbreak of this war, ideals became facts. Every detail proposed to the Delevingne Committee was put into practice.' 'Resettlement, The End of Workmen's Compensation', *Lancet*, 18 Nov. 1944, p. 666.
4. Freidson, 'Foreword' to Glenn Gritzer and Arnold Arluke, *The Making of Rehabilitation* (Berkeley, 1985), p. xx.
5. Jones, 'Foreword' to P. C. Varrier-Jones, 'Village Settlements', *Cripples' J.*; 5, (1928), p. 18. On Papworth, see Linda Bryder, *Below the Magic Mountain* (Oxford, 1988), *et passim*.
6. J. N. Meachen, *A Short History of Tuberculosis* (1936), quoted in E. Brieger, *After-Care and Rehabilitation. Principles and Practice* (Cambridge, International Union Against Tuberculosis. Committee for After-Care and Rehabilitation, (1937)), which sees 'after-care as a social-hygienic and social-economic problem' (p. 49). Arnold Wilson admitted to being greatly influenced by Brieger's publication and to having his notion of rehabilitation shaped by it: *Royal Commission on Workmen's Compensation, Minutes of Evidence* (1939/40), p. 354. See also P. Angrove, 'The Rehabilitation of the Tuberculous', *Rehab. Rev.*, 1 (1927), pp. 55–9.
7. V. Warren Low, 'Workmen's Compensation and the Surgeon', *Lancet*, 23 October, 1926, pp. 849–50.

8. See Sir William Bennett, *The Present Position of the Treatment of Simple Fractures of the Limbs . . . to which is appended a summary of the opinions and practice of about three hundred surgeons* (1900), p. 23, and Arbuthnot Lane's opening address at the BMA Annual Meeting, 'Section on Surgery', *BMJ*, 30 July 1910, p. 1059.

9. *Cripples' J.*, 2 (1926), p. 242. The editorial drew upon the simultaneously published autobiographical account, 'The making of a Cripple' which, together with Agnes Hunt's 'A Plea for the Adult Cripple' in volume 3 (1926), pp. 89–94, was later published separately as *Fag-ends*.

10. *Cripples' J.*, 5 (1929), p. 381. Whitley, 'What America is doing for Her Civil and Industrial Cripples', *Cripples' J.*, 4 (1927/8), pp. 115–22, 193–204 and 311–16.

11. See, for example, D. McCrae Aitken, 'The Injured and Crippled in Relation to Economic Employment and the Application of Insurance', *Cripples' J.*, 5 (1929), pp. 213–18; Dame Agnes Hunt, 'The Training of Cripple Boys and Girls from 14 to 25 Years of Age', *Cripples' J.*, 5 (1929), pp. 151–3.

12. PRO:ED/50/171. The reply from H. Ramsbotham at the Board of Education indicates that they had received from Peto a memo on 'Difficulties at Present Existing in Securing Satisfactory Occupational Training for Victims of Industrial Accidents' prepared by the CCCC in 1930 for submission to the Accident Officer's Association, a body representing the insurance companies which was formed in 1906 to regulate workmen's compensation premiums.

13. Contained in PRO:PIN/12/50. See also his letter to *The Times*, 1 Sept.1937.

14. Bywaters' memorandum of 12 July 1932 contained in PRO:ED/50/171.

15. The latter can be traced to 19 Sept.1935 when Thomas Carnwath (Senior MO) and W. A. McLaughlin of the Ministry of Health conferred with Home Office officials R. R. Bannatyne and G. R. A. Buckland, and R. C. Douglas of the government training centre, with regard to Peto's letter to Major Glinn on the subject of persons suffering from industrial, road and other accidents: see PRO:ED/50/173. On the Leatherhead Training College, see Arnold Wilson and Hermann Levy, *Workmen's Compensation*, vol. 2 (Oxford, 1941), p. 222.

16. Obituary, *Br. J. Indust. Med.*, 5 (1948), p. 236; and the obituary by Reginald Watson-Jones in *JBJS*, 31B (1949), pp. 130–1. On the dinner at the Reform Club and the King's reaction to what was said to be the first ever film on the rehabilitation of the injured, see *JBJS*, 30B (1948), p. 3.

17. The Cunard Papers at the University of Liverpool reveal little on Maitland. He does not appear to have sought a rehabilitation centre for Cunard employees, and he is known to have favoured an accident hospital for Liverpool. There is some evidence that he helped Gissane establish the Birmingham Accident Hospital.

18. The meeting took place on 19 Dec. 1930. A report of the discussion and the submitted documents are contained in 'Orthopaedic After Care, 1932–47': GLRO:PH/HOSP/1/66.

19. A. A. Eagger, *Venture in Industry: the Slough Industrial Health Service, 1947–1963* (1965), p. 22.

20. See 'Injuries in Industry: progress of modern methods,' *The Times*, 19 Mar. 1937, p. 9.

21. In 1942 his nomination for Honorary Membership in the BOA was rejected in favour of his taking out Full Membership: BOA Minutes. 9 July 1942. Griffiths was a member of the BMA's Industrial Health Committee (for whom he drafted a memorandum on 'Suggestions for a Scheme for Co-operation Between the Voluntary Hospitals & Industry' in Dec.1940: PRO:MH 76/157). He was also on the wartime Medical Advisory Committee on Rehabilitation and became a founding member of the British Council for Rehabilitation. In 1948, in the company of Ernest Bevin (then Secretary of State), George Isaacs (Minister of Labour), Aneurin Bevan (Minister of Health), James Griffiths (Minister of National Insurance), George Starauss (Minister of Supply), Hugh Gaitskell (Minister of Fuel and Power), and the presidents of the Royal Colleges, Griffiths delivered the MacKenzie Industrial Health Lecture to the 9th International Congress on Industrial Medicine (*JBJS*, 30B (1948), p. 395). In 1949 he was knighted for his efforts in this field. See obituaries in *BMJ* and *Lancet*, 27 May 1961, pp. 1545 and 1176–7, respectively.

22. In the discussion on Moore's 'Avoidable Wastage in Connection with Industrial Injuries', *Liverpool Medico-Chirurg. J.*, 41 (1933), pp. 19–50 at p. 43 (this was Moore's only publication). That Maitland 'stimulated and encouraged' Moore is claimed in the obituary by Watson-Jones in *JBJS*, 34B (1952), p. 708.

    Moore, one-time house-surgeon to Sir William Thorburn, graduated from Manchester in 1904 and went into general practice. He served in a tented hospital on Cape Helles during the war and afterwards was employed by the Ministry of Pensions in Darlington. He became an Associate Member of the BOA in 1934, at which time he was a member of the BMA Fracture Committee. During the Second World War, on Watson-Jones's recommendation, he was appointed Consultant Adviser on Rehabilitation to the Royal Air Force and Special Consultant in Rehabilitation to the Royal Navy. Information on Moore and the hospital at Crewe is contained in PRO:RAIL/410/1347; 'Crewe Rehabilitation Hospital' in PRO:ED/50/171; and the report of the visit by Dr Bywaters to 'Crewe Railway Hospital' on 4 Mar. 1935 in PRO:ED/50/173. See also the obituary in *BMJ*, 16 Aug. 1952, pp. 397–8.

23. The post entailed treating some 1,200 accident victims a year, examining the 1500 or so annual candidates for company service, acting as a witness in court (for the company) in workmen's compensation cases, as well as teaching first aid and attending to railway accidents in the Crewe district. The hospital itself, Moore reported to company headquarters in 1922, was 'the worst equipped and dirtiest he had ever come across' and required a good 'shaking down'. Moore might not have been quite so critical had he known that highly qualified candidates had been 'tumbling over themselves' for this attractive £950-a-year post, as Lord Knutsford wrote to Sir William Thorburn, the LMS Company's consulting surgeon. Moore seems to have secured the post partly because he had formerly been one of Thorburn's house-surgeons at the Manchester Royal Infirmary, but mainly, because his wife was a fully

qualified nurse (trained at Oswestry) who could therefore serve as the hospital's matron and conveniently share with the MO the tied accommodation. On the origins of the hospital and the candidates for the post, see PRO:RAIL/410/1347 and PRO:ED/50/171.

24. Moore, 'Avoidable Wastage', p. 19.

25. *3rd Annual Report, LMS Railway Hospital*, 31 Dec. 1929, extracts in Moore, 'Observations on the After-Treatment of Industrial Casualties' as reported in GLRO:PH/HOSP/1/66.

26. In 1922 Sir William Thorburn, the consulting surgeon to the company, had suggested to them the possibility of using the Town Hospital, but had also cautioned against it, arguing that the company hospital 'gives much greater freedom to the M.O. in dealing with serious cases and obviates the possibility of difficulties or friction with the staff of the Town Hospital'. Thorburn also pointed out that '"Orthopaedic" treatment, baths, massage, gymnastics, &c. would save much time in recovery from many injuries'. It further occurred to him that 'the new workmen's councils may press for such facilities in the near future': letter dated 20 Jan. 1922 in PRO:RAIL/410/1347. See also GLRO:PH/HOSP/ 1/66 and PRO:RAIL/1007/555.

27. 'Avoidable Wastage', p. 24.

28. 'Orthopaedic After-Care': GLRO:PH/HOSP/1/66.

29. 'Avoidable Wastage', p. 21.

30. I am grateful to Robert Roaf, emeritus Professor of Orthopaedics at Liverpool, for this information (communication 30 Mar. 1990). For career details on Watson-Jones prior to 1927, see his 'Application and Testimonials' for the honorary assistant surgeonship at the LRI, BOA archives. In 1940 he published the textbook *Fractures and Other Bone and Joint Injuries*, which is still being reprinted. He was subsequently the editor of the British volumes of *JBJS* and president of the BOA (see obituary in *BMJ*, 26 Aug. 1972). In 1943 he was appointed director of the Orthopaedic and Accident Department of the London Hospital, and in that same year, in company with a TUC delegation, he was a member (with Rock Carling and G. Gordon-Taylor) of a mission to Russia sponsored by the British Council (see his 'Russian Surgeons and Russian Surgery', *BMJ*, 28 Aug. 1943, p. 276, and Walter Citrine, *Two Careers: a second volume of autobiography* (London, 1967), pp. 167–8, 171). He was regarded as a radical; however, when it came to implementing the NHS, by which time his private practice in London was large and world-famous, he declared: 'We want freedom from medical control, and that freedom demands private practice. I saw the abolition of such freedom in Russia, and it has meant the end of medical progress in that country': 'The Consultant's Vote', *BMJ*, 7 Feb. 1948, pp. 264–7, at p. 266.

31. This was mentioned in Bywaters' report, and is alluded to in Rowley Bristow's comments on H. E. Moore's paper (p.38). Robert Jones had suggested this to the Minister of Health, Neville Chamberlain, in his letter of January 1927, adding that the acquired infirmaries should be linked to the teaching hospitals and universities. Quoted in Watson, *Life of Jones*, pp. 255–7.

32. Charles Hill, the secretary of the BMA, wrote to William Allen Daley

(Chief MO for the LCC) on 18 April 1934, while the BMA Fracture
Committee was sitting: 'You deal with the larger problem of orthopaedic
cases in general and had [sic] in mind particularly the administrative
problems encountered as a result of the variety of agencies at work. The
Fracture Committee dealing with a smaller, a clear-cut problem, is
approaching its problem in a missionary spirit with perhaps a lesser
consideration for administrative problems': GLRO:PH/HOSP/1/72. See
also Wilson and Levy, *Workmen's Compensation*, pp. 213–14.

33. There are fascinating insights into the process of negotiation in
GLRO:PH/HOSP/1/66. See also PRO:ED/50/173.

34. See obituary in *BMJ*, 21 Nov. 1959, p. 1102, and Iain D. Levack and
H. A. F. Dudley (eds), *Aberdeen Royal Infirmary* (1992), pp. 139–40.
Along with Gissane, H. B. W. Morgan and Watson-Jones, he was also
on the BMA's Committee on the Rehabilitation of Disabled Persons
(1945–58).

35. On Nicoll and Berry Hill, see William Waugh, *The Development of
Orthopaedics in the Nottingham Area* (Nottingham, 1988), pp. 160–9.

36. On Sidney Alan Stormer Malkin (1892–1964), who went to Nottingham
in 1923, see H. Platt, 'The Special Orthopaedic Hospital Past and
Future' (the first Malkin lecture), *J. Roy. Col. Surg. Edin.*, 21 (Mar.
1976), pp. 67–74; and Waugh, *Orthopaedics in Nottingham*.

37. Quoted in Waugh, *Orthopaedics in Nottingham*, p. 162.

38. E. A. Nicoll, 'Rehabilitation of Injured Miners', *News Letter of the
Central Council for the Care of Cripples*, no. 4, Oct.1940, page proofs in
PRO:MH/76/157. See also, Nicoll, 'Rehabilitation of the Injured', *BMJ*,
5 Apr. 1941, pp. 501–6. Nicoll told the Ministry of Health in 1941 that
there were about 100 000 men in the Nottingham coalfield and a yearly
accident rate of 18 000: PRO:MH/76/117.

39. 'Miners' Welfare Commission: rehabilitation of injured miners', *JBJS*,
31B (1949), p. 487.

40. Quoted in Waugh, *Orthopaedics in Nottingham*, pp. 163–4. The Centre
did not, however, become a part of the orthopaedic scheme of the EMS,
despite much protest; in 1941 only 10 of its 50 beds were occupied by
miners, the others by maternity cases. See PRO:MH/76/117 and
MH/76/157.

41. Until 1949 there were no official statistics for accidents and diseases
sustained in industry. The figure of £12.5 million comes from Political
and Economic Planning, *British Health Services* (1937), p. 65, where
other Home Office statistics are cited. Mr R. S. Hudson, Parliamentary
Secretary to the Ministry of Health in 1937, stated that 'in 1934 industry
was responsible for nearly 400,000 cases of injury, which cost over
£4,000,000 in compensation, while the hospitals treated no fewer than
1,300,000 accidents of which over 200,000 were cases of fracture': *The
Times*, 19 Mar. 1937, p. 9. Some of the most useful statistics for this
period are contained in John L. Williams, *Accidents and Ill-Health at
Work* (1960).

42. *The Times*, 25 Apr. 1945, paraphrased in A. R. Thompson, 'Recent
Advances in Physical Medicine', in F. Bach (ed.), *Recent Advances in
Physical Medicine* (1950), p. 403.

43. Quoted in Waugh, *Orthopaedics in Nottingham*, p. 163.

44. Minutes of the University of Liverpool, Medical Faculty's Board of Orthopaedic Studies, 'Report of the Sub-Committee on Rehabilitation', 9 Nov. 1937: University of Liverpool Archives, S 3359.

45. Miller, 'Late Rehabilitation of the Injured: a survey of seven years' experience in an industrial clinic', *BMJ*, 22 Aug. 1942, pp. 209–11 (whence all subsequent quotations). See also *Br. J. Indust.Med.*, 3 (1946), p. 54.

46. To a degree, what was observed in America in this respect also applied in Britain: 'patients cling faithfully to their own medical adviser. They often feel that their doctors are thoroughly competent to treat any and all types of injuries, regardless of severity': R. N. Gray, 'Disability and Cost of Industrial Fractures' (read before the AOA in 1927), *JBJS*, 10 (1928), p. 35.

47. 'Rehabilitation, Physiotherapy and Orthopaedics', *J.Chartered Soc. Physiotherapy*, 29 (1944), p. 72; Watson-Jones, 'Fractures and Joint Injuries', quoted in Thomas M. Ling, and C.J.S. O'Malley, *Rehabilitation after Illness and Accident* (1958), pp. 73–4.

48. Ogilvie, 1945, reprinted in *Surgery: orthodox and heterodox* (Oxford, 1948), p. 220. For Watson-Jones on Plato, see his 'Surgical Rehabilitation', *Bull. de l'Académie Suisse des Sciences Médicales*, 2 (1946/7), pp. 235–48 at pp. 244–5; and idem, *Surgery is Destined to the Practice of Medicine: Hunterian Oration* (1959).

49. Cited in Ling and O'Malley, *Rehabilitation*, p. 18.

50. See the comments of Walter Citrine at the BOA meeting at Nottingham in September 1942, reported in *JBJS*, 25 (1943), p. 218. See also D. McCrae Aitken, 'The Injured and Crippled in Relation to Economic Employment and the Application of Insurance', *Cripples' J.*, 5 (1929), p. 217; Helen Jones, 'The Politics of Health at Work in Britain, 1900–1950', PhD thesis, University of London, 1986; and Raymond Munts, *Bargaining for Health: labor unions, health insurance, and medical care* (Madison, 1967), p. vi *et passim*. According to Hermann Levy (*War Effort and Industrial Injuries*, Fabian Society Tract no. 253 (1940), p. 10): 'Lump sums may be popular, in the sense in which large doses of morphia are popular with persons suffering pain; it does not follow in either case that the sufferer's judgment of what is best for him is correct.' This forgets that workers' concerns with wages, hours, and conditions of employment *are* interests in the determinants of health.

51. See Bevin's letter to Sir David Munro (Industrial Health Research Board) of 12 Mar. 1934: Bevin Papers, Churchill College, Cambridge.

52. *JBJS*, 25 (1943), p. 217.

53. On the cuts see Pat Thane, *The Foundations of the Welfare State* (1982), p. 192. British orthopaedists maintained that, so far as malingering existed, it was the result of inadequate medical services which left discharged patients quite incapable of a speedy return to work – in other words, as Watson-Jones claimed, malingerers were 'made . . . not born'. Although professional self-interest can be read into such statements, British orthopaedic surgeons noticeably did not follow the example of some American orthopaedists in claiming expertise in the detection of

malingering. Since backache was alleged to be the scrimshanker's favoured complaint, orthopaedists stood to gain in any case. On malingering, see Donald C. Norris, 'Malingering', *Br. Ency. Med. Practice*, 8 (1938), pp. 354–67; and sources cited in K. Figlio, 'How does Illness Mediate Social Relations? Workmen's compensation and medico-legal practices, 1890–1940,' in P. Wright and A. Treacher (eds), *The Problem of Medical Knowledge* (Edinburgh, 1982), pp. 174–224.

54. GFTU, *Report of Conference on Institutional Treatment of Fractures, 7 October 1936 at Onward Hall, Deansgate, Manchester* (1936).

55. Robert Hyde, the founder in 1918 of the Industrial Welfare Society, in ibid., p. 5. Hyde also wrote a letter on 'Treatment of Fractures' to *The Times*, 18 Mar. 1935, p. 8, endorsing the reform of insurance legislation.

56. See Alice Prochaska, *History of the General Federation of Trade Unions, 1899–1980* (1982), esp. p. 192.

57. See Ch. 9 above; see also the TUC pamphlet *John Smith has an Accident* (1933).

58. See Samuel James Woodall, *Manor House Hospital* (1966), which discusses Morgan on pp. 145–7; see also press cuttings in the hospital's archives.

59. See Ray Earwicker, 'A Study of the BMA–TUC Joint Committee on Medical Questions, 1935–1939', *J. Social Policy*, 8 (1979), pp. 335–56.

60. Earwicker, p. 340 *et passim*; on Smyth (1882–1966), who was also on Bevin's Factory Welfare Advisory Board during the Second World War (to whom he submitted a Memorandum on Fracture Treatment and Rehabilitation (PRO:LAB/14/425)), see Frank Honigsbaum, *The Division in British Medicine* (1979), pp. 337–8. Bevin, besides being involved with the Industrial Orthopaedic Society, had been one of the eight members of the Industrial Health Research Board of the Medical Research Council from 1932, and a member of the Empire Rheumatism Council. In Sept. 1937 he addressed the BMA on the need for cooperation with the unions in developing an industrial medical service: 'The Wider Issues of Health Legislation', *BMJ*, 25 Sept. 1937. pp. 610–12. See also Alan Bullock, *The Life and Times of Ernest Bevin*, vol. 1 (1967), pp. 602–3.

61. Earwicker, pp. 342ff.

62. BMA, 'Joint Committee of BMA and TUC (1936–9)', 3 vols, BMA archives.

63. See *BMJ*, 18 Dec. 1937, pp. 367–71.

64. 'Joint Committee of BMA and TUC', BMA archives, p. 7.

65. The BMA Fracture Committee thought it advisable 'to make some effort to increase the numbers of employers who run their own insurance': *BMJ*, suppl., 16 Feb. 1935, p. 61.

66. 'Appendix D' to the Memorandum of Evidence by the TUC to the *Report of the Royal Commission on Workmen's Compensation* (1939–40), pp. 445–8; TUC General Council's Report to the Blackpool Congress, 1938, paragraphs 88–103. See also *BMJ*, suppl., 18 Dec. 1937, pp. 367–71. The *Royal Commission* was prompted by Sir Samuel Hoare, the Home Secretary, who felt that there was strong support in the House of Commons for reform.

67. Elmslie was the only surgeon on the committee. His illness and death on 26 July 1940 contributed to the circumstances that eventually led to the suspension of the commission without report. See P. W. J. Bartrip, *Workmen's Compensation in Twentieth Century Britain: law, history and social policy* (Aldershot, 1987), p. 171.

68. 'Assumption B' insisted that 'rehabilitation must be continued from the medical through the post-medical stage till the maximum of earning capacity is restored and that a service for this purpose should be available for all disabled persons who can profit by it irrespective of the cause of their disability': Beveridge, *Social Insurance and Allied Services*, 1942, para. 438. For the political context of Beveridge's *Report*, see Paul Addison, *The Road to 1945* (1977), ch. 8, and José Harris, *William Beveridge: a biography* (Oxford, 1977), ch. 16.

69. Honigsbaum, *Division in British Medicine*, p. 329, and idem, 'Medical Politics and the State: the part played by the medical profession in the development of the national health service', *Bull. Soc. Social Hist. Med.*, no. 29 (1981), pp. 20–1.

70. Earwicker, 'BMA-TUC Joint Committee'. I am grateful to Charles Webster for clarification on this issue.

71. A. F. Young, *Industrial Injuries Insurance: an examination of British policy* (1964), p. 103. Also, as Wilson and Levy noticed, provision for medical rehabilitation, 'to which the B. M. A. attached such importance', found no place in the Workmen's Compensation Acts (1925–34) Amendment Bill that was defeated in 1938: *Workmen's Compensation*, vol. 2, pp. 208–9. See also Bartrip, *Workmen's Compensation*, chs 8–10.

72. In the report of the BOA meeting at Nottingham, Sept. 1943: *JBJS*, 25 (1943), p. 218. Citrine, also present, remarked on the need to overcome conservatism on the part of both the medical profession and the workforce.

73. The MWC was created in 1920 as a result of the Coal Mines Regulation Act, and upon the recommendation of the Sankey Commission of 1919. It had central and district administrations and derived its income from a levy of 1*d* per tonne of coal raised. Until the 1930s, when pit head baths were promoted by the Commission, most of its income was spent on miners' convalescent homes. As late as 1937, health matters took up only 6.3 per cent of the grant money. See 'History of the Miners Welfare Fund with Special Reference to the Principles which have governed its administration', in Mines Department, Miners' Welfare Fund, Departmental Committee of Inquiry (1931), *Report to the Secretary for Mines*, presented 1933, pp. 1–24. Cmd.4236. See also the Annual Reports of the MWF.

74. MWC, *16th Annual Report* (1937), p. 60; Waugh, *Orthopaedics in Nottingham*, p. 39; *16th A.R.*, p. 58; and Manchester Royal Infirmary, *Annual Report for 1938*, p. 21.

75. See, for example, Harold Balme, 'Some Problems of Rehabilitation', *BMJ*, 10 July 1943, p. 47; H. A. T. Fairbank, 'Notes on Rehabilitation', PRO:MH/76/157; and R. E. Matkin, 'Rehabilitation: an ambiguous term and unfulfilled ideal', *Rehab. Lit.*, 46 (1985), pp. 314–20.

76. Henry Kessler in an address to the British Council for Rehabilitation, *JBJS*, 32B (1950), p. 439.

77. See Howard A. Rusk and Eugene J. Taylor, *New Hope for the Handicapped: the rehabilitation of the disabled from bed to job* (New York, 1949).

78. They complained, for example, that the TUC's submission to the Hetherington Commission had concerned itself with rehabilitation only in relation to fractures. *Workmen's Compensation*, vol. 2, pp. 234–5. They also thought that 'for the treatment of fractures less specialised "industrial" knowledge and experience is perhaps needed than for other "industrial" injuries involving loss of a limb or eye or hand or leg'.

79. PRO:MH/76/509. Platt and Robert Ollerenshaw (Manchester's other 'pure' consultant in orthopaedics) were involved in advising on the project.

80. Hermann Levy, *War Effort and Industrial Injuries*, Fabian Society Tract, no. 253 (1940)).

81. Joan Anderson, *A Record of Fifty Years' Service to the Disabled 1919–1969 by the Central Council for the Disabled* [1970], p. 25.

82. 'Provision of Specialist Services', section 4: 'The Orthopaedic Service' in PRO:MH/80/24.

11 The Phoney War

1. Richard Titmuss, *Problems of Social Policy (History of the Second World War, United Kingdom Civil Service)*, ed. W. K. Hancock (1950), p. 480.

2. While observing that by 1944 the number of hospitals employing rehabilitation methods had doubled over 1943 and that the daily average of patients receiving rehabilitation had risen to 31 000, Titmuss also noted that of 457 hospitals surveyed in 1944, only 131 were classified as 'Grade A', i.e., as 'possessing and using facilities for all forms of rehabilitation'; 136 possessed limited facilities, and 190 had no facilities at all (ibid., pp. 479–80).

3. Bevan, Foreword to *The Road to Health – The Story of Medical Rehabilitation* (Ministry of Health, 1947), p. 2. See also Ann Carr, 'Rehabilitation and Resettlement' in *Health and Social Welfare 1945–1946* (1947), pp. 43–8.

4. See Stephen J. Watkins, 'Occupational Health Services – Part of the Health Care System?', MSc thesis, University of Manchester, 1982; C. N. Moss, 'Rehabilitation and Occupational Medicine', *J. Occupat. Med.*, 16 (1974), pp. 81–5; and T. A. Lloyd Davies, 'Whither Occupational Medicine', *Proc. Roy. Soc. Med.*, 66 (1973), pp. 818–21. For America, see Edward D. Berkowitz, 'The Federal Government and the Emergence of Rehabilitation Medicine', *The Historian*, 43 (1981), pp. 530–45.

5. (G. Tomlinson), *Report of the Inter-Departmental Committee on the Rehabilitation and Resettlement of Disabled Persons*, 1942–3. Cmd 6415.

6. H. Osmond-Clarke and J. Crawford Adams, 'Orthopaedic Surgery: general review', in Zachary Cope (ed.), *Surgery (History of the Second World War)* (1953), pp. 234–70, at p. 234.

7. E. J. Crisp, 'Physical Medicine for Foot-strain', in Cope, *Surgery*, pp. 723–5 at p. 723. '[A] third to a half or even more of the cases referred to an Orthopaedic Centre for an opinion were "feet"', lamented H. A. T. Fairbank: 'Rehabilitation of Orthopaedic Cases in the Emergency Medical Services', in Cope, *Surgery*, pp. 271–80 at p. 278. On McIndoe, see Hugh McLeave, *McIndoe: plastic surgeon* (1961), and *DNB.*; on the lavish attention to burns during the war see also Morris Fishbein, *Doctors at War* (1945; reprinted Freeport, New York, 1972), p. 17.

8. Fairbank: 'Rehabilitation of Orthopaedic Cases', p. 278.

9. Osmond-Clarke and Adams, 'Orthopaedic Surgery', p. 237. See also John B. Coates and M. Cleveland (eds), *Orthopaedic Surgery in the European Theater of Operations* (Washington, DC, 1956).

10. Osmond-Clarke and Adams, 'Orthopaedic Surgery', p. 237.

11. Burgess, *Little Wilson and Big God: being the first part of the confessions of Anthony Burgess* (Harmondsworth, 1988), p. 244.

12. William Heneage Ogilvie insisted that the traumatic injuries of war were quite unlike those experienced in civilian life: it is 'a fallacy', he maintained, ' that . . . war surgery and traumatic surgery are synonymous. They are not. War surgery is traumatic surgery applied under conditions of war.' In 'General Introduction: Surgery in War-Time', in Cope (ed.), *Surgery*, p. 3.

13. See Jim Fyrth, *The Signal Was Spain* (1986), pp. 148ff; ' "Trueta's Message" ' in J. Trueta, *Surgeon in War and Peace* (1980), pp. 265–79; and 'Minor Injuries in Civil Bombardment – Dr Trueta's Address', *BMJ*, 16 Dec. 1939, pp. 1197–9. However, Trueta's major therapeutic contribution – his closed plaster treatment of wounds – was no sooner taken up by orthopaedists than it was forced to be abandoned (or radically modified) as a result of changes in the tactical circumstances of the war, and by the introduction of sulphonamides and, later, penicillin. See J. C. Scott, 'Closed Plaster Treatment of Wounds', in Cope, *Surgery*, pp. 280–7; William Heneage Ogilvie, 'The Surgery Of War Wounds: a forecast (1940) and a retrospect (1945)' in his *Surgery: orthodox and heterodox* (1948), pp. 204–11; Osmond-Clarke and Adams, 'Orthopaedic Surgery', pp. 237–43; and *Army Medical Department Bulletin*, Suppl. No.22, May 1945, section 'Wound Treatment Before 1943'.

14. J. A. MacFarlane, 'Wounds in Modern War', *JBJS*, 24 (1942), pp. 739–52 at p. 739.

15. Ibid., p. 744.

16. Cited in St J. D. Buxton, 'Prevention of Accidents and Limitation of Injury', *Lancet*, 5 Jan. 1946, p. 27.

17. See Ministry of Health, Emergency Medical Services Instruction, Part I: *Medical Treatment and Special Centres* (1943), pt 3: 'Orthopaedic'; and 'Rehabilitation of War Injuries', PRO:MH/76/157.

18. Rowley Bristow, 'Some Surgical Lessons of the War', *JBJS*, 25 (1943), pp. 524–34 at p. 524.

19. Fairbank, 'Rehabilitation of Orthopaedic Cases', p. 272; and see idem., 'Rehabilitation of the Injured in this War and the Last', *Lancet*, 29 July 1944, pp. 131–4.

20. A. S. Blundell Bankart was appointed Regional Orthopaedic Consultant for Section V of London, and S. A. S. Malkin for Region III, Nottingham.
21. Bristow, 'Surgical Lessons', p. 526.
22. Platt, 'The Place of Orthopaedics in Medical Education and in the Regional Hospital Services [BOA Address]', *Lancet*, 1945, reprinted in Platt, *Selected Papers*, pp. 32–41 at p. 34. Cf. Ernest Hey Groves, 'Hero Worship in Surgery', *BMJ*, 22 Feb. 1930, pp. 321–6, and Wilfred Trotter, 'The Commemoration of Great Men' in his *Collected Papers* (1941), pp. 15–31.
23. EMS Instruction, p. 5.
24. See Ministry of Health, 'Ministry of Health, Consultant Advisers on the organisation of hospitals in war time', 7 Mar. 1939 (copy in Platt Papers). Williams, was Dean of University College Medical School and the editor of Christopher Heath's popular text *Minor Surgery and the Treatment of Fractures* (22nd edn, 1940).
25. On Platt's appointment, which was made on the £500 per annum salary scale for full-time appointments, see PRO:MH/76/157. He looked after most of the orthopaedic organization in the north of the country.
26. See Neville M. Goodman, *Wilson Jameson: architect of national health* (1970), p. 112; 'Nuffield Hospital Trust: papers of Sir William Jameson, CMO, Ministry of Health', PRO:MH/77/24; J. V. Pickstone, *Medicine and Industrial Society* (Manchester, 1985), p. 312; and Daniel Fox, *Health Policies, Health Politics* (Princeton, 1986), p. 97.
27. See Osmond-Clarke and Adams, 'Orthopaedic Surgery', pp. 235–6; Watson-Jones, 'Organisation of an Accident Service', in his *Fractures and Joint Injuries* (4th edn, 1955), vol.2, pp. 1039–48, esp. at p. 1040; *Br. J. Indust. Hygiene*, 2 (1945), pp. 123–4; and 'Rehabilitation After War Injuries', especially 'Extracts from a Report on the Organisation of Orthopaedic and Fracture Services in R.A.F. Hospitals' in PRO: MH/76/157.
28. Perkins (1892–1979) was replaced in 1940 by another London orthopaedist, St John Dudley Buxton (1891–1981), Fairbank's successor at King's College Hospital: see the latter's obituary in *JBJS*, 63B (1981), pp. 449–50.
29. 'The Organisation of War Orthopaedic Services', 2 June 1940, in Platt Papers.
30. On both, see BOA Minutes: 3 Jan. 1941.
31. 'Ministry of Health, Emergency Hospital Service, Orthopaedic Scheme, Introduction', September 1940. (Copy in Platt Papers.)
32. Ibid. The main Royal Navy orthopaedic unit was at the RN Auxiliary Hospital, Sherborne, Dorset: see, EMS Instruction 403, 27 Jan. 1943.
33. 'EMS. Orthopaedic Scheme. Introduction', 26 Aug. 1940.
34. Bristow, 'Surgical Lessons', p. 525.
35. Ibid., pp. 525–6. A 'long and lively discussion' on the organization of orthopaedics in the Army and the EMS took place at the executive meeting of the BOA on 3 Jan 1941, a result of which were letters to both the administrations applauding progress but calling for much more.

36. The movement toward coordination within the EMS is reflected, for example, in EMS Instruction 406, 'Service of Consultants', 23 Feb. 1943.
37. Titmuss, *Problems of Social Policy*, pp. 477–8.
38. Superimposed on these criteria was that of vulnerability to enemy attack. Thus Fracture Hospitals 'A' were those where experienced surgeons were supposed to be available, where liability to enemy attack was minimal, where all types of fractures could be treated, and where there existed long-term inpatient facilities; 'B'-category fracture hospitals were those suitable only for ambulant cases requiring inpatient treatment for a few days. The 'C'-category were outpatient clinics 'to provide for the treatment and supervision of fracture cases in civilians following their discharge from hospital'. See EMS Instruction, Part I: *Medical Treatment at Special Centres* (1943), p. 8; Platt, 'Treatment of Fractures in E.M.S. Patients', 6 Mar. 1941, in Platt Papers; and Platt, 'Essentials of a Fracture Department B', 1 Aug. 1941, Platt Papers.
39. Titmuss, *Problems of Social Policy*, p. 477.
40. H. Balme, 'Organisation of Occupational Therapy', in Cope, *Surgery*, pp. 729–32 at p. 730. See also P. Bauwens, 'Medical Electrology in Wartime', in ibid., pp. 725–9.
41. Robert Stanton Woods, 'Physical Medicine' in C. L. Dunn (ed.), *The Emergency Medical Services (History of the Second World War)* (1952), pp. 366–87 at p. 366. See also 'Physical Medicine' in Ministry of Health, *National Health Service: the development of consultant services* (1950), pp. 15–16; Francis Bach, *Recent Advances in Physical Medicine* (1950); and 'History of the Archives – a journal of ideas and ideals', *Arch. Phys. Med. and Rehab.*, 50 (1969), pp. 6–42. Woods (1877–1954) was physician to the London Hospital in charge of the Physical Medicine department there from 1911; president of the RSM section on Physical Medicine from 1932; president of the International Congress on Physical Medicine in 1936 and from 1936 to 1946 consultant adviser to the Ministry of Health. See obituary in *BMJ*, 27 Nov. 1954, pp. 1295–6, 1362.
42. See 'Rehabilitation: suggestions for scheme for occupational physiotherapy': PRO:MH/76/321, which contains letters from various intermediaries and a memorandum by Smart on 'Physical Medicine in War Time'. Smart was also among those in 1940 who called for pit-head and other local clinics, and who expressed the hope that these would become 'the nucleus for a nation-wide physiotherapeutic scheme'. Cited in Hermann Levy, *War Effort and Industrial Injuries*, Fabian Society Tract, no. 253 (1940), p. 16.
43. See 'Advisory Committee on Physical Medicine', PRO:MH/76/326 and MH/76/117. Rhaiadr Jones was formerly House-Surgeon to Robert Jones at the Royal Southern Hospital and subsequently the MO at Leasowe (until 1919) before going into commerce. On the advice of Robert Jones and at the request of Agnes Hunt, he undertook the organization of the Derwen Cripples' College of which he subsequently became the Hon. Consultant Superintendent. He was also on the Executive of the CCCC, and had a hand in organizing the Stanmore branch of the RNOH: see his curriculum vitae in PRO:MH/76/157.

44. See the letter of Rhaiadr Jones to Professor F.R. Fraser, Acting Director-General of the EMS, 7 July 1942, complaining about the circular on 'Rehabilitation' which had been drafted by the Advisory Committee on Physical Medicine without his consultation: PRO:MH/76/117. The failure to ask Jones's opinion reflects the view of Fairbank and Stanton Woods that Jones lacked the necessary general experience for the rehabilitation of the wounded, having 'no real knowledge of . . . massage, electrical treatment, gymnasia &c.': See letters in PRO:MH/76/157. Platt and Girdlestone were also disappointed with Jones's appointment, having themselves sought to appoint E.S. Evans, the Consultant to the Cripples' Training College at Leatherhead: see letter from Fairbank to Fraser, 19 Sept. 1940, in PRO:MH/76/157.

45. Stanton Woods, 'Physical Medicine', p. 372; Titmuss, *Problems of Social Policy*, p. 470; and see 'Accidents to key workers in vital war factories', EMS Instruction 401, 11 Jan. 1943; EMS Instruction 437, 19 Nov. 1943: 'Fracture and Other Injury Cases Amongst Certain Industrial and Civil Defence Workers'; and EMS Suppl. 28, 'Preparing the Sick and Injured for Return to Work', copy in EMS – Instructions and Memos, 1939–45 in GLRO:PH/War/2/4.

46. See PRO:MH/76/117 (letter of 28 Feb. 1941); and MH/76/322.

47. Stanton Woods, 'Physical Medicine', pp. 372–6; Titmuss, *Social Policy*, p. 479; and Balme, 'Occupational Therapy', p. 730. The Society of Physiotherapists was also active in promoting the use of private physiotherapy clinics. See Levy, *War Effort*, p. 16.

48. See Gerald Larkin, *Occupational Monopoly and Modern Medicine* (1983), ch. 4: 'Physiotherapy'; and, for the American experience, Glenn Gritzer and Arnold Arluke, *The Making of Rehabilitation* (Berkeley, 1985), ch 5: 'The Rediscovery of Rehabilitation, 1941–1950'.

49. See Berkowitz, 'Federal Government and Rehabilitation'. A major American influence was the Baruch Committee on Physical Medicine, established in 1943, on which see also Gritzer and Arluke, *Rehabilitation*, pp. 94ff.; and *Report of the Baruch Committee on Physical Medicine* (n.p., April 1944). The American Congress of Physical Therapy was established in 1921; the American Academy of Physical Medicine held its 16th annual meeting in 1939.

50. BOA, 'Memorandum on Fracture Services', dated 2 Sept. 1941, para. 6 (BOA Archives). As during the Great War, there was also a shortage of trained orthopaedists. In Scotland in 1940 this led the Department of Health to import and pay the salaries of nine orthopaedic surgeons and 22 specially trained nurses from Canada to work at the Hairmyers Hospital in Lanarkshire. See 'Report of the Committee Appointed to Organize Canadian Orthopaedic Unit for Scotland', Central Council of the Canadian Red Cross Society, 8 Jan. 1942. *Canadian Red Cross Minute Book no. 2*, CRC Archives, Toronto. I am grateful to David Cantor for this reference.

51. See J. Cyriax and E.H. Schiotz, 'Manipulation: by whom?' in their *Manipulation: Past and Present* (1975), ch. 5; cf. F.J. Gaenslen in his AOA Presidential Address on 'The Role of Physical Therapy in

Orthopaedic Surgery', *JBJS*, 18 (1936), pp. 559–65 at p. 559: 'It . . . behooves us as a group to recognize not only the virtues but also the limitations of physical therapy, in order that his important branch of therapy may be included in our program of treatment.'

52. Stanton Woods, 'Physical Medicine', p. 373.

53. (Tomlinson), *Report on Rehabilitation*, p. 3.

54. Contained in BOA Minutes 1 Dec. 1942.

55. See George Riddoch and H. J. Seddon (of the Wingfield-Morris Orthopaedic Hospital), 'Surgery of Peripheral Nerve Injuries', in Cope, *Surgery*, pp. 517–33 at p. 517.

56. See Bristow, 'Surgical Lessons', p. 528.

57. On 'segregation within segregation' in relation to fracture treatment – 'so successfully adopted for peripheral nerve injuries' – see Platt's version of Girdlestone's 'Memorandum re Fracture Hospital E.M.S. to D.G. E.M.S.', 2 May 1941 (in Platt papers).

58. Platt, 'Fracture Department, Region IX', 1 Aug. 1941, Platt Papers.

59. See Gissane, 'Discussion on the Relationship of Orthopaedic Surgery to Traumatic Surgery', *JBJS*, 31B (1949), p. 633, and Arnold Wilson and Hermann Levy, *Workmen's Compensation*, vol. 2 (Oxford, 1941), pp. 203, 213–14, 234–5. Gissane was appointed to the BOA's Sub-committee on Fractures in January 1944 at the time when the accuracy of its title was being questioned. BOA Minute. 28 Jan. 1944.

60. G. R. Girdlestone, 'A Regional Orthopaedic and Accident Service', *BMJ*, 23 Apr. 1949, pp. 720–22; and J. Trueta, *Gathorne Robert Girdlestone* (Oxford, 1971), p. 79. Girdlestone, himself, appears not to have worked in the Oxford service. C. Max Page was in charge, succeeded by Trueta.

61. BOA Minutes, 14 Dec. 1944.

62. Draft of the 'Memorandum on Accident Services', 2 Jan. 1942 (BOA Archives); cf. printed version, p. 4.

63. This is also apparent in the postwar paper by the London-based orthopaedic surgeon, St J.'D. Buxton, 'Prevention of Accidents and Limitation of Injury', *Lancet*, 5 Jan.1946, pp. 27–9.

64. 'Accident Services and "traumatology" ', BOA, Memorandum on Accident Services, 1943, pp. 3–4.

65. F. D. Dickson in his presidential address to the AOA in 1936: *JBJS*, 18 (1936), pp. 263–9. In 1941 it was claimed by an American reviewer of the first *Year Book of Industrial and Orthopedic Surgery* (Chicago, 1940) that 'probably no branch of surgery has developed to a greater extent in scope, progress, and recognition of its importance than traumatic surgery' (*JBJS*, 23 (1941), p. 201). Reflecting this market trend, the *Year Book* was retitled in 1947 *Orthopedic and Traumatic Surgery*.

66. Gissane was insistent, however, that accident services and accident hospitals should be under the control of persons with extensive orthopaedic training; that they 'must not become segregated'; and must be connected to teaching hospitals. See the discussion on 'Orthopaedic Surgery and Traumatic Surgery' at the annual meeting of the BOA, as reported in *BMJ*, 31 Dec. 1949, p. 1524, and *JBJS*, 31B (1949), pp. 633–5.

67. In *JBJS*, 31B (1949), p. 634. 'The real solution', he thought, lay in a better distribution of orthopaedic surgeons throughout the country. Only when this had been achieved could traumatology 'be separated from the practice of general surgery and from that of general practitioners "with a flair".'
68. See BOA Minutes, 2 Jan. 1942, item 6 – alterations in the wording of the Memo, after 'a long and animated debate'.
69. MS in Platt Papers, dated 8 Mar. 1946.
70. Sir Frederick Menzies in 1941, quoted in Charles Webster, *The Health Services Since the War*: vol. 1: *Problems of Health Care, The National Health Service before 1957* (1988), p. 83; see also on the 'Regional Hospital Service', pp. 262–70, and PRO:MH/80/24.
71. A. S. M. MacGregor (MOH for Glasgow) in the memorandum on 'The Future Development of the Hospital Service', 24 Apr. 1938, p. 7: PRO:MH/80/24. In line with this 'regionalist' planning was the Royal Commission on Local Government in Tyneside, which recommended a multi-purpose regional authority to replace 16 separate authorities.
72. Girdlestone, 'A Regional Orthopaedic and Accident Service', *BMJ*, 23 Apr. 1949, p. 722. See also 'Regional Orthopaedics [review of Girdlestone's *A Regional Orthopaedic Service*, 1948]', *BMJ*, 23 Oct. 1948, p. 751; Trueta, *Surgeon in War and Peace*, p. 142; and 'Traumatic and Orthopaedic Surgery', in *NHS Consultant Services* (1950), para. 167–71.
73. The teaching hospitals, moreover, did not come under the administration of the Regional Health Authorities. They retained their autonomy and had their own boards of managers.
74. See Webster, *NHS*, Table IV, p. 266.
75. Discussion on the 'Memorandum on Accident Services', BOA Minutes, 25 Sept. 1942.
76. 174 out of 228: cited in Walpole Lewin, 'Medical Staffing of Accident and Emergency Services: a report prepared on behalf of the Joint Consultants Committee and Presented in April 1978', BMA, 1978. The BOA in 1959 perceived accident services as a failure, and insisted that their organization 'be undertaken by the state as a quasi-military operation': 'Memorandum on Accident Services', *JBJS*, 41B (1959), pp. 458–63 at p. 463. For further insights into the postwar history of accident services, see Nuffield Provincial Hospitals' Trust, *Casualty Services and Their Setting: a study in medical care* (Oxford, 1960), and the BOA's *Casualty Departments: The Accident Committee* (July 1973).
77. 'Report of the Fracture Sub-Committee', BOA Minutes. 14 Dec. 1944.

12 An End to 'Adolescence'

1. By 1960 less than one-quarter of all patients in orthopaedic hospitals were children: 'The Orthopaedic Hospital', appended to BOA Minutes, 14 Apr. 1961. *The Orthopaedic Yearbook* currently devotes only one chapter in twelve to paediatrics.
2. Among assistant, associate and full professors of medical specialisms in America in 1983–4, orthopaedists were among the second- and third-highest paid, the mean salary for full professors being $142 800,

surpassed only by cardiovascular surgeons ($170 500) and plastic surgeons ($154 900). See R. G. Petersdorf, 'Medical Schools and Research: Is the tail wagging the dog?', *Daedalus*, 115 (1986), pp. 99–118 at p. 108.

3. W. R. Bristow, 'Injuries of Peripheral Nerves in Two World Wars [the Robert Jones Lecture]', *Br. J. Surg.*, 34 (1947), pp. 333–48.

4. William Waugh, *John Charnley: the man and the hip* (1990).

5. Cited in J. Trueta, *Surgeon in War and Peace* (1980), pp. 214–15.

6. See *JBJS*, 37A (1955), pp. 622–7, and the editorial on 'Experimental Surgery' in 36B (1954), pp. 1286–90. Among the research pursued was bone transplant, the use of frozen tendons for grafting, and chronic arthritis. The *Journal of Orthopaedic Research* was started in 1982. For expenditure on research in orthopaedics in the USA in 1953, see A. R. Shands, Jr, 'The Orthopaedic Research Foundation', *JBJS*, 36A (1954), p. 1287.

7. *J. Clinical Orthop.*, 4 (1954), pp. 99–114 at p. 99; cf. *JBJS*, 34A (1952), special issue on 'Chemotherapy in Bone and Joint Tuberculosis'.

8. The prewar careers of Leo Mayer and Fred Albee exemplify such aspirations. See Mayer, 'Orthopaedic Surgery in the USA', *JBJS*, 32B (1950), p. 472; and W. A. Rogers, 'Leo Mayer: an appreciation', *Bull. Hosp. Joint Dis.*, 21 (1960), pp. 81–5 at p. 82; on Albee, see Leo Mayer, 'Reflections on Some Interesting Personalities in Orthopaedic Surgery During the First Quarter of the Century', *JBJS*, 37A (1955), pp. 374–83 at p. 381; H. Osmond-Clarke, 'Half a Century of Orthopaedic Progress', *JBJS*, 32B (1950), p. 645; and Albee, *Autobiography* (1950).

9. Clarence Starr, AOA presidential address, *JOS*, 2 (1920), pp. 381–9 at pp. 386–7. See also P. C. Colonna, 'The Evolution of Physiological Concepts in Bone Surgery', *JBJS*, 38A (1956), pp. 1169–74.

10. Robert Osgood, presidential address, *JOS*, 3 (1921), p. 266.

11. Baer, AOA address, 1924, *JBJS*, 6 (1924), pp. 508–9. Much of Baer's own research was on the use of maggots in wound healing: see W. S. Baer, 'The Treatment of Chronic Osteomyelitis with the Maggot (Larva of the Blow Fly)', *JBJS*, 13 (1931), pp. 438–75; and W. Robinson, 'Stimulation of Healing in Non-healing Wounds by Allantoin occurring in Maggot Secretions and of Wide Biological Distribution', *JBJS*, 17 (1935), pp. 267–71.

12. See Charles V. Heck (compiler), *Fifty Years of Progress in Recognition of the 50th Anniversary of the AAOS* (Chicago, 1983); and De Forest P. Willard, 'The [AOA] President's Address', *JBJS*, 17 (1935), pp. 531–5 at p. 533.

13. See, for example, Jones's praise for the work on the aetiology of tuberculosis in children which, according to him, 'emanated from Edinburgh under the direction of Sir Harold Stiles, assisted by [John] Fraser and Mr. Mitchell'. With 'scientific methods and infinite labour [Stiles's team] traced the infections from cows to children, and were able to correlate their clinical and pathological findings': Jones, 'The Cameron Lecture on the Necessity of Orthopaedic Training: its relation to the prevention and cure of deformities. Delivered at the University of Edinburgh', *BMJ*, 5 Feb. 1921, pp. 181–6 at p. 181.

14. See above, Ch. 3 An associate of Fairbank at King's College recalled that Fairbank's no-touch technique was meticulous: 'we scrubbed for ten minutes by the clock and wore dry gloves . . . woe betide any assistant who took up a swab in his fingers': H. L. C. Wood, June 1982, letter in BOA archives.

15. Robert Jones, 'The Orthopaedic Outlook in Military Surgery', *BMJ*, 12 Jan. 1918 (offprint) p. 2.

16. See Hey Groves, 'A Surgical Adventure: an autobiographical sketch', reprinted from *Bristol Medico-Chirurg. J.*, 50 (1933), p. 10, and idem, *On Modern Methods of Treating Fractures* (Bristol, 1916). Hey Groves's research was conducted at University College, London; like Tubby, who conducted research into the properties of flavine and its derivatives, Hey Groves had previously studied in Germany. For Tubby's work, see his *A Consulting Surgeon in the Near East* (1920) p. 120.

17. Clinical research was poorly catered for in Britain in the 1920s, in large part because the Medical Research Council (established as an independent body after the First World War) became increasingly the preserve of basic scientists, especially of physiologists and biochemists. See David Hamilton, '"Too Difficult for Doctors": British attitudes to clinical research in the early twentieth century', paper presented to the Society for the Social History of Medicine/British Society for the History of Science conference, Manchester, July 1985. Cf. Sir Christopher Booth, 'Clinical Research', in Joan Austoker and L. Bryder (eds), *Historical Perspectives on the Role of the MRC* (Oxford, 1989), pp. 205–41.

18. Bristow, 'Orthopaedic Surgery – Retrospect and Forecast', *Lancet*, 6 Nov. 1937, pp. 1061–4 at p. 1063. The quotation from Fairbank is from his 'Some Recent Advances in the Treatment of Fractures', *Post-Graduate Med. J.*, 13 (1937), pp. 341–3.

19. Quoted in Daniel Fox, 'The National Health Service and the Second World War: the elaboration of consensus', in Harold Smith (ed.), *War and Social Change: British Society in the Second World War* (Manchester, 1986), p. 51.

20. Ministry of Health and Department of Health for Scotland, *Report of the Inter-Departmental Committee on Medical Schools* (HMSO, 1944), ch. 3. See also Robert Milnes Walker, *Medical Education in Britain* (1965), pp. 23ff., and Sir George Pickering, *Quest for Excellence in Medical Education, a personal survey* (Oxford, 1978), pp. 10ff.

21. Francis M. McKeever, 'The Need for Basic Investigation in Orthopaedic Surgery', *JBJS*, 35A (1953), pp. 285–8 at p. 287. The first chair of orthopaedic surgery in the British Isles was at Trinity College, Dublin, in 1925 (see *JBJS*, 7 (1925), p. 740); Girdlestone's Chair in Oxford was the first in Britain. There are now 18 chairs of orthopaedic surgery in Britain.

22. S. A. S. Malkin, 'The Scientific Approach to Orthopaedic Surgery [presidential address to the BOA , 1948]', *JBJS*, 31B (1949), pp. 5–9 at pp. 6–7. Cf. Sir Herbert Seddon, BOA presidential address, 1961, 'The Scientific Surgeon, cited in J. N. Wilson, 'The Unscientific Surgeon', *J. Roy. Soc. Med.*, 77 (1984), pp. 281–4.

23. Waugh, *Charnley*, p. 38 and pp. 59–62. In 1955 the American orthopaedist, D. C. Riordan could report after a visit to Britain that

'there has been a considerable swing away from the conservative treatment which was advocated for so long by British orthopaedists. Everywhere we went we saw patients being treated by open operations with the removal of abscesses and necrotic debris': 'Report of Travels of the Exchange Fellows', *JBJS*, 37A (1955), p. 1096.

24. Robert Roaf at a BOA meeting in Liverpool in 1955, *JBJS*, 37B (1955), p. 720, quoted in Waugh, *Charnley*, p. 57.

25. Platt, 'The Place of Orthopaedics in Medical Education and in the Regional Hospital Services', *Lancet*, 1945, reprinted in his *Selected Papers* (Edinburgh, 1963), pp. 32–41, at p. 34.

26. Rosemary Stevens, *Medical Practice in Modern England* (New Haven, 1966), p. 111: 'Table 1: Distribution of Specialists by Specialty: England and Wales, 1938–1964' is drawn from the questionnaire that A. Bradford Hill sent out in 1947 for evidence for the *Spens Report on Remuneration of Consultants and Specialists* (1948). For a breakdown of these figures for 1949–55, see Charles Webster, *The Health Services Since the War:* vol. 1: *Problems of Health Care, the National Health Service before 1957* (1988), p. 310. By 1978 orthopaedic surgeons constituted 17.7 per cent of all surgical specialties in England and Wales: Department of Health and Social Security, *Orthopaedic Services: waiting time for out-patient appointments and in-patient treatment. Report of a working party to the Secretary of State for Social Services*, Chairman, R. B. Duthie (HMSO, 1981), p. 11.

27. Sasha Lewis, 'The Impact of World War I On the Development of Neurosurgery as a Speciality', BSc dissertation, University of Manchester, 1991, p. 20. Of the 576 Fellows who died between 1952 and 1964, whose main period of training would have been between 1900 and 1910, 71 (12.3 per cent) claimed specialist interest in orthopaedics. 'General surgery' tops both lists, accounting for 52.2 per cent (425) in the first sample period, and 39.2 per cent (226) in the second.

28. See BOA Minutes, 28 Jan. 1944 and 11 May 1944. In the USA at this time, 60 per cent of orthopaedists had specialty board certificates: Rosemary Stevens, *American Medicine and the Public Interest* (New Haven, 1971), pp. 543–4. NHS 'consultants' were recognized by regional panels in 1948.

29. The BMA Orthopaedic Group Committee in July 1939 estimated that there were then approximately 200 orthopaedic surgeons of more or less senior status. BMA Archives. This nearly accords with the NHS figures for 1949 given in Stevens, *Medical Practice in Modern England*, and in Webster, *The Health Services Since the War*.

30. Goodenough Report, Appendix H.

31. See, for example, the record of his visit to Leeds of 13 June 1941 in Platt Papers.

32. J. C. Scott (a Canadian who worked under Girdlestone at Oxford), BOA Minutes, 28 May 1942. The shortage of trained orthopaedic surgeons led the Ministry of Health in 1943 to arrange for fortnightly practical training courses in orthopaedic units and fracture clinics for general surgeons and general practitioners. See EMS Instruction, nos 391 and 404.

33. Partly, it seems, as a result of the influence of George Gask. See James Calnan, *The Hammersmith: [1935–1985] the first fifty years of the Royal Postgraduate Medical School at Hammersmith Hospital* (Lancaster/ Boston, 1985), pp. 22, 92, 169. E. Laming Evans of the RNOH became Consulting Orthopaedic Surgeon to the Hammersmith in 1935, but it was only in 1946 that the British Postgraduate Medical Federation of the University of London established the Institute of Orthopaedics at the RNOH with H. Jackson Burrows as its first Dean. See Osmond-Clarke, 'Half a Century', p. 637. In the 1930s Hey Groves and Watson-Jones were among those to offer special short courses on orthopaedics at the school.

34. See 'Classification of Students', University of Liverpool Archives, ref. 5490.

35. 'Remarks on Orthopaedic Surgery in Relation to Hospital Training', reprinted from *BMJ*, 20 Nov. 1929, p. 2; see also idem, 'The Necessity of Orthopaedic Training', p. 183; and Tubby, *Advancement*, p. 131.

36. In London in the 1920s clinical sessions in orthopaedics were held at St Bartholomew's Hospital, under Elmslie; at the London Hospital under Openshaw; at St Thomas's, under Bristow; and at the Westminster Hospital, under Ernest Rock Carling. See 'Meeting of London Fellowship of Medicine', *JOS*, 2 (1920), p. 436. See also, Robert Ollerenshaw, 'Clinical Lecture on Orthopaedic Cases at the Salford Royal Hospital', reprinted from the *Clinical J.*, 16 Sept. 1914 (these were a part of the Manchester Hospitals' Post-Graduate Clinics). Lectures in orthopaedics were also available at many of the London and provincial voluntary hospitals (dating from as far back as 1861 at St Bartholomew's Hospital), and it became common policy for honorary hospital appointments to be combined with university lectureships in order that income could be derived from the latter. See: 'Appointment of [Orthopaedic] Surgeon [Radcliffe Infirmary, Oxford]', 23 May 1929, in Girdlestone Correspondence, BOA Archives. (The Lectureship was worth £600 p. a.) See also Girdlestone to Jones, 11 Apr. 1923, quoted in J. Trueta, *Gathorne Robert Girdlestone* (Oxford, 1971), p. 46.

37. See Minutes of Board of Orthopaedic Studies, Liverpool, 16 Jan. 1939 and 8 Feb. 1940, and Minutes of the Medical Faculty Board, 10 Nov. 1939. Bristow also refers to the problem in his 'Retrospect and Forecast', p. 1061.

38. H. Jackson Burrows, 'The Orthopaedic Department', *St Barts Hosp. J.*, 69 (1965), pp. 355–9 at p. 357.

39. Indeed, it was not uncommon for those who had practised only orthopaedics during the war in the armed services to enter general surgical practice upon demobilization. Charnley's senior colleague at Cairo during the war, Clifford Brewer, was an orthopaedic surgeon in charge of No. 5 Orthopaedic Unit, who returned to Liverpool to become a general surgeon. Waugh, *Charnley*, p. 24.

40. Cited in Iain D. Levack and H. A. F. Dudley (eds), *Aberdeen Royal Infirmary* (1992), p. 139. The matter was compounded by the fact that acute bony injuries continued to be treated at the ARI by general surgeons until 1953.

41. Platt, 'The Place of Orthopaedics in Medical Education and in the Regional Hospital Services [BOA Address],' *Lancet*, 1945, reprinted in Platt, *Selected Papers*, pp. 32–41, p. 35 and, slightly different, in the report in *JBJS*, 28 (1946), p. 194.

42. Jones, 'The Orthopaedic Outlook in Military Surgery', *BMJ*, 12 Jan. 1918, (offprint), p. 2. Cf. Charles Painter, *Yearbook of Industrial and Orthopaedic Surgery* (Chicago, 1941), p. 3: 'Today the lure of surgery of bones and joints may be said almost to rival that of the abdomen in its attraction for the young surgeon. Much of the enthusiasm for this type of surgery has arisen since the advent of bone grafting in its multitudinous ramifications and the newer methods of treating certain fractures.'

43. Platt, 'The Place of Orthopaedics in Medical Education', p. 35

44. 4 Oct. 1938, BMA Archives.

45. See Neville M. Goodman, *Wilson Jameson: architect of national health* (1970), p. 112.

46. As Charnley encountered orthopaedics in Manchester around 1934, it consisted of 12 lectures from general surgeons, and 12 lecture-demonstrations by Platt and H. Osmond-Clarke at the Royal Infirmary and R. Ollerenshaw at the Salford Royal Hospital. Waugh, *Charnley*, p. 12.

47. Platt, 'Medical Education', p. 4; see also details in his 'The Evolution and Scope of Modern Orthopaedics' in Platt (ed.), *Modern Trends in Orthopaedics* (1950), pp. 1–4.

48. *JBJS*, 28 (1946), p. 194.

49. Ibid., p. 194; BOA Minutes, 1 Jan. 1942; and *JBJS*, 32B (1950), p. 132. In his 'Medical Education' (p.38) Platt wrote:

    We have now to decide whether there is any place for a special degree or diploma in orthopaedics in the comprehensive scheme of training to which we are already committed; and also whether, in the long run, specialist diplomas are in the interests of surgical science as a whole. Opinions on these questions are necessarily divided. It should be stated, however, that the present attitude of the Council of the Royal College of Surgeons of England is one of opposition to any modification of its Fellowship, except for the fully differentiated specialisms of ophthalmology and otolaryngology. Moreover, it is held that the Fellowship of the College or of the sister Colleges, followed by an adequate period of training, renders unnecessary the creation of further special diplomas or degrees.

50. The latter was especially so in the USA. See Stevens, *American Medicine and the Public Interest*, pp. 326–7.

51. On the appointment of the orthopaedic surgeon George Perkins as Professor of Surgery in the University of London, St Thomas's Hospital Medical School, see *JBJS*, 30B (1948), p. 206.

52. 'Training in the Major Surgical Specialties', discussion document at the working party on the training of surgeons, International Federation of Surgical Colleges, Rome, 1963 – typescript in Platt Papers, p. 2.

53. Ibid., p. 5.

54. 'Medical Education', p. 2.
55. Robert Osgood, *The Evolution of Orthopaedic Surgery* (St Louis, 1925), p. 69. Watson-Jones shared these ideals with Platt, as the title of his 1959 Hunterian Oration reveals: *Surgery is Destined to the Practice of Medicine*. See also H. Jackson Burrows, 'An Even Keel in Orthopaedics', *JBJS*, 35B (1953), pp. 321–3.
56. Platt, 'Orthopaedics in Medical Education and in Regional Hospital Services', *JBJS*, 28 (1946), p. 194.
57. Sir Sheldon Dudley, 'Naval Experience in Relation to a National Health Service', *Lancet* 29 July 1944, p. 134. See also William Heneage Olgivie on specialization in 'A Surgeon's Life [the Rutherford Morison Lecture, 1948]' reprinted in his *Surgery: heterodox and orthodox* (1948), pp. 35–49.
58. For sources, see David Armstrong, *The Political Anatomy of the Body: medical knowledge in Britain in the twentieth century* (Cambridge, 1983), pp. 61ff. See also Fielding Garrison on 'the mental staleness and ennui which result from narrow specialism': *Introduction to the History of Medicine* (4th edn, Philadelphia, 1929), p. 8 (previous editions do not contain these remarks).
59. Goodenough Report, p. 148 ('Educational problems created by specialization').
60. See N. T. A. Oswald, 'A Social Health Service Without Social Doctors', *Soc. Hist. Med.*, 4 (1991), pp. 295–315.
61. Goodenough Report, p. 164.
62. See BOA Report on 'The Orthopaedic Hospital', 1961, p. 3 (contained in BOA Minutes, 14 Apr. 1961); and *BMJ*, 10 Oct. 1970, p. 113.
63. Platt, 'Evolution', p. 21. See also Norman Capener, 'Forty Years On', *JBJS*, 40B (1958), pp. 615–17 at p. 617.
64. Platt, 'Organisation of Orthopaedic Services in a Large City', *Med. Officer*, 14 Jan. 1939, pp. 15–16.
65. See Steve Sturdy, 'The Political Economy of Scientific Medicine: science, education and transformation of medical practice in Sheffield, 1890–1922', *Med.Hist.*, 36 (1992), p. 146ff.; Frank Honigsbaum, *The Division in British Medicine: a history of the separation of general practice from hospital care, 1911–1968* (1979); and Daniel Fox, *Health Politics, Health Policies* (Princeton, 1986), pp. 108–9.
66. J. V. Pickstone has observed: 'the fact that scarcely anyone before Aneurin Bevan had expected local authorities to lose control of their hospitals, meant that no-one had stressed the deep and inevitable connections between hospitals, welfare, domiciliary services and public health': 'Psychiatry in District General Hospitals: history, contingency and local innovation in the early years of the National Health Service', in J. V. Pickstone (ed.), *Medical Innovations in Historical Perspective* (1992), pp. 185–99 at p. 190.
67. Watson-Jones, 'Resettlement,The End of Workmen's Compensation', *Lancet*, 18 Nov. 1944, p. 666.
68. Ministry of Labour and National Services, *Report of the Committee of Inquiry on the Rehabilitation, Training and Resettlement of Disabled Persons* (HMSO, 1956), Cmd 9883, p. 10. See also paras 16, 40, 42 and

63. The Committee took evidence from the BOA, the CCCC, the British Council for Rehabilitation, the British Rheumatism Council, and the Red Cross among others.

69. H. Osmond-Clarke, *Accident Services Review Committee of Great Britain & Ireland, Interim Report*, BOA [1961], p. 5.

70. Platt, 'The Special Orthopaedic Hospital Past and Future', *J. Roy. Coll. Surg. Edin.*, 21 (1976), pp. 64–74 at p. 72. In an interview with the author (14 Dec. 1983) he confessed that the first ten years of the NHS 'went swimmingly, because it was just a continuation of the old system, with the hospitals not completely secure in their financial backing'.

BIBLIOGRAPHY

1. Manuscript sources
2. Reports and official publications
3. Orthopaedic sources
   (i) Journals
   (ii) Selected secondary sources
4. Non-orthopaedic sources
   (i) Journals
   (ii) Selected secondary sources

1. MANUSCRIPT SOURCES

Cambridge

Churchill College: Bevin Papers.

Liverpool

University Archives: Minutes of the Faculty's Board of Orthopaedic Studies.
Liverpool Medical Institution: Hugh Owen Thomas Papers; Robert Jones Papers.
Liverpool Public Library: Bickerton Collection, biographical files.
Royal Liverpool Children's Hospital: Minutes of the Executive Board, 1880–1920; *Annual Reports*, 1900–19. Heswall Branch, admission registers, 1899–1933.

London

British Medical Association: BMA–TUC Joint Committee (1936-39); Committee on the Rehabilitation of Disabled Persons (1945–58); Orthopaedic Group Committee Minutes (1938–44); Minutes of the Committee on the Treatment of Fractures, 1910–12.
British Orthopaedic Association: G. R. Girdlestone/Robert Jones Correspondence; Minutes of the Executive Council meetings, 1918–59; H. A. T. Fairbank, Boer War letters; H. Platt and W. R. Bristow, 'The Log of Vienna', 26–29 September 1929.
Greater London Record Office: Public Health files; Welfare Department records; General files; Hospital files.
London Hospital: Material relating to the Accident Relief Society, 1844–1860; *Annual Reports*, 1890–1914.
Manor House Hospital: Operations Register, 1933–42; Minute Books of the Management Committee, from 1917; Collection of press cuttings (2 vols).

Public Record Office, Kew: Board of Education files [ED]; Ministry of Labour files [LAB]; Ministry of Health files [MH]; Ministry of Pensions and National Insurance files [PIN]; Railways files [RAIL]; War Office files [WO].

Royal College of Surgeons, Library: Hugh Owen Thomas, notebook, letter book, and casebook; Evan Thomas and Hugh Owen Thomas, letter book; Minutes of Council, 1918; Minutes of the Committees for Temporary Purposes, 1907–22.

Royal National Orthopaedic Hospital: Minute Books of City Orthopaedic Hospital; Minute Books of National Orthopaedic Hospital; Minute Books of (Royal) Orthopaedic Hospital; Minute Books (1885–1915), Kensington National Home for Crippled Boys.

Wellcome Institute, Contemporary Medical Archives: Alfred Tubby: typescript diary and autobiography, c. 1915–17.

Manchester

University of Manchester, John Rylands Library: Harry Platt Papers; Typescript, Annual Reports, and other material relating to the Greengate Dispensary and School, 1904–59.

University of Manchester, Wellcome Unit for the History of Medicine: Transcripts of interviews with Harry Platt and Norman Roberts.

Cripples' Aid Society: Minute Books of the Crippled Children's Help Society, 1906–46.

Oxford

Bodleian Library: University Registry Archives: material concerning Girdlestone's appointment to the Chair of Orthopaedics; H. A. L. Fisher Papers: Girdlestone correspondence; J. Johnson Collection: box files on the Great War.

Shrewsbury

Shropshire County Record Office: Annual Reports of Baschurch Convalescent Home; Material relating to Robert Jones and Agnes Hunt Orthopaedic Hospital.

2. REPORTS AND OFFICIAL PUBLICATIONS (chronological)

Charity Organization Society. *Report of the Sub-Committee Appointed to Consider the Means By Which the Abuses of the Out-Patient Departments of General Hospitals May Best Be Remedied*, 1870.

Report from the Select Committee of the House of Lords on Metropolitan Hospitals, together with . . . Minutes of Evidence, 1890. Cmd 392.

Second and Third Report from the Select Committee of the House of Lords on Metropolitan Hospitals, together with . . . Minutes of Evidence, 1891. Cmd 457.

Charity Organization Society. *The Epileptic and Crippled Child and Adult*, 1893.

Committee on Defective and Epileptic Children (Education Department). *Report of the Departmental Committee on Defective and Epileptic Children.* Volume II: *Minutes of Evidence, Appendices, etc, presented to both Houses of Parliament,* 1898. Cmd 8747.

Departmental Committee on the Ambulance Service in the Metropolis. *Report of the Ambulance Committee.* Volume I: *Report and Appendices,* 1909. Cmd 4563; Volume II: *Minutes of Evidence with Index,* 1909. Cmd 4564.

Board of Education. *Annual Reports of the Chief Medical Officer,* 1909–35.

City of Birmingham Education Committee. *Report of a Special Sub-Committee of Enquiry Concerning Physically-Defective Adults and Children, Presented to the Education Committee, 27 October 1911.*

King Edward's Hospital Fund for London. *Report of Committee on Out-Patients,* July 1912.

British Medical Association. 'Report on the Treatment of Simple Fractures'. *British Medical Journal,* 30 November 1912, pp.1505–41.

Leasowe Open-Air Hospital for Children, Liverpool. *Annual Reports,* 1913–35.

Royal Commission on Mining. Reports of the Minutes of Evidence on the Second Stage of the Enquiry, 1919. Cmd 360. [Chairman, Lord Sankey].

Ministry of Health. *An Outline of the Practice of Preventive Medicine. A Memorandum Addressed to the Minister of Health by Sir George Newman,* 1919. Cmd 363.

Department Committee on Workmen's Compensation. Minutes of Evidence, 1920. Cmd 816. [Holman Gregory Report].

Ministry of Health. Consultative Council on Medical and Allied Services. *Interim Report on the Future Provision of Medical and Allied Services,* 1920. Cmd 693. [Dawson Report].

Reports by the Joint War Committee and the Joint War Finance Committee of the British Red Cross Society and the Order of St. John of Jerusalem in England on Voluntary Aid Rendered to the Sick and Wounded at Home and Abroad and to British Prisoners of War, 1914-1919, with appendices. London, HMSO, 1921.

Miners' Welfare Fund. *Annual Reports,* 1921–38.

Ministry of Health. *Voluntary Hospitals' Committee, Interim Report,* 1921. Cmd 1206. [Cave Report].

King Edward's Hospital Fund for London. *Ambulance Case Disposal Committee: Report of a Special Committee.* London, 1924.

Ministry of Health. *Voluntary Hospitals' Committee, Final Report,* 1928. Cmd 1335. [Cave Report].

Royal Commission on National Health Insurance, Report, 1928. Cmd 2596.

British Medical Association. 'Report of the Committee on Fractures'. *British Medical Journal,* supplement, 16 February 1935, pp. 53–62.

Political and Economic Planning. *Report on the British Health Services.* London: P.E.P., 1937.

Interim Report of the Inter-Departmental Committee on the Rehabilitation of Persons Injured by Accidents. London: HMSO, 1937. [Chairman, Malcolm Delevinge].

Report by the Departmental Committee on Certain Questions Arising under the Workmen's Compensation Act, 1938. Cmd 5657. [Chairman, Charles Stewart].

Final Report of the Inter-Departmental Committee on the Rehabilitation of Persons Injured by Accidents. London: HMSO, 1939. [Chairman, Malcolm Delevingne].

Voluntary Hospitals' Committee for London. *Organised Fracture Services for London: Report by Fracture Sub-Committee.* June 1939.

Royal Commission on Workmen's Compensation, Minutes of Evidence, 1939–40. Cmd 6588. [Chairman, Sir Hector Hetherington].

Social Insurance and Allied Services, 1942. Cmd 6404. [Beveridge Report] Also published by Macmillan, New York, 1942.

Report of the Inter-Departmental Committee on the Rehabilitation and Resettlement of Disabled Persons, 1942–3. Cmd 6415. [Chairman, George Tomlinson].

British Orthopaedic Association. *Memorandum on Fracture and Accident Services Committee,* 1943.

Ministry of Health. *Report of the Inter-Departmental Committee on Medical Schools.* London: HMSO, 1944. [Goodenough Report].

Ministry of Health. *A National Health Service, 1944.* Cmd 6502.

Ministry of Health. *Hospital Survey: The Hospital Services of the North-Western Area.* London: HMSO, 1945. Written by E. Rock Carling and T. S. McIntosh.

Ministry of Health. *National Health Service: The Development of Consultant Services.* London: HMSO, 1950.

Ministry of Labour and National Services. *Report of the Committee of Inquiry on the Rehabilitation, Training and Resettlement of Disabled Persons,* 1956, Cmd 9883.

Nuffield Provincial Hospitals' Trust. *Casualty Services and Their Setting: A Study in Medical Care.* Oxford: NPHT, 1960.

British Orthopaedic Society. *Accident Services Committee of Great Britain and Ireland, Interim Report,* 1961. Compiled by H. Osmond-Clarke.

British Medical Association. *Medical Staffing of Accident and Emergency Services: A Report Prepared on Behalf of the Joint Consultants Committee and Presented in April 1978.* Written by Walpole Lewin.

3. ORTHOPAEDIC SOURCES

(i) Journals

American Journal of Care for Cripples; Official Organ of the Federation of Associations for Cripples and the Welfare Commission for Cripples, edited by D. C. McMurtrie, 1914–19.

American Journal of Orthopaedic Surgery, 1902–19

Archives of Physical Medicine and Rehabilitation, 1920–

British Journal of Children's Diseases, 1904–44

British Journal of Industrial Medicine, 1944–

Cripples' Journal, 1925–1930

Journal of Bone and Joint Surgery, 1922–

Journal of Orthopaedic Surgery, 1919–22

Physical Medicine and Rehabilitation, 1920–
Physiotherapy. The Journal of the Chartered Society of Physiotherapy, 1915–
Recalled to Life, 1917–18, retitled *Reveille*, 1918–19
Rehabilitation Literature, 1940–
Rehabilitation Review: Devoted to the Restoration and Employment of the Disabled, 1926–38
Society for the Study of Diseases in Children, Annual Reports, 1900–1908
Transactions of the American Orthopaedic Association, 1887–1902
Transactions of the British Orthopaedic Society, 1896–99
Year Book of Industrial and Orthopaedic Surgery, ed. Charles Painter, 1940–7; subsequently titled the *Year Book of Orthopaedics, Traumatic and Plastic Surgery*

(ii) Selected secondary sources on orthopaedics
(reflective, historical, biographical and bibliographical only)

Abt, Henry Edward. *The Care, Cure, and Education of the Crippled Child, . . . a complete bibliography of literature bearing on this subject; and a complete directory of institutions and agencies engaged in this work*. Elyria, Ohio: International Society for Crippled Children, 1924.
Aitken, D. McCrae. *Hugh Owen Thomas, His Principles and Practice*. London: Oxford University Press, 1935.
Albee, Fred H. *A Surgeon's Fight to Rebuild Men: An Autobiography*. London: R. Hals, 1950.
Anderson, Joan. *A Record of Fifty Years' Service to the Disabled, 1919–1969 by the Central Council for the Disabled*. London: CCD, 1970.
Anderson, Roy N. *The Disabled Man and his Vocational Adjustment: A Study of the Types of Jobs Held by 4,404 Orthopaedic Cases in Relation to the Specific Disability*. New York: Institute for the Crippled and Disabled, 1932.
Barker, Sir Herbert. *Leaves From My Life*. London: Hutchinson, 1927.
Bick, Edgar M. *Source Book of Orthopaedics*. Baltimore: Williams & Wilkins, 1937.
Bristow, W. Rowley. 'Orthopaedic Surgery – Retrospect and Forecast'. *Lancet*, 6 November, 1937, 1061–4.
Brown, Thornton. *The American Orthopaedic Association: A Centennial History*. N.p.: American Orthopaedic Association, 1987.
Buxton, St J. D. 'Sir Thomas Fairbank'. Special issue of the *Journal of Bone and Joint Surgery*, 38B (1956).
Central Council for the Care of Crippled Children. *Directory of Orthopaedic Institutions, Voluntary Organisations and Official Schemes for the Welfare of Cripples*. London: CCCC, 1935.
Cholmeley, John A. *History of the Royal National Orthopaedic Hospital*. London: Chapman & Hall, 1985.
Cohen, J. 'Orthopaedics'. In J. Walton, P. B. Brown and R. B. Scott, eds. *Oxford Companion to Medicine*. Oxford: Oxford University Press, 1986, 953–61.
Cooter, Roger. 'The Meaning of Fractures: Orthopaedics and the Reform of British Hospitals in the Inter-war Period'. *Medical History*, 31 (1987), 306–31.

Cyriax, James and Eiler H. Schiotz. *Manipulation Past and Present, With an Extensive Bibliography*. London: Heinemann, 1975.

Eastwood, W. J. 'Orthopaedics: Old and New'. *Liverpool Medico-Chirurgical Journal*, 45 (1937), 186–94.

Freiberg, A. H. 'Orthopaedic Surgery in the Light of its Evolution'. *Journal of Bone and Joint Surgery*, 15 (1933), 279–301.

Gibney, Robert. *Gibney of the Ruptured & Crippled*, edited by Alfred R. Shands, Jr. New York: Appleton-Century-Crofts, 1969.

Girdlestone, G. R. 'The Robert Jones Tradition. A lecture given to the staff of the Wingfield-Morris Orthopaedic Hospital, Oxford'. Printed for private circulation, 1947.

Goldthwait, Joel E. *The Division of Orthopaedic Surgery in the A.E.F.* Norwood, Mass.: Plimpton Press, 1941.

Goodwin, George M. *Russell A. Hibbs, Pioneer in Orthopaedic Surgery, 1869–1932*. New York: Columbia University Press, 1935.

Gressmann, C. *Tradition und Fortschritt in der Orthopädie*. Frankfurt: Georg Thieme Verlag, 1985.

Harvey, A. McGehee. 'Orthopedic Surgery at Johns Hopkins'. *Johns Hopkins Medical Journal*, 150 (1982), 221–45.

Heck, Charles V. (compiler). *Fifty Years of Progress in Recognition of the 50th Anniversary of the American Academy of Orthopaedic Surgeons*. Chicago: American Academy of Orthopaedic Surgeons, 1983.

Hey Groves, Ernest. 'A Surgical Adventure: An Autobiographical Sketch'. Reprinted from the *Bristol Medico-Chirurgical Journal*, 50 (1933).

Hunt, Agnes G. *The Story of Baschurch*. Oswestry: Woodhall, n.d.

——. *This is My Life*. London: Blackie, 1938.

Heritage of Oswestry: The Origins and Development of the Robert Jones and Agnes Hunt Orthopaedic Hospital (1900 to 1975). Published by the Hospital, 1975.

Jones, A. Rocyn. 'The Evolution of Orthopaedic Surgery in Great Britain'. *Proceedings of the Royal Society of Medicine*, 31 (1936–7), 19–26.

——. 'A Review of Orthopaedic Surgery in Britain'. *Journal of Bone and Joint Surgery*, 38B (1956), 27–45.

Jones, Robert. 'The Cripple: A Retrospect and a Forecast'. *Cripples' Journal*, 5 (Oct. 1928), 5–18.

——. 'Lady Jones Lecture on Crippling Due to Fractures: Its Prevention and Remedy'. *British Medical Journal*, 16 May 1925, 909–13.

——. 'Orthopaedic Surgery'. *Encyclopaedia Britannica*. 1922 supplement to the 11th edition covering the period 1911–21.

——. 'A Chronological List of Sir Robert Jones's Contributions to Surgical Literature'. *Journal of Bone and Joint Surgery*, 39B (1957), 212–17.

Keith, Arthur. *Menders of the Maimed: The Anatomical and Physiological Principles Underlying the Treatment of Injuries to Muscles, Nerves, Bones and Joints*. London: Hodder & Stroughton, 1919.

Kessler, Henry H. *The Crippled and the Disabled: Rehabilitation of the Physically Handicapped in the United States*. New York: Columbia University, published PhD thesis, 1935.

Le Vey, David. *The History of Orthopaedics: An Account of The Study and Practice of Orthopaedics From the Earliest Times to the Modern Era.* Carnforth, Lancashire and Park Ridge, New Jersey: Parthenon, 1990.

——. *The Life of Hugh Owen Thomas.* Edinburgh: E. & S. Livingstone, 1956.

Little, Ernest Muirhead. 'Orthopaedics Before Stromeyer'. In *The Robert Jones Birthday Volume.* London: Oxford University Press, 1928, 1–26.

Lorenz, Adolf. *My Life and Work: The Search for a Missing Glove.* New York: Scribners, 1936.

MacAlister, Charles J. *The Origin and History of the Liverpool Royal Southern Hospital with Personal Reminiscences.* Liverpool: W. B. Jones, 1936.

——. *The Origin and History of the Royal Liverpool Country Hospital for Children at Heswall from the Time of its Inception (1895–1898) and Foundation (1899) to the Year 1930.* Campden, Gloucestershire: Alcuin Press, 1930.

McMurtrie, Douglas. *Bibliography of the Education and Care of Crippled Children.* New York: Published by the author, 1913.

Mayer, Leo. 'Orthopaedic Surgery in the United States of America'. *Journal of Bone and Joint Surgery,* 32B (1950), 461–569.

——. 'Reflections on Some Interesting Personalities in Orthopaedic Surgery During the First Quarter of the Century'. *Journal of Bone and Joint Surgery,* 37A (1955), 374–83.

Mercer, Sir Walter. 'Some Edinburgh Pioneers in Orthopaedic Surgery.' *Annals of Royal College of Surgeons of England,* 26 (1960), 339–61.

Orr, H. Winnett. *On the Contributions of Hugh Owen Thomas of Liverpool, Sir Robert Jones of Liverpool and London, John Ridlon, M.D. of New York and Chicago to Modern Orthopedic Surgery.* Springfield, Illinois: Charles C. Thomas, 1949.

Osgood, Robert. *The Evolution of Orthopaedic Surgery.* St Louis: C. V. Mosley, 1925.

Osmond-Clarke, H. 'Half a Century of Orthopaedic Progress in Great Britain'. *Journal of Bone and Joint Surgery,* 32B (1950), 620–75.

Paul, John R. *A History of Poliomyelitis.* New Haven: Yale University Press, 1971.

Platt, Harry. 'The Care of the Physically Handicapped Child'. *Manchester University Medical Gazette,* 45 (1964), 4–11.

——. 'The Evolution of the Treatment of Fractures'. *Manchester University Medical School Gazette,* 17 (1938), 56–62.

——. 'Orthopaedic Surgery in Boston'. *Medical Chronicle,* 58 (March 1914), 473–9.

——. *Selected Papers.* Edinburgh: E. & S. Livingstone, 1963.

Pritchard, D. G. *Education and the Handicapped, 1760–1960.* London: Routledge, 1963.

Rogers, William A. *et al.,* eds. 'Half a Century of Progress in Orthopaedic Surgery'. Special issue of *The Journal of Bone and Joint Surgery,* 32B (1950), 451–778.

—— . 'Dr. Leo Mayer: An Appreciation'. *Bulletin of the Hospital for Joint Diseases,* 21 (1960), 81–5.

Ruttimann, Beat. *Wilhelm Schulthess (1855–1917) und die Schweizer Orthopädie seiner Zeit*. Zurich, Schulthess Polygraphischer, 1983.

Schleichkorn, Jay. *'The Sometime Physician': William John Little, Pioneer in Treatment of Cerebral Palsy and Orthopedic Surgery (1810–1894)*. Farmingdale, New York: Published by the Author, 1987.

Shands, A. R., Jr. *The Early Orthopaedic Surgeons of America*. St Louis, Missouri: C.V. Mosby, 1970.

Thornton, J. L. 'Orthopaedic Surgery at St. Bartholomew's Hospital, London'. *St. Bartholomew's Hospital Journal*, 59 (1955), 195–204.

Trueta, Josep. *Gathorne Robert Girdlestone*. Oxford: Oxford University Press, 1971.

——. *Trueta: Surgeon in War and Peace, the Memoirs of Joseph Trueta, M.D.* London: Gollancz, 1980.

Tubby, Alfred H. *The Advance of Orthopaedic Surgery*. London: Lewis, 1924.

Valentin, Bruno. *Geschichte der Orthopädie*. Stuttgart: Thieme, 1961.

Watson, Frederick. *The Life of Sir Robert Jones*. London: Hodder & Stoughton, 1934.

Waugh, William. *The Development of Orthopaedics in the Nottingham Area*. Nottingham: Harlow Wood Orthopaedic Hospital, 1988.

——. *John Charnley, The Man and the Hip*. London: Springer-Verlag, 1990.

West, John L. *The Taylors of Lancashire: Bonesetters and Doctors, 1750–1890*. Walkden, Worsley: H. Duffy, 1977.

Whitley, J. 'What America is Doing for Her Civil and Industrial Cripples'. *Cripples' Journal*, 4 (1927/8), 115–22, 193–204, 311–16.

Whitman, Royal. 'The Emancipation of Orthopaedic Surgery'. *Proceedings of the Royal Society of Medicine*, 36 (May 1943), 327–9.

——. 'A Review of the Evolution of the Orthopaedic Branch of Surgery in New York City'. *Journal of Bone and Joint Surgery*, 29 (1947), 250–3.

Woodall, Samuel J. *The Manor House Hospital, A Personal Record*. London: Routledge, 1966.

4. NON-ORTHOPAEDIC SOURCES

(i) Journals

Archives of Pediatrics
Bristol Medico-Chirurgical Journal
British Journal of Children's Diseases
British Journal of Industrial Medicine
British Journal of Surgery
British Medical Journal
Charities Annual and Register
Charity Organization Review
Child
Engineering
Hospital
Hospitals and Charities Annual

Irish Journal of Medical Science
Journal of the American Medical Association
Journal of Industrial Hygiene
Lancet
Liverpool Medico-Chirurgical Journal
Manchester University Medical School Gazette
Medical Annual
Medical Directory
Medical Officer
Medical Press and Circular
Medical Record
Medical Times and Gazette
Newcastle Medical Journal
New York Medical Journal
Proceedings of the Royal Society of Medicine
Practitioner
St Bartholomew's Hospital Journal
Transactions of the Hunterian Society
Westminster Review

(ii) Selected non-orthopaedic secondary sources

Abel-Smith, Brian. *The Hospitals, 1800–1948*. London: Heinemann, 1964.
Abrams, P. 'The Failure of Social Reform: 1918–1920'. *Past and Present*, 24 (1963), 43–64.
Addison, Paul. *The Road to 1945: British Politics and the Second World War*. London: Quartet, 1977.
Aldcroft, Derek H. *The British Economy*, vol. 1: *The Years of Turmoil, 1920–1951*. Brighton: Wheatsheaf, 1986.
Armstrong, David. *Political Anatomy of the Body: Medical Knowledge in Britain in the Twentieth Century*. Cambridge: Cambridge University Press, 1983.
Austoker, Joan and L. Bryder, eds. *Historical Perspectives on the Role of the MRC: Essays in the History of the Medical Research Council of the United Kingdom and its Predecessor, the Medical Research Committee, 1913–1953*. Oxford: Oxford University Press, 1989.
Ayers, Gwendoline M. *England's First State Hospitals and the Metropolitan Asylums Board, 1867–1930*. London: Wellcome Institute, 1971.
Bartrip, Peter W. J. *Workmen's Compensation in Twentieth Century Britain: Law, History and Social Policy*. Aldershot: Gower, 1987.
Bartrip, Peter W. J. and S. B. Burman. *The Wounded Soldiers of Industry: Industrial Compensation Policy 1833–1897*. Oxford: Oxford University Press, 1983.
Bateman, Donald. *Berkeley Moynihan, Surgeon*. London: Macmillan, 1940.
Behlmer, George K. *Child Abuse and Moral Reform in England, 1870–1908*. Stanford: Stanford University Press, 1982.
Berkowitz, Edward D. 'The Federal Government and the Emergence of Rehabilitation Medicine'. *The Historian*, 43 (1981), 530–45.

Berridge, Virginia. 'Health and Medicine'. In F.M.L. Thompson, ed., *The Cambridge Social History of Britain, 1750–1950*, vol. 3: *Social Agencies and Institutions*. Cambridge: Cambridge University Press, 1990, pp. 171–242.

Beveridge, William. *Voluntary Action: A Report on Methods of Social Advance*. London: Allen & Unwin, 1948.

Bickerton, Thomas H. *A Medical History of Liverpool from the Earliest Days to the Year 1920*. London: John Murray, 1936.

Bonner, Thomas N. *The Kansas Doctor: A Century of Pioneering*. Lawrence, Kansas: University of Kansas Press, 1959.

Bosanquet, Helen. *Social Work in London 1869 to 1912: A History of the Charity Organisation Society*. London: John Murray, 1914.

Bowman, A.K. *The Life and Teaching of Sir William Macewan: A Chapter in the History of Surgery*. London: William Hodge & Co, 1942.

Brand, Jeanne L. *Doctors and the State: The British Medical Profession and Government Action in Public Health, 1870–1912*. Baltimore: Johns Hopkins University Press, 1965.

Bryder, Linda. *Below the Magic Mountain: A Social History of Tuberculosis in Twentieth-Century Britain*. Oxford: Clarendon Press, 1988.

Bullock, Alan. *The Life and Times of Ernest Bevin*. vol. 2: *Minister of Labour 1940–45*. London: Heinemann, 1967.

Burrow, James G. *Organized Medicine in the Progressive Era: The Move Toward Monopoly*. Baltimore: Johns Hopkins University Press, 1977.

Bynum, W.F. *et al.*, eds. *The Emergence of Modern Cardiology*. *Medical History*, Supplement No. 5 (1985).

Cantor, David. 'The Aches of Industry: Philanthropy and Rheumatism in Inter-war Britain'. In Jonathan Barry and Colin Jones, eds. *Medicine and Charity in Western Europe Before the Welfare State*. London: Routledge, 1991.

——. 'The Contradictions of Specialization: Rheumatism and the Decline of the Spa in Inter-War Britain'. In Roy Porter, ed., *The Medical History of Waters and Spas*. *Medical History*, Supplement No. 10 (1990), 127–44.

——. 'Cortisone and the Politics of Empire, 1918–1955'. *Bulletin of the History of Medicine* (forthcoming).

Cartwright, Frederick. *The Development of Modern Surgery from 1830*. London: A. Barker, 1967.

Chandler, Alfred D., Jr. *The Visible Hand: The Managerial Revolution in American Business*. Cambridge, Mass.: Harvard University Press, 1977.

Clapesattle, Helen. *The Doctors Mayo*. New York: Garden City Publishing Co., 1943.

Cliff, Kenneth Stephenson. *Accidents: Causes, Prevention and Services*. London: Croom Helm, 1984.

Cooter, Roger. 'Bones of Contention? Orthodox Medicine and the Mystery of the Bone-setter's Craft'. In W.F. Bynum and Roy Porter, eds. *Medical Fringe and Medical Orthodoxy, 1750–1850*. London: Croom Helm, 1987, 158–73.

——. 'Medicine and the Goodness of War'. *Canadian Bulletin of Medical History*, 7 (1990), 147–59.

——, ed. *In the Name of the Child: Health and Welfare, 1880–1940*. London: Routledge, 1992.

——. 'War and Modern Medicine'. In W. F. Bynum and Roy Porter, eds. *Encyclopedia of the History of Medicine*. London: Routledge (forthcoming).

Cope, Sir Zachary, ed. *Surgery*. (*History of the Second World War: Surgery*.) London: HMSO, 1953. 2 vols.

Cox, Alfred.*Among the Doctors*. London: Christopher Johnson, 1950.

Craig, W. J. *John Thomson: Pioneer and Father of Scottish Paediatrics, 1856–1926*. Edinburgh: E. & S. Livingstone, 1968.

Cushing, Harvey. *From a Surgeon's Journal 1915–1918*. London: Constable, 1936.

——. 'The Society of Clinical Surgery in Retrospect'. *Annals of Surgery*, 169 (1969), 1–9.

Davidson, Maurice. *The Royal Society of Medicine: The Realization of an Ideal, 1805–1955*. London: Royal Society of Medicine, 1955.

Davis, Loyal. *Surgeon Extraordinary: The Life of J. B. Murphy*. London: G. G. Harrop & Co., 1938.

Digby, Anne and Nick Bosanquet. 'Doctors and Patients in an Era of National Health Insurance and Private Practice, 1913–1938'. *Economic History Review*, 41 (1988), 74–94.

Dunn, C. L, ed. *The Emergency Medical Services*. (*History of the Second World War*.) London: HMSO, 1952. 2 vols.

Dwork, Deborah. *War is Good for Babies and Other Young Children: A History of the Infant and Child Welfare Movement in England, 1898–1918*. London: Tavistock, 1987.

Earwicker, Roy. 'A Study of the BMA–TUC Joint Committees on Medical Questions 1935–1939'. *Journal of Social Policy*, 8 (1979), 335–56.

Eckstein, Harry. *The English Health Service: Its Origins, Structure and Achievement*. Cambridge, Mass.: Harvard University Press, 1959.

——. *Pressure Group Politics: The Case of the British Medical Association*. London: Allen & Unwin, 1960.

Eder, Norman R. *National Health Insurance and the Medical Profession in Britain, 1913–39*. New York: Garland, 1982.

Ehrenreich, Barbara and Deirdre English. *For Her Own Good: 150 Years of the Experts' Advice to Women*. London: Pluto Press, 1979.

Eksteins, Modris. *Rites of Spring: The Great War and the Birth of the Modern Age*. New York: Archor Books, 1989.

Emsley, Clive *et al.*, eds. *War, Peace and Social Change in Twentieth Century Europe*. Milton Keynes: Open University Press, 1989.

English, Peter C. *Shock, Physiological Surgery, and George Washington Crile: Medical Innovation in the Progressive Era*. Westport, Conn.: Greenwood Press, 1980.

Fee, Elizabeth and Roy M. Acheson, eds. *A History of Education in Public Health*. Oxford/New York: Oxford University Press, 1991.

Finney, J. M. T. *A Surgeon's Life: The Autobiography of J. M. T. Finny*. New York: Putnam's sons, 1940.

Fox, Daniel M. *Health Policies, Health Politics: The British and American Experience, 1911–1965*. Princeton: Princeton University Press, 1986.

Fraser, Derek. *The Evolution of the British Welfare State: A History of Social Policy Since the Industrial Revolution*. 2nd edition. London: Macmillan, 1984.

Fulton, John F. *Harvey Cushing: A Biography*. Springfield, Illinois: C.C. Thomas, 1946.

Fussell, Paul. *The Great War and Modern Memory*. Oxford: Oxford University Press, 1975.

Gilbert, Bentley B. *The Evolution of National Insurance in Britain: The Origins of the Welfare State*. London: M. Joseph, 1966.

Goodman, Neville M. *Wilson Jameson: Architect of National Health*. London: Allen & Unwin, 1970.

Gore, John F. *Sydney Holland: Lord Knutsford, A Memoir*. London: John Murray, 1936.

Granshaw, Lindsay. *St. Mark's Hospital, London: A Social History of a Specialist Hospital*. London: King Edward's Hospital Fund, 1985.

Granshaw, Lindsay and Roy Porter, eds. *The Hospital in History*. London: Routledge, 1989.

Green, David G. *Working-Class Patients and the Medical Establishment: Self-help in Britain from the Mid-Nineteenth Century to 1948*. London: Gower/Temple Smith, 1985.

Gritzer, Glenn and Arnold Arluke. *The Making of Rehabilitation. A Political Economy of Medical Specialization, 1890–1986*. Berkeley: University of California Press, 1985.

Guthrie, D., ed. *The Royal Edinburgh Hospital for Sick Children 1860–1960*. Edinburgh: E. & S. Livingstone, 1960.

Halpern, Sydney A. *American Pediatrics: The Social Dynamics of Professionalism, 1880–1980*. Berkeley: University of California Press, 1988.

Harris, José. *William Beveridge: A Biography*. Oxford: Clarendon Press, 1977.

Hazlett, T. Lyle and William W. Hummel. *Industrial Medicine in Western Pennsylvania, 1850–1950*. Pittsburgh: University of Pittsburgh Press, 1957.

Hodgkinson, Ruth. *The Origins of the National Health Service: Medical Services and the New Poor Law, 1834–1871*. London: Wellcome Institute, 1967.

Holland, Sydney. *In Black and White*. London: E. Arnold, 1926.

Honigsbaum, Frank. *The Division in British Medicine: A History of the Separation of General Practice from Hospital Care, 1911–1968*. London: Kogan Page, 1979.

——. *Health, Happiness and Security: The Creation of the National Health Service*. London: Routledge, 1989.

——. *The Struggle for the Ministry of Health, 1914–1919*. London: Ball, 1970.

Howell, Joel. '"Soldier's Heart": The Redefinition of Heart Disease and Speciality Formation in Early Twentieth-Century Great Britain'. In W.F. Bynum *et al.*, eds. *The Emergence of Modern Cardiology. Medical History*, Supplement No. 5, 1985, 34–52.

Hunter, Donald. *Health in Industry*. Harmondsworth: Penguin, 1959.

Johnson, Paul Barton. *Land Fit for Heroes: The Planning of British Reconstruction 1916–1919*. Chicago: University of Chicago Press, 1968.

Jones, Helen. 'Employers' Welfare Schemes and Industrial Relations in Inter-War Britain'. *Business History*, 25 (1983), 61–75.

Keynes, Geoffrey. *The Gates of Memory*. Oxford: Oxford University Press, 1981.

Klein, Rudolf. *The Politics of the National Health Service*. London: Longman, 1983.

Kunitz, Stephen J. 'Efficiency and Reform in the Financing and Organization of American Medicine in the Progressive Era'. *Bulletin of the History of Medicine*, 55 (1981), 497–515.

Larkin, Gerald. *Occupational Monopoly and Modern Medicine*. London: Tavistock, 1983.

Lawrence, Christopher. 'Incommunicable Knowledge: Science, Technology and the Clinical Art in Britain 1850–1914'. *Journal of Contemporary History*, 20 (1985), 503–20.

——. 'Moderns and Ancients: The "New Cardiology" in Britain 1880–1930'. In W. F. Bynum *et al.*, eds, *The Emergence of Modern Cardiology. Medical History*, Supplement No. 5, 1985, 34–52.

——, ed. *Medical Theory, Surgical Practice: Studies in the History of Surgery*. London: Routledge, 1992.

Layton, T. B. *Sir William Arbuthnot Lane: An Enquiry into the Mind and Influence of a Surgeon*. Edinburgh: E. & S. Livingstone, 1956.

Learmonth, James Rognväld. *The Thoughtful Surgeon*. Edited by Donald M. Douglas. Glasgow: University of Glasgow Press, 1969.

Levy, Hermann. 'The Economic History of Sickness and Medical Benefit Since the Puritan Revolution'. *Economic History Review*, 14 (1944), 135–62.

Lewis, Jane. *The Politics of Motherhood: Child and Maternal Welfare in England, 1900–1939*. London: Croom Helm, 1980.

——. *What Price Community Medicine? The Philosophy, Practice and Politics of Public Health Since 1919*. Brighton: Harvester, 1986.

Long, Diana and Janet Golden, eds. *The American General Hospital: Communities and Social Contexts*. Ithaca, New York: Cornell University Press, 1989.

Lowe, Rodney. 'The Erosion of State Intervention in Britain, 1917–1924'. *Economic History Review*, 31 (1978), 270–86.

Lyons, J. B. *The Citizen Surgeon: A Biography of Sir Victor Horsley, FRS, FRCS, 1857–1916*. London: Downay Ltd, 1966.

Macpherson, W. G. *et al. Medical Services in the History of the Great War*. London: HMSO, 1922–31. 12 vols.

Martin, Franklin H. *The Joy of Living: An Autobiography*. Long Island: Doubleday, Doran & Co., 1933. 2 vols.

Marwick, Arthur. *The Deluge: British Society and the First World War*. London: Macmillan, 1965.

——. 'The Impact of the First World War on British Society'. *Journal of Contemporary History*, 3 (1968), 51–63.

——. 'The Labour Party and the Welfare State in Britain, 1900–48'. *American Historical Review*, 73 (1967), 380–403.

Maulitz, Russell and D. Long, eds. *Grand Rounds: One Hundred Years of Internal Medicine*. Philadelphia: University of Pennsylvania Press, 1988.

Mayo, Charles W. *Mayo: The Story of My Family and My Career*. New York: Doubleday, 1968.

Morgan, Kenneth and Jane. *Portrait of a Progressive: The Political Career of Christopher, Viscount Addison*. Oxford: Clarendon Press, 1980.

Morman, Edward T., ed. *Efficiency, Scientific Management, and Hospital Standardization: An Anthology of Sources.* New York: Garland, 1989.

Moscucci, Ornella. *The Science of Woman: Gynaecology and Gender in England, 1800–1929.* Cambridge, Cambridge University Press, 1990.

Mosse, George L. *Fallen Soldiers: Reshaping the Memory of the World Wars.* New York: Oxford University Press, 1990.

Mowat, Charles Loch. *Britain Between the Wars, 1918–1940.* London: Methuen, 1956.

——. *The Charity Organization Society 1869–1913: Its Ideas and Work.* London: Methuen, 1961.

Munts, Raymond. *Bargaining for Health. Labor Unions, Health Insurance and Medical Care.* Madison: University of Wisconsin Press, 1967.

Navarro, Vicente. *Class Struggle, the State and Medicine: An Historical and Contemporary Analysis of the Medical Sector in Great Britain.* Oxford: Martin Robertson, 1978.

Neuburger, Max. *The Doctrine of the Healing Power of Nature Throughout the Course of Time.* Trans. L. J. Boyd. New York: printed, 1942. (German edition, 1926.)

Newman, George. *The Building of a Nation's Health.* London: Macmillan, 1939.

Newsholme, Arthur. *Fifty Years in Public Health: A Personal Narrative.* London: Allen & Unwin, 1935.

——. *The Last Thirty Years in Public Health: Recollections and Reflections on My Official and Post-Official Life.* London: Allen & Unwin, 1936.

——. *Medicine and the State: The Relations Between the Private and Official Practice of Medicine with Special Reference to Public Health.* London: Allen & Unwin, 1932.

Ogilvie, William Heneage. *Surgery: Orthodox and Heterodox.* Oxford: Blackwell Medical Publications, 1948.

Parry, Noel and José. *The Rise of the Medical Profession.* London: Croom Helm, 1976.

Perkin, Harold. *The Rise of Professional Society: England Since 1880.* London: Routledge, 1989.

Pickering, George. *Quest for Excellence in Medical Education: A Personal Survey.* Oxford: Oxford University Press, 1978.

Pickstone, John V., ed. *Medical Innovations in Historical Perspective.* London: Macmillan, 1992.

——. *Medicine and Industrial Society: A History of Hospital Development in Manchester and Its Region, 1752–1946.* Manchester: Manchester University Press, 1985.

Pinker, Robert. *English Hospital Statistics, 1861–1938.* London: Heinemann, 1966.

Plowden, William. *The Motor Car and Politics in Britain 1896–1970.* Harmondsworth: Penguin, 1971.

Pollard, Sidney. *The Development of the British Economy, 1914–1980.* London: Edward Arnold, 1983.

Prochaska, Frank. *Philanthropy and the Hospitals of London: The King's Fund, 1897–1990.* Oxford: Clarendon Press, 1992.

——. *Women and Philanthropy in Nineteenth Century England.* Oxford: Clarendon Press, 1980.

Rabinbach, Anson. 'The Body Without Fatigue: A Nineteenth Century Utopia'. In Seymour Drescher, D. Sabeen and A. Sharlin, eds, *Political Symbolism in Modern Europe.* New Brunswick: Transaction Books, 1982, 42–62.

——. *The Human Motor: Energy, Fatigue, and the Origins of Modernity.* Berkeley: University of California Press, 1992.

Reverby, Susan. *Ordered to Care: The Dilemma of American Nursing, 1850–1945.* New York: Cambridge University Press, 1987.

Rivett, Geoffrey. *The Development of the London Hospital System, 1823–1982.* London: King Edward's Hospital Fund, 1986.

Rooff, Madeline. *Voluntary Societies and Social Policy.* London: Routledge, 1957.

Rose, Nikolas. *The Psychological Complex: Psychology, Politics and Society in England, 1869–1939.* London: Routledge, 1985.

Rosen, George. *A History of Public Health.* New York: MD Publications Inc., 1958.

——. *The Specialization of Medicine with Particular Reference to Ophthalmology.* New York: Froben Press, 1944.

——. *The Structure of American Medical Practice, 1875–1941,* edited by Charles Rosenberg. Philadelphia: University of Pennsylvania Press, 1983.

Rosenberg, Charles. *The Care of Strangers: The Rise of America's Hospital System.* New York: Basic Books, 1987.

Rosner, David. *A Once Charitable Enterprise: Hospitals and Health Care in Brooklyn and New York 1885–1915.* New York: Cambridge University Press, 1982.

Rosner, David and Gerald Markowitz. 'The Early Movement for Occupational Safety and Health, 1900–1917'. In J.W. Leavitt and R.N. Numbers, eds. *Sickness and Health in America.* 2nd edition, revised. Madison: University of Wisconsin Press, 1985, 507–21.

Rosner, David and Susan Reverby, eds. *Health Care in America: Essays in Social History.* Philadelphia: Temple University Press, 1979.

Rothstein, William G. *American Medical Schools and the Practice of Medicine, A History.* New York: Oxford University Press, 1987.

Sand, René. *The Advance of Social Medicine.* London: Staples Press, 1952.

Schivelbusch, Wolfgang. *The Railway Journey: The Industrialization of Time and Space in the 19th Century.* Leamington Spa: Berg, 1986.

Searle, Geoffrey R. *The Quest for National Efficiency: A Study in British Politics and Social Thought, 1899–1914.* Oxford: Blackwell, 1971.

Sharpey-Schäfer, E.A. 'The Relations of Surgery and Physiology'. *British Medical Journal,* 27 October 1923, 739–45.

Shepherd, J.A. *Lawson Tait, The Rebellious Surgeon (1845–1899).* Lawrence, Kansas: Coronado Press, 1980.

Smith, Clement A. *The Children's Hospital of Boston.* Boston/Toronto: Little, Brown & Co., 1983.

Smith, F.B. *The People's Health, 1830–1910.* London: Croom Helm, 1979.

——. *The Retreat of Tuberculosis, 1850–1950.* London: Croom Helm, 1988.

Smith, Harold L., ed. *War and Social Change: British Society in the Second World War*. Manchester: Manchester University Press, 1986.

Starr, Paul. *The Social Transformation of American Medicine*. New York: Basic Books, 1982.

Steedman, Carolyn. *Childhood, Culture and Class in Britain: Margaret McMillan, 1860–1931*. London: Virago, 1990.

Stevens, Rosemary. *American Medicine and the Public Interest*. New Haven: Yale University Press, 1971.

——. *In Sickness and in Wealth: American Hospitals in the Twentieth Century*. New York: Basic Books, 1989.

——. *Medical Practice in Modern England: The Impact of Specialization and State Medicine*. New Haven: Yale University Press, 1966.

Stone, J. E. *Hospital Organization and Management (Including Planning and Construction)*. 3rd edition. London: Faber & Faber, 1939.

Sturdy, Steve. 'The Political Economy of Scientific Medicine: Science, Education and the Transformation of Medical Practice in Sheffield, 1890–1922'. *Medical History*, 36 (1992), 125–59.

——. 'From the Trenches to the Hospitals at Home: Physiologists, Clinicians and Oxygen Therapy, 1914–1930'. In J. V. Pickstone, ed., *Medical Innovations in Historical Perspective*. London: Macmillan, 1992, 104–23, 234–45.

Summers, Anne. *Angels and Citizens: British Women as Military Nurses*. London: Routledge, 1988.

Tanner, William E. *Sir W. Arbuthnot Lane: His Life and Work*. London: Baillière, 1946.

Thane, Pat. *The Foundations of the Welfare State*. London: Longman, 1982.

Titmuss, Richard. *Essays on 'The Welfare State'*. 2nd edition. London: Allen & Unwin, 1958.

——. *Problems of Social Policy (History of the Second World War, United Kingdom Civil Services)*, edited by W. K. Hancock. London: HMSO, 1950.

Trombley, Stephen. *Sir Frederick Treves: The Extra-Ordinary Edwardian*. London: Routledge, 1989.

Vogel, Morris. *The Invention of the Modern Hospital: Boston 1870–1930*. Chicago: Chicago University Press, 1980.

Walker, Robert Milnes. *Medical Education in Britain*. London: Nuffield Provincial Hospitals' Trust, 1965.

Wangensteen, Owen H. and Sarah D. Wangensteen. *The Rise of Surgery From Empiric Craft to Scientific Discipline*. Folkestone, Kent: Dawson, 1978.

Watson, Francis. *Dawson of Penn*. London: Chatto & Windus, 1950.

Wear, Andrew, ed. *Medicine in Society: Historical Essays*. Cambridge: Cambridge University Press, 1991.

Webster, Charles, ed. *Biology, Medicine and Society, 1840–1940*. Cambridge: Cambridge University Press, 1981.

——. 'Conflict and Consensus: explaining the British health service'. *Twentieth Century British History*, 1 (1990), 115–51.

——. *The Health Services Since the War*, vol. 1: *Problems of Health Care, The National Health Service Before 1957*. London: HMSO, 1988.

——. 'Healthy or Hungry Thirties'. *History Workshop Journal*, 13 (1982), 110–29.

Weindling, Paul, ed. *The Social History of Occupational Health.* Beckenham, Kent: Croom Helm, 1985.

Whiteside, Noel. 'Counting the Cost: Sickness and Disability Among Working People in An Era of Industrial Recession, 1920–39'. *Economic History Review,* 40 (1982), 228–46.

Williams, Gertrude, *The State and the Standard of Living.* London: P. S. King, 1936.

Williams, John L. *Accidents and Ill-Health at Work.* London: Staples Press, 1960.

Wilson, Arnold and Hermann Levy. *Workmen's Compensation.* Oxford: Oxford University Press, 1939, 1942. 2 vols.

Winter, J. M. *The Great War and the British People.* London: Macmillan, 1986.

Wohl, Anthony S. *Endangered Lives: Public Health in Victorian Britain.* London: Dent & Sons, 1983.

Wright, Peter and Andy Treacher, eds. *The Problem of Medical Knowledge: Examining the Social Construction of Medicine.* Edinburgh: Edinburgh University Press, 1982.

Young, A. F. *Industrial Injuries Insurance: An Examination of British Policy.* London: Routledge & Kegan Paul, 1964.

Zelizer, Viviana. *Pricing the Priceless Child: The Changing Social Value of Children.* New York: Basic Books, 1985.

INDEX

Individual institutions are located by town or country, except in the case of London, where no location is given. Societies and legislation are British unless stated otherwise.

Abbott, E. G. (1870–1938) 280n83
abdominal surgery 47, 124, 243, 302n127, 335n92
Accident Assurance Company 82
Accident Officers'
    Association 313n71, 327n17, 336n12
Accident Relief Society 85
accident services 188, 211, 229–33, 241, 349n76
accidents
    as cause of crippling 68
    and childbirth 60
    and compensation 83; see also insurance companies; workmen's compensation
    conceptualization 79, 80, 181, 282n1
    on docks 87–8
    domestic 189, 330n51
    and employers 102
    and historiography 8
    hospital provision 83, 87–9, 188, 191–3
    increases 211–212
    prevention 79, 86
    public 307–8n21
    v. public health 79
    railway 79, 81, 337n23
    reporting 86
    and rest treatment 20
    road traffic 148, 189–90, 330n51
    statistics on 191, 282n1, 339n41
    street 81, 90, 99, 189–90
    at work 8, 81–5, 140–1, 189, 205–6, 221, 307n21, 308n21, n30, 330n51
    victims 85–7

Adams, William (1820–1900) 26, 256n20, 257n34, 260n64, 263n28
Addison, Christopher 278n69
Adkins, Ryland 318n33
Aetna Life Insurance
    Company 313n66
after-care
    clinics 153, 167, 178
    concept and practice 38, 58, 77, 91, 121, 134, 138, 147, 157, 183, 200–1, 205, 207
air-raids 221, 223, 224, 225, 226, 227, 228
Aitken, David McCrae
    (1876–1954) 128, 300n93, 322n72
Albee, Fred 139, 142, 143, 144, 263n20, 350n8
Albert Dock Hospital see Royal
    Victoria
Alexander, William
    (1844–1919) 62–3, 276n48
Alexandra Hip Hospital 21, 54, 259n56
Alexandra Therapeutic
    Institute 305n149
Alton, Lord Trealor's Cripples'
    Home and Hospital 56, 60, 68, 69, 75, 173
Amar, Jules 123
ambulances 82, 94–9, 188
American Academy of
    Orthopaedics 235
American Academy of Physical
    Medicine 347n49
American Civil War 97, 98
American College of Surgeons 50, 129, 270n102, 308n29

American Congress of Physical
Therapy 347n49
*American Journal for the Care of
Cripples* 71
American Medical Association 34,
129, 139, 140, 141, 327n22
American Orthopaedic
Association 26, 46, 77, 81, 127,
135, 136, 171, 179, 235
and BOS 36–9
and First World War 107, 129
foreign members 42, 258n44,
263n25
formation 34–5,
survey (1910) 43
symposium (1909) 70
American Orthopaedic
Institution 255n8
American Pediatric Society 44
American Public Health
Association 140
American Railway Surgeons'
Association 140–1
American Surgical Association 49
amputations 20, 68, 84, 110–11,
112, 284n24
amputees 295n30
anaesthesia 18, 33
anatomy 35, 47, 121, 124, 236, 237
Andry, Nicolas (1659–1742) 11–12,
14, 51
Anglesey 23
antibiotics 235
antisepsis 18, 19, 23, 109
antivivisection 21
Argyll, Duchess of 59
Armour, Theodore R.W. 300n93
Army Council of Consultants 116
Army Medical Advisory
Board 116, 133, 296n56
Army Medical Council 115
Army Medical School 130
Army Medical Service
First World War 106, 114, 115,
116, 123, 125
Second World War 222, 224, 225,
230
Army Orthopedic Advisory Board,
USA 129

Arnold, Thomas 59
arthritis 39, 350n6
artificial limbs 123, 145
asepsis 19, 20, 31, 51, 110, 113,
236
Ashby, Henry 42, 44, 64–5
Ashton-under-Lyne 88
asphyxia 121
Association for the Improvement of
London Workhouse
Infirmaries 92
Association of Industrial
Officers 204
Association of Surgeons of Great
Britain and Ireland 268n79,
270n104
Atholl, Duchess of 152
Austrian National Insurance
Company 191, 192

backache 238, 341n53, 304n137
bacteriology 28, 35, 52, 165
Baer, William Stevenson
(1872–1931) 235, 265n52,
270n106
Baltimore 43, 301n110
Johns Hopkins University Medical
School 41, 44, 49, 126, 235,
238, 266n62, 268n88,
301n106
Bankart, A. S. Blundell
(1879–1951) 128, 265n46,
303n134, 345n20
Barclay, A. E. 182
Barker, Herbert 119–20
Barnardo's 56, 274n31
Barrow-in-Furness 88
Baruch Committee on Physical
Medicine 347n49
Baschurch, Salop Convalescent
Home for Women and
Children 72, 74
Bath 185, 187
Bauer, Louis 261n70
Beavan, Margaret (1877–1931) 58,
59, 156, 275n36, 320n60
Bennett, William Edward
(1865–1927) 128

Benson, Godfrey Rathbone *see* Charnwood

Beresford-Jones, A. B. (1881–1974) 304n147

Berlin 46, 71, 88, 263n26

Bernal, J. D. 246

Berry Hill Rehabilitation Centre 204, 207, 208–9, 210

Bevan, Aneurin (MP) 218, 237, 355n66, 337n21

Beveridge Report 146, 214

Beveridge, William 146, 172, 218

Bevin, Ernest 209, 211, 212, 213, 216, 228, 337n21

Bielefeld, Germany 63

Biesalski, Konrad (1868–1930) 71, 72, 107, 156, 312n55

Bigelow, Henry Jacob 13

Bigg, Henry Heather (1826–81) 12, 255n14

Birmingham 49, 58, 64, 66, 278n68, 294n25, 300n93, 332n62
    Accident Hospital 193, 231, 336n15
    Children's Hospital 41
    Cripples' Union 76, 277n55
    Education Committee 64, 67, 68, 141
    Queen's General Hospital 94, 332n68
    Royal Orthopaedic and Spinal Hospital 255n10, 261n1

Black, Arthur 156

blood poisoning 18, 109

Bloomsbury 13, 14

Board of Control 157

Board of Education 57, 58, 59, 61, 66, 152
    and adult cripples 202–3
    funding for cripples 68–70, 157
    and the National Scheme for cripped children 156–9, 161

Board of Industrial Medicine and Traumatic Surgery, USA 308n29

Board of Trade 157, 278n75

Boards of Guardians *see* Poor Law

Bodington, George 74

body metaphors 48, 93, 122, 200

Boer War 57, 107, 109, 111, 114

Böhler's clinic *see* Vienna Accident Hospital

Böhler, Lorenz 191–3, 310n45, 331n55

Bolshevism 142

bone disease 42, 68; *see also* tuberculosis; spinal deformities

bone-grafting 52, 263n20, 354n42

bonesetting 20, 21, 23, 24–5, 38, 86

Bonney, Victor 246

boots, surgical 56, 118

Boston 98, 99, 126, 130, 182, 183, 185, 268n88, 323n80
    Carney Hospital 39
    Children's Hospital 41, 174, 265n52
    Industrial School for Crippled and Deformed Children 70
    Orthopaedic Institution 255n8

bow leg 15, 16, 41, 80, 256n20

Bowlby, Anthony 259n56, 270n103, 300n92

braces 14, 34, 44, 56, 142

Brackenbury, Henry 186, 326n16

Brackett, Elliot G. (1860–1943) 39, 41, 129, 181, 255n13

Bradford, Edward H. (1848–1926) 41, 57, 70

Brander, William 134, 204, 205, 332n67

Briant, Frank (MP) 189

Bricheteau, Isidore 12

Brieger, Gert 21

Brighton 117, 312n57

Bristol 36, 50, 177, 194, 242, 300n93, 303n130

Bristow, Walter Rowley (1883–1947) 125, 128, 132, 156, 170, 184, 191, 221, 223, 224, 225, 229, 236, 237, 247, 274n32, 328n30, 338n31, 353n36

British Association for Dermatology 148

British Association for Radiology and Physiotherapy 135, 321n71

British Association of Surgeons 49

British Council for
 Rehabilitation 337n21,
 343n76, 356n68
British Electro-Therapeutic
 Society 46
British Expeditionary Forces 223
British Hospitals' Association 94,
 95, 97, 148, 180, 232, 233,
 327n17
British Journal of Surgery 50,
 283n12
British Medical Association 17, 25,
 49, 146, 167, 180, 257n33,
 327n17
 and Charity Organization Society
 joint committee 65
 children's medicine section 45
 Committee on the Rehabilitation
 of Disabled Persons 339n34
 Committee/Report on Fractures
 (1912) 80, 141, 283n11
 Committee/Report on Fractures
 (1935) 183–4, 186, 194, 197,
 198, 204, 207, 214, 326n16,
 333n69,n77, 337n22
 Industrial Health
 Committee 337n21
 meetings: (1884) 36; (1919) 139;
 (1925) 185
 and motor vehicles accidents 190
 and private practitioners 213
 Orthopaedic Group
 Committee 243, 321n71
 Orthopaedic Section 329n33
 and salaried service 316n6,
 334n80
 and Trades Union Congress 214
British Medical Journal 21, 44, 77,
 82, 84, 95, 96, 106, 132, 153, 168,
 170, 184, 186
British Orthopaedic
 Asssociation 43, 108, 125, 136,
 150, 178, 181, 184, 191, 193, 204,
 207, 211, 222, 268n79, 270n107,
 327n17, 331n55
 and accident services 248
 and BMA–TUC alliance 213–15
 constitution 235, 238
 divisions 223

 founding 50, 106, 131, 132,
 fractures sub-committee on 230,
 348n59
 meetings (1943) 340n50, 342n72;
 (1949) 231
 membership 153, 240–1
 Memorandum on Accident
 Services (1943) 230–33, 244
 Memorandum on Education 244
 Memorandum on Rehabilitation
 (1942) 228–9
 organization 128
 and Rehabilitation Committee
 (1956) 356n68
British Orthopaedic Society 35–39,
 40, 45, 69, 106
British Postgraduate Medical
 Federation 353n33
British Rheumatism
 Council 356n68
British Social Science
 Association 94
Broadstairs 174
Brodie, Benjamin 270n106
Brompton Chest Hospital 173
bronchitis 175, 216
Brown, Buckminster
 (1819–90) 255n13
Brown, John Bull 255n13
Bruce, Harold W. (MO) 332n67
Bryant, Thomas 26
Buckland, G. R. A. 336n15
Buller, Dame Georgina 172
Burdett, Henry (1847–1920) 17, 93,
 94–5, 97, 99, 100, 102, 127,
 302n127, 329n42
Bureau of Mines, USA 140
Burgess, Anthony 220
Burnett, Napier 156
Burrows, H. Jackson 353n33
Butterworth, John J. (MOH) 161
Buxton, St John Dudley
 (1891–1981) 345n28
Bywaters, Muriel 203

Calais 112
Callender, George (1830–79) 20, 21,
 24
Camus, Jean 122–3

Canadian Hospital, Ramsgate 118
Cancer Act (1939) 197
carbolic spray 22
Cardiff 50, 127, 186, 300n93,
    329n39, 332n68
cardiology, 'new' 48, 120
Carling, Ernest Rock
    (1877–1960) 244, 335n92,
    353n36, 338n30
Carnwath, Thomas (MO) 204, 205,
    207, 336n15
Carrel, Alexis 111
Carshalton, St Mary's Hospital for
    Children 178
Carter, William (1870–1907) 63
Casualty Clearing Stations 83, 167,
    121, 111, 112, 122
'casualty' patients, defining 90; *see
    also* accidents; air-raids
Cave, Lord 187
Central Association for the Care of
    the Mentally Deficient 156
Central Council for Infant and Child
    Welfare, 156
Central Council for the Care of
    Cripples 155–7, 158, 159, 161,
    164, 172, 201, 202, 203, 210,
    346n43, 356n68
Chadwick, Edwin 92
Chailey, Sussex 56, 59, 73, 178
Chamberlain, Austen 145
Chamberlain, Neville 160, 311n47,
    338n31
Chance, Arthur 270n103, 300n93
Charing Cross Hospital 29, 188,
    256n26, 290n101
charity 'abuse' 92, 94; *see also*
    outpatient departments
Charity Organization Society
    (COS) 55–8, 59, 65, 73, 76, 90,
    91, 92, 95, 275n44
Charnley, John (1911–82) 234, 238,
    239, 354n46
Charnwood, Lord
    (1864–1945) 297n58, 318n30
Chartered Society of Massage and
    Medical Gymnastics 151,
    305n150
chemotherapy 235, 319n47

Chesterfield 208
Cheyne, W. Watson 51
Chicago 26, 49, 98, 130
child
    health and welfare 7, 53–5, 70,
        176, 198,
    labour 55, 79
    psychology 53, 65
Children's Bureau, USA 71
children's homes and hospitals 7,
    40, 41–2, 51
children's hospitals
    appointment of orthopaedic
        surgeons 265n46
    and BOA members 153
    growth of surgery in 61–2
    image 42
    v. orthopaedic hospitals 40
    patient turn-over 62
children's medicine and
    surgery 40–43
children, sub-acute chronically
    ill 61–4
chiropodists 251n15
Church of England's Children's
    Union of Waifs and Strays
    Society 156, 160
Chutro, Dr Pedro 122
Citrine, Walter 212, 213, 215,
    338n30, 340n50, 342n72
City Orthopaedic Hospital 13, 14,
    17, 37, 40, 264n42
Clark, Robert Veitch
    (MOH) 318n39, 320n60
cleft palate 61, 42
clergy 160
clinics
    private 49
    tuberculosis, 77
    see also after-care; fracture;
        hospitals; infant;
        orthopaedics; *and under
        individual institutions*
club foot 13, 15, 16, 17, 18, 36, 61
coal industry 86, 138, 207, 215, 233,
    282n1
Coal Mines Act (1911) 307n8
Coalmasters' Association 210
Cochrane, W. A. 323n80, 327n17

Codivalla, Alessandro
(1861–1912) 235
Coleshill, Warwickshire 69
Colonia, New Jersey 142
communications 92, 93, 98, 111, 194
community care *see* social medicine
company hospitals 102, 150
Cone, Sydney M. 301n110
Conference Board of Physicians in Industry, USA 140
Congress of Orthopaedic Surgery (Berlin, 1908) 71
conscription 115, 124
conservative surgery 19–23, 24, 36, 39, 48, 76, 81, 110, 143, 144, 167, 169, 170, 238
consultants, hospital, 49, 233
and EMS 237
honorariums 102, 230, 335n91
pay structures 155
'pure' 326n6, 343n79
qualifications 3, 242
in surgery 206
traditional routine 182
trainees 131
see also specialists
continuity in patient care, concept and practice 62, 82, 100, 113, 114, 121, 134, 160, 183, 188, 197, 206, 228, 230, 248
contractures 80
corporatism 6–7, 103, 122, 218, 252n21
Council of National Defense, USA 125, 129
Coventry 221
Crewe 204, 205, 209, 291n113
Crile, George 47, 49, 51
'cripple', defining 66, 71
Crippled Children's Charter (1918) 59
crippled children
'discovery' 53, 54–9
and the medical profession 61, 236
commercial exploitation 60
home care 58
legislation 70, 156

in London 67, 178
'made not born' 165, 185
schooling *v.* medicine 60–1
and surgery 60
statistics 66, 67, 156, 176
welfare and funding
agencies 54–55, 57, 58–9, 157–61; *see also* Board of Education
Cripples Home and Industrial School for Girls, Marylebone 54
Cripples' Journal 164, 201, 202
cripples
adult 40, 140, 172, 201–2
industrial, salvage of 202
surveys 66
training and employment 176, 203
crippling, causes 54, 67–8, 320n60
Crutch and Kindness League 55
Cunard Steamship Company 203, 336n17
curative medicine *v.* public health 166, 167, 168, 198
curative workshops 118–19, 127
Cushing, Harvey 41, 49, 170, 262n7, 302n115
Cyclopaedia of the Diseases of Children (1899) 80
Cyriax, James 239, 322n71

Dakin, Henry D. 111
Daley, William Allen (MO) 338–9n32
Danforth, Murray S. 126–7, 262n7, 328n31
Dawson Report 166–9, 188
Dawson, Lord Bertrand 38, 122, 133, 233, 246, 315n6
deformities 18, 39, 80; *see also* spinal deformities
Delevingne Reports/ Committees 184, 186, 196, 197, 203, 207, 208, 213, 224, 226, 229–30, 326–7n17, 335n3
Delpech, Jacques 12, 13, 14
Dendy, Mary 65
Department of Labor, USA 140

Depression: (1920s) 160; (1930s) 175
Derby 208
Derwen Cripples College 346n43
diet 14, 21, 44, 170, 173, 259n56, 320n60
diphtheria 33, 166
Disabled Persons (Employment) Act (1944) 214
disabled soldiers 150, 154, 311n54
employment appeal 145
political fears of 119, 133
see also rehabilitation
dislocations 16, 39, 80
divisions of labour in medicine and welfare 2–3, 48, 92, 121
dockers 87–8, 93–4
Dockers' Union 212
doctor/patient relations 47, 91, 199, 211, 287n65; *see also* whole-person medicine
Douglas, W. R. 182, 183
Dreadnought Seamen's Hospital, Greenwich 88, 97
dresserships 242, 244
Dublin 50
Dudley, Sheldon 246
Dundee 195
Dunn, Naughton (1884–1939) 128, 207, 261n1, 300n93

East Grinstead Plastic Surgery Centre 219
East India Company Docks 88
East London Hospital for Children 44
Eastwood, W. J. 314n80, 331n57
Eccles, W. McAdam 328n25, 333n69
Edinburgh 20, 24, 42, 131, 300n93, 303n130, 350n13
Chalmers Hospital 47
Cripples and Invalid Children's Aid Society 64
Princess Margaret Rose Hospital 43
Royal Hospital for Sick Children 41, 43, 47, 61, 64
Education Act (1907) 65–6

Education and Care of Crippled Children, *Bibliography* (1913) 71
education, compulsory 53, 55; *see also* schools
'efficiency', concept and practice 142, 47–8, 51, 92–3, 103, 120–4; *see also* scientific management
'effort syndrome' 121
Eichholz, Alfred 274n25,n34, 324n103
electrotherapy 20, 118, 125
Elementary Education Acts 55, 57, 59, 152
Elliot, Walter (MP) 315n84, 334n83
Elmslie, Reginald Cheyne (1878–1940) 65, 67, 156, 172, 128, 132, 214, 261n1, 263n31, 274n32, 297n62, 300n93, 353n36
Emergency Medical Service (EMS) 193, 221–3, 225, 227, 229–30, 237
Empire Rheumatism Council 322n71, 341n60
Employers' Liability Act (1880) 102
Engineering 138
epilepsy 54, 61, 63, 112
Erichsen, John Eric 26, 260n67
Evans, E. Laming (1871–1945) 128, 300n93, 353n33
Evatt, G. J. H. 98, 99
Evelina Hospital for Sick Children 35
excision of joints 21, 24, 25, 26, 62, 266n55
exercise 14, 28
Exeter, St Loyes College for the Disabled 172
experiments on humans 239
expertise, in medicine and welfare 47–8, 92, 108; *see also* fractures
extension-movement 25

Fabian state centralism 57
Factory Inspector's Report (1885) 82
factory legislation 54

Factory Welfare Advisory
  Board 341n60
Fairbank, H. A. T. (1876–1961) 29,
  156, 221, 223, 236, 265n46,
  270n103, 294n21, 331n62,
  347n44, 351n14
fatalism *v.* curative optimism 63, 72
'fatigue' 33, 47
Federal Board for Vocational
  Education, USA 142
Federal Vocational Rehabilitation
  Act (1920), USA 142, 306n5
Federated Employers' Insurance
  Association 149, 180
Federation of British Industry 180
feeble-mindedness 64–5
feet cases 16, 118, 127, 219, 221,
  256n20; *see also* club foot
Fergusson, William (1808–77) 20,
  21, 170, 287n60
Finch, Ernest 269n98,n100, 326n6
First World War
  American Expeditionary
    Forces 129
  American doctors 127
  aid and dressing stations 96, 106,
    111
  expenditure on Poor Law
    premises 304n143
  locomotor injuries 107, 110
  in medical literature 105
  and military manpower 113, 114,
    119, 124
  Military Massage Service 135
  Military Orthopaedic
    Centres 105, 117, 127, 133–
    4, 164, 303n130
  and military orthopaedics 106,
    108, 122
  recruits 118, 140
  special hospitals 112
  and orthopaedic training 130–1
  see also disabled soldiers; hospitals;
    orthopaedics
first aid 81, 82, 97, 98, 290n100
Fisher Education Act (1918) 59,
  156
Fisher, Alfred George
  Timbrell 322n71

Fisher, H. A. L. 156, 158
'fitness'
  among workers, concept 103, 142
  see also 'labour'; 'fatigue'
Foucault, Michel 12
Fox, Daniel 155
Fox, Robert Fortesque (1858–1940),
  135
fracture
  apparatus 293n13
  'experts' 80, 186, 196, 197
  'surgeon', the term 335n90
  wards 80
fracture clinics 183, 192–7, 244
  v. accident/trauma services 207,
    230, 231, 233
  v. rehabilitation services 148, 149
fracture treatment 22, 33
  and GPs 185–6
  in general surgery 134
  in homes 185
  internal fixation 52, 187, 283n11
  mending-time 285n30
  in 19th century 84–5
  in Royal Navy 246
  and technology 187
  textbooks on 185
  traditional 284n26
  in USA, pre-1914 80–1
  wartime courses in 241
fractures 106, 112, 194
  compound 18, 84
  compound comminuted 139
  femur 124, 295n32
  and First World War 108, 109,
    110–11
  and medical education 186
  non-united 39, 183
  Pott's 189
  as proportion of accident
    cases 339n41, 331n62
  simple 85
Franco-Prussian War 97, 110
Fraser, John 266n57, 350n13
Fraser, Sister Teresa 73, 77
Freiberg, Albert 139
Freidson, Eliot 200, 250n8
fresh air 20, 44, 56, 72, 74, 170,
  320n604; *see also* open-air

friendly societies 8, 85, 86, 89, 292n116
Furley, John 97, 283n14

Galsworthy, John 297n58, 311n51
Gamgee, Sampson 257n37
Garrison, Fielding 123, 355n58
gas poisoning *see* asphyxia
Gask, George E. 183, 185, 186, 333n69, 353n33
Gauvain, Henry (1878–1945) 60, 169, 170, 174, 178, 201, 319n47
Geddes, E. C. 306n5
General Federation of Trade Unions 211–12
General Medical Council 131–2
general practitioners (GPs)
  and charity 'abuse' 90, 91
  and Dawson Report 166–9, 188
  and education 185, 246
  and fracture treatment 80, 85, 185–6, 328n31
  and Manchester Ship Canal 100
  and middle classes 328n31
  and National Health Insurance 146, 186, 206
  and orthopaedics/orthopaedists 36, 77, 115, 131, 150, 143, 177, 179, 184–5, 247
  part-time in British industry 144
  and rehabilitation 210
  and specialism 4, 45, 252n15
  status 2
  and workmen's compensation 147
  USA 39, 131, 328n31
general surgeons 3, 80, 121
  and bone and joint work 18, 37, 150, 205–6, 334n83
  'new wave' 46–51
  v. orthopaedists 6, 17, 29, 127, 131–3, 142, 144, 183, 184, 186
George, David Lloyd 68, 157, 306n5
geriatrics 166
germ theory 19, 22, 70, 257n37
Germany 13, 63, 71–2, 107, 123; accident insurance 140

open-air TB sanatoria 74
orthopaedic lectures 267n64
germicide 111
Gibney, Virgil P. (1847–1927) 34, 35, 38, 39, 262n7, 279n83
Gillespie, R. D. 148
Gillette, Arthur J. (1864–1921) 70, 77, 262n7
Girdlestone, Gathorne R. (1881–1950) 76, 132, 143, 154, 155, 156, 170, 171, 177, 221, 230, 232–3, 244, 255n13, 300n93, 301n106, 303n134, 351n21
Gissane, William (1898–1981) 193, 197, 230, 231, 325n4, 332n62, 336n15, 339n34
Glasgow 51, 96
  Children's Hospital 41, 164, 165, 170
  Erskine Hospital 295n29
  Orthopaedic and Rheumatic Clinic 315n84
  Orthopaedic Clinic 150–1
  Physiotherapy and Rehabilitation Clinic 315n84
  Royal Infirmary 285n29
  Victoria Infirmary 207
Goldthwait, Joel (1867–1961) 39, 41, 43, 107, 113, 114, 119, 122, 125, 126, 129, 142, 262n7, 323n80
Goodenough Committee/Report on Medical Education 238, 244, 246–7
Goodwin, T. H. John 125, 130, 133
Gordon-Watson, Charles 224
Granshaw, Lindsay 18
Gray, A. M. Henry 50, 264n31, 270n103, 300n89
Grenfell, Wilfred 116
Griffiths, Hugh Ernest (1891–1961) 204, 207, 208, 211, 214, 217, 331n62
Grimké, Dr Theodore (1817–86) 61
Groves, Ernest William Hey (1872–1944) 134, 148, 149, 177, 186, 193, 194, 195, 236, 300n93, 327n17, 333n69n75, 353n33
Guild of Good Samaritans 55

Guild of the Brave Poor Things 55, 59
Guilds of Help 57
Guy's Hospital 132, 261n1, 263n31, 266n55, 270n106, 283n11, 284n26, 294n25
  Massage Department 258n43
  Physical Exercise Department 303n135
gynaecology 49, 62, 264n31, 267n68, 304n137

haemorrhoids 33
Haldane, R. B. 114
Halsted, William 21, 22, 41, 49
Hammer, Dr W. H. 67
Harrison, Reginald 97
Hart, Ernest 95, 96
Harvard 130
  Infantile Paralysis Commission Clinic 323n86
  Medical School 266n52
Harveian Society of London 28
health and welfare, legislation 69, 248
heart disease 67, 112
heliotherapy 227
Helvétius, Claude 12
Henderson, Melvin 43, 327n22
hereditarianism 52, 57
Hetherington Royal Commission on Workmen's Compensation (1939–40) 214
Hill, Charles 213, 338n32
Hill, Leonard 318n30
Hilton, John (1805–78) 20, 21, 24
hips 22, 42, 62, 234, 260n57, 264n35
Hodge, John (MP) 120
Hoffa, Albert 71
holistic philosophy 21, 39, 48, 166, 167, 170, 200, 211, 229, 238, 245–46
  v. fragmentation in patient care 91
  see also integration; whole-person medicine
Holland, C. Thurston 36, 76
Holland, Sydney *see* Knutsford

Holman Gregory Committee on Workmen's Compensation in 1919 146
Holmes, Timothy 18, 62
Home Office 99, 149, 157, 203, 278n75, 336n15, 339n41
homoeopathy 28, 256n16, 258n43
Honigsbaum, Frank 196
Horsley, Victor 46, 47, 48, 49, 50, 283n12
Hospital 94
Hospital and Home for Incurable Children, Maida Vale 54
Hospital Officers' Association 290–1n101
Hospital Saturday Funds 86–7, 94
Hospital Sunday Funds 86, 94
Hospitals and Charities Annual 94, 98
Hospitals' Association *see* British Hospitals' Association
hospitals
  accident 191–3
  'bloodless' 22
  casualty 100
  casualty departments 84
  cottage 2, 185, 189, 205
  and cross-infection 83
  and First World War 83, 106, 111, 112, 114, 116, 121, 126
  in health planning 247
  v. home 83, 89, 146
  hutted 169
  and industrialists 89
  infectious disease 42, 69
  in medical history 8, 93
  municipal 188, 194, 198, 206, 338n31
  open-air 77, 167, 169, 171, 177
  outpost 99
  orthopaedic 14–17, 40, 153–5, 167, 176, 177, 179, 265n50
  pay beds/patients 2, 39, 83, 94, 195, 287n60, 317n22, 321n67, 333n80
  pre-payment schemes 89, 195
  private 47
  reconstruction, USA 129
hospitals (*cont.*)

specialist 3, 17, 48, 91
specialty departments and
    clinics 3, 91
territorial 111, 112, 155
for traumatic surgery and
    rehabilitation 193
tuberculosis 173
uniform accounting 94
workers' contribution
    schemes 86–7, 182
see also accidents; children's
    hospitals; First World War;
    outpatient departments;
    Second World War; *and under
    names of hospitals*
hospitals, voluntary
finances/funding 64, 89, 92, 94,
    189
in London 95–9
lay governors 93, 182, 195, 237
rationalization and reform 50,
    64, 83, 95, 99
and the state 92, 194–6
teaching 48, 155, 166, 179, 184,
    196, 232, 247
transformation in late 19th
    century 88–9, 286n50
House of Commons 116, 119, 184,
    202, 203, 209, 341n66
Howard Ambulance 97, 98
Howard, Dr Benjamin 95, 96, 97
Howell, B. Whitechurch 300n93
Huddersfield 83
Hunt, Agnes (1867–1948), 72–3, 75,
    77, 201, 255n13, 320n60
Hunter, John 20, 299n81
Hyde, Robert 341n54
hydrotherapy 20, 118, 227,
    305n149
hygiene 44, 109, 173

immobilization 20, 25
Imperial College 114, 295n31
individualist philosophy 56–7, 93
Industrial Fatigue Research
    Board 141
industrial health and the National
    Health Service 248

Industrial Health Research
    Board 340n51, 341n60
Industrial Hygiene, Section of
    American Public Health
    Association 140
Industrial Injuries Act (1946) 209,
    214
industrial injuries
and compensation, legislation 80
market 144, 231
and manpower 'wastage' 205
and medical history 8
and orthopaedics 139ff
and statistics 79, 141
and surgery 150, 310n45
see also accidents; rehabilitation;
    workmen's compensation
industrial medicine 141, 144,
    308n27, 310n45
Industrial Medicine and Surgery,
    Advisory Committee,
    USA 141
Industrial Orthopaedic Society 151,
    212, 341n60
industrial rehabilitation
    centres 197
Industrial Welfare Society 341n54
Infant and Maternal Welfare
    Clinics 157, 169
infantile paralysis 16, 44, 60, 61, 67,
    68, 173, 88, 307n21; *see also*
    poliomyelitis
Injured Soldiers (Parliamentary)
    Committee 120, 135
Institute of Orthopaedics 353n33
insurance companies 147–9, 190–2,
    205, 213, 313n71, 341n55; *see
    also* workmen's compensation
integration of health care, concept
    and practice 38, 100, 121, 123,
    157, 166, 200, 216, 228, 231, 239,
    245; *see also* holistic philosophy
Inter-Allied Conference on
    Rehabilitation 306n5
intercommunication, principle and
    practice 98, 99, 102, 166
Internal Medicine, USA 9
International Conference on Cripples
    (The Hague, 1931) 202

International Congress on Industrial
  Medicine (London,
  1948) 337n21
International Congress on Physical
  Medicine 346n41
International Congress on School
  Hygiene (London, 1907) 65
International Federation of Surgical
  Colleges 354n52
International Health Exhibition
  (London, 1884) 290n101
International Hygiene Exposition
  (Dresden, 1911) 280n86
International Medical Congress
  (Berlin, 1890) 46, 261n68
International Medical Congress
  (London, 1881, 1913) 26, 30,
  46, 51
International Medical Congress for
  Industrial Accidents and
  Occupational Diseases
  (Budapest, 1928) 310n45
International Society for Crippled
  Children 272n14, 280n85
Invalid Children's Aid
  Association 58, 60, 75, 156,
  160, 161, 178
Invalid Transport Corps 97
Ireland 333n76
iron and steel industry 88
Isaacs, George (MP) 337n21
Italy 70, 107, 250n4, 306n5

Jacobi, Abraham 44, 70
Jefferson, Geoffrey 183, 229,
  297n60, 335n93
John Groom's Crippleage 55
Johns Hopkins *see* Baltimore
Johnson, Robert W. Jr 143,
  301n106
joint diseases/injuries 26, 33, 61,
  106; *see also* excision;
  tuberculosis
Joint Parliamentary Advisory
  Committee 59, 156
Jones, Ernest 47
Jones, Rhaiadr 227, 347n44,
Jones, Robert (1857–1933), 38, 42,
  43, 72, 84, 88, 138, 139, 142, 143,

144, 150, 164, 165, 168, 171, 173,
  201, 206, 222, 224, 270n103,
  314n80, 315n82
early career 28–34
and First World War 109ff
and fracture treatment 88, 184–5,
  186, 187
as general surgeon 33–4, 38
and Macalister 32, 62–4
and the Manchester Ship
  Canal 100–3
and modern orthopaedics 29, 35
and the new surgical
  cohort 46–51
National Scheme 77, 153–9, 167,
  177, 232
and orthopaedic education 132,
  241–2
and Oswestry 74–6
private practice 31, 102
and professionalization in
  orthopaedics 35–40, 46
and RNOH 178–9
and SSDC 44–5
Jones, Thomas (1848–1900) 42, 43
*Journal of the American Medical
  Association* 34, 132, 139
*Journal of Bone and Joint
  Surgery* 171, 234
*Journal of Clinical
  Orthopaedics* 235
Journal of Orthopaedic Surgery 127,
  139, 171
Joynson, William 290n101

Keith, Arthur (1866–1955) 121
Kent and Canterbury
  Hospital 285n29, 304n147
Keogh, Alfred (1857–1936) 114,
  115, 116, 117, 118, 130, 135, 137,
  222, 294n25, 299n84, 315n82
Kerr, James (1861–1941) (SMO) 66
Kerr, James Rutherford
  (1878–1942) 150
Kessler, Henry 310n43
Keynes, Geoffrey 327n24
Kimmins, Dame Grace
  (1870–1954) 58–9, 73

King's College Hospital 330n46, 345n28, 351n14
King's Fund 94–5, 188–9
King's Roll 298n77, 312n55
knock knee 80, 256
Knutsford, Viscount (Sidney Holland, 1855–1931) 93–4, 98, 112, 298n76, 337n23

Labour Party 8, 165, 166, 187–8, 196, 212, 214, 316n8; *see also* Trades Union Congress
Labour Party Advisory Committee on Public Health 212, 298n77, 316n8, 321n64
labour policy *v.* welfare 212
'labour', reconceptualization 33, 103, 138, 141, 149, 292n120; *see also* 'fitness'
Lambeth Board of Guardians 189
Lanarkshire Medical Practitioners' Union 210
Lanarkshire Orthopaedic Association 207, 210
Lancashire 66, 196, 334n83
Lancashire and Cheshire Miners' Permanent Relief Fund 306n7, 314n80, 315n82
Lancashire and Cheshire Miners' Welfare Committee 216
Lancashire County Council 161, 318n35, 319n42
Lancet 14, 19, 90, 138, 139, 144, 147, 183, 189, 191, 194, 201, 246, 270n106
Lane, Arbuthnot 106, 270n103, 270n106, 283n11, 283n12
Lansbury, George (MP) 212
Lawrence, Christopher 8, 18
League of Hearts and Hands 55
Leasowe Open-Air Hospital 58, 69, 173, 346n43
Leatherhead
    Cripples' Training College 203, 336n15, 347n44
    Lord Roberts's Workshops 145
    Queen Elizabeth's Training College for the Disabled, 172

Leeds 25, 26, 46, 161, 242, 269n98, 270n104, 352n31
Leeds General Infirmary 124, 257n33, 260n67
Letchworth 203
Levick, G. Murray (1877–1956) 125, 274n32
Levy, Hermann 216, 340n50
Lewis, Thomas 48
light work 86, 147, 208, 210
light-therapy clinics 175
limb-fitting 110, 122
Lister, Joseph 18, 19, 20, 24, 36, 51, 70
Listerians/Listerism 19–22, 84, 257n30,n37
lithotomy 28
Little, Ernest Muirhead (1854–1935) 46, 128, 294n21
Little, William John (1810–94) 13, 14, 17, 18, 28, 29, 46, 51
Littlewood, Harry 124
Liverpool 23, 28, 30, 43, 58, 62, 96, 179, 221, 300n93
    Alder Hey Poor Law Infirmary 114, 117, 118, 127, 142, 144, 154, 221, 296n48, 303n130, 305n150
    ambulance services 290n94
    Children's Convalescent Home, West Kirby 63, 66
    Country Hospital for Chronic Diseases of Children (Heswall) 62, 63, 64, 66, 69, 77
    Daily Post and Mercury 63
    Hospitals' Commission 334n80
    Hospital for Diseases of the Chest and . . . Women and Children 88
    Infirmary for Children 63
    Kyrle Society for Children 58, 76
    Maghull Colony for Epileptics 63
    v. Manchester 245
    Margaret Beavan Hospital 273n25
    Medical Institution 25, 33, 36, 205, 261n74
    Medical Missionary Society 62

Nelson Street Clinic 23, 24, 26, 31, 34, 49, 76, 84, 102, 141, 191, 262n7
Northern Hospital 87, 290n101
orthopaedic education and training in 132, 242
Royal Infirmary 36, 63, 64, 206, 260n66, 276n50, 327n24
Royal Insurance Company 206
Royal Southern Hospital 13, 31, 32, 34, 36, 44, 58, 62, 63, 75, 84, 87, 102, 258n43, 276n50, 326n10, 346n43
'school' of orthopaedics 76, 171, 178, 223
School of Tropical Medicine 63
Society for the Care of Invalid Children 275n36
Stanley Hospital 31, 33, 88, 102, 276n50
University 124, 127, 132, 242, 267n64
Victoria Settlement 58
Western General Territorial Hospital 113
workhouse 272n9
Llanelli 213
Lloyd, Eric (1892–54) 265n46
local authorities 6, 56–7, 77, 144, 157–8, 161, 165, 168–9, 197, 233, 247
local education authorities 59, 66, 156, 157, 176, 279n75
Local Government Act (1929) 157, 161, 194
Local Government Board 57, 69, 78, 157, 158, 159
London Ambulance Service 96
London and Provincial Vehicle Workers' Union 212
London Clinic for Injuries 322n71
London Clinic of Physiotherapy 301n97
London County Council (LCC) 58, 64, 93, 180, 181, 193, 194, 203, 330n51
hospitals 134, 195, 204, 331n62, 334n80
Ambulance Service 99, 188

and health of schoolchildren 66–7, 278n67
Education Committee 65, 156
London Fire Brigade 99
London Hospital 14, 37, 88, 90, 94, 96, 261n1, 283n9, 286n52, 291n111, 333n77, 338n30, 346n41, 353n36,
London Midland and Scottish (LMS) Railways Co. 204–5, 313n65, 327n17
London School Board 59
London School of Tropical Medicine and Hygiene 310n45
Longmore, Thomas 97–8
Lonsdale, E. F. 256n20
Lord, John Prentiss 327n22
Lorenz, Adolf (1854–1946) 16, 22, 262n7
Lovett, Robert (1859–1924) 41, 262n7, 282n6, 323n86
Lucas-Championnière, J. M. M. 239, 283n11
lump-sum settlements 145, 147, 340n50
Lynn-Thomas, John (1861–1939) 50, 270n103, 300n89,n93, 318n30

Macalister, Charles (1860–1943) 32, 44, 45, 62, 63, 64, 273n25
Macewen, William (1848–1924) 19, 41, 51, 235, 257n36, 260n64, 295n29
MacGregor, A. S. M. (MOH) 349n71
McIndoe, Archibald 219
Mackenzie, Forbes 77
Mackenzie, James 48, 299n81
McKenzie, Robert Tait (1867–1938) 135
MacKenzie, William Colin 306n5
McLaughlin, W. A. 336n15
McMurray Thomas Porter (1888–1949) 128, 132, 221, 300n93, 331n56
McMurtrie, Douglas 71
MacNalty, A. S. 318n31
McNeill, Ronald (MP) 296n55

MacVeagh, Jeremiah (MP) 298n76
Maghull *see* Liverpool
Maida Vale Hospital for Nervous
    Diseases 265n46
Maitland, Thomas Gwynne
    (1875–1948) 203–4, 205, 207
Makins, George 51, 123, 133,
    270n103, 300n92
malingering 211
Malkin, Sidney Alan Stormer
    (1892–1964) 208, 311n47,
    345n20
malnutrition 175; *see also* diet
Manchester 58, 98, 99, 211, 221,
    234, 300n93
    accidents in 330n51–2
    air pollution 161
    ambulance services 290n94
    Ancoats Hospital 181, 182–3,
        194, 206
    Band of Kindness 55
    Christie Hospital 182
    Corporation 100
    Cripple Children's Help
        Association 275n37
    cripples 66–7
    Docks 216
    Grangethorpe House, 183, 297n60
    Guardian 103
    Heaton Park 304n149
    Hospitals' Post-Graduate
        Clinics 353n36
    Joint Hospitals' Advisory
        Board 198, 247
    Medical School Gazette 204
    military medicine 183
    orthopaedics in 242, 354n46
    Pendlebury Children's
        Hospital 42, 44, 61, 62, 64
    Royal Infirmary 42, 82, 83, 87,
        196, 215, 266n56, 284n20,
        337n23, 354n46
    School Board 65
    schoolchildren 175
    Ship Canal 31, 100–3, 120, 121,
        138
    Swinton House 65
    University 42, 182, 244
    workhouse 272n9

Manchester and Salford Crippled
    Children's Help Society 55, 61
Manchester and Salford Sanitary
    Association 61
manipulative techniques 20, 37,
    228; *see also* massage;
    physiotherapy
Mann, Tom 212
Manners, Lady Diana 106
Manoel, King of Portugal 117–18,
    222
Manor House Hospital, Golders
    Green 107, 150–1, 205, 212
Mansfield General Hospital 208
Marnock, John 124
Marsh, Frederick Howard
    (1839–1915) 20–21, 22, 24, 25,
    26, 62, 327n24
Martin, Franklin 50, 125, 127, 129,
    141, 270n102
Marx, Karl 93, 239
Massachusetts 70, 313n66
Massachusetts General
    Hospital 39, 43, 127, 181,
    188–9, 287n65, 327n22
massage 14, 20, 25, 28, 135, 227,
    230, 239, 283n11, 305n150,
    321n71, 338n26
Maternity and Child Welfare Act
    (1918) 77
maternity health services 77, 166,
    176, 180, 198, 213, 215
Mayer, Leo (1884–1972) 31, 139,
    262n7, 294n18, 316n11, 350n8
Mayo Clinic *see* Rochester, Minn.
Mayo, Charles (1865–1939) 49
Mayo, William (1861–1939) 33, 38,
    46, 49, 51, 262n7
means testing 94, 194
mechanical apparatus/therapy 14,
    20, 21, 22, 24, 38, 88
Medical Attendance Organization
    Committee 90
medical education 91, 166, 185–6,
    197, 237–48, 266n63; *see also*
    hospitals, voluntary;
    orthopaedics
medical gymnastics 14, 20, 22, 227,
    228, 258n43, 338n26

medical market, Britain *v.* USA 2, 3, 39
Medical Officers of Health (MOsH) 5, 57, 58, 69, 157, 168, 169, 176, 177, 179, 198, 247, 316n8
Medical Officers of Schools Association 65
medical profession
  and cripples' homes 60
  metropolitan/provincial split 46, 48-9
  referral system 3, 131, 188
  remuneration, from lecturing 353n36
  structure 2-4, 39, 148
  and salaried service 111, 127, 247, 315-16n6, 321n64, 335n91; *see also* BMA
  and subdivision of labour process 191
  traditionalism 144
  see also BMA; consultants; doctor/patient relations; general surgeons; general practitioners; medical education; orthopaedics; orthopaedists; professionalization; specialization; state; surgery
medical records 82, 121, 123, 127, 183, 187, 287n65
Medical Research Council (MRC) 229, 236, 341
medical schools *see* medical education
mendicity 56, 272n13
meniscectomies 132
Mennell, James B. (1880-1957) 283n11, 297n62, 321n71
mental handicap 12, 55, 61, 64-5, 66, 157, 277n61
Menzies, Frederic 207, 333n80, 335n90
Metropolitan Asylums Board (MAB) 92, 96, 97
Middlesbrough 86
Middlesex Hospital 303n134

Midland Colliery Owners' Mutual Indemnity Co. Ltd 208, 327n17
militarism 97, 99
milk 167, 320n60
Miller General Hospital 88, 294n21
Miller, Alexander (1904-59) 207, 209, 210, 211, 217
Miners' Welfare Commission 208, 209, 210, 215-16
Mineworkers' Union 210
Ministry of Fuel and Power 209
Ministry of Health 149, 152, 154, 157, 158, 159, 161, 162, 165, 168, 169, 177, 180, 197, 203, 204, 217, 221, 222, 224, 226, 227, 232, 322n71, 327n17
  and Ministry of Labour 223
  and orthopaedic treatment of industrial injury 206
  Medical Advisory Committee on Rehabilitation 337n21
  Post Graduate Medical Education Committee 303n129
Ministry of Labour 145, 180, 202-3, 219, 227, 228, 311n54, 312n55, 327n17
Ministry of Mines 216, 306n8
Ministry of Munitions 137, 138
Ministry of Pensions 118, 131, 132, 145, 150, 155, 157, 159, 180, 225, 296n44, 297n58, 311n51, 314n80, 327n17
Mitchell, Alexander (1881-1953) 255n13
Mitchell, Arthur 124, 350n13
'modernist' ideology 33, 47, 52; *see also* scientific management
Montessori, Maria 275n37
Montgomery, William Percy (1867-1911) 266n56
Montreal 185, 328n31
  McGill University 304n149
Montreal General Hospital 327n22
Moore, Benjamin 318n30
Moore, Harold Ettrick (1878-1952) 204-6, 211, 214
Morgan, H. B. W. (MP) 212, 213, 339n34

Morgan, Montague Travers 204
Morison, Rutherford
    (1853–1939) 46, 47, 50, 325n4
Morley, John 182, 183, 335n92
Mott, Valentine 12–13, 14
Moynihan, Berkeley
    (1865–1936) 46, 47, 50, 115,
    116, 117, 122, 123, 124, 133, 170,
    185, 190, 243, 270n103, 270n104,
    318n30
municipalism 93, 98, 100, 103, 160,
    168
Munro, David 340n51
Murphy, J.B. (1857–1916) 49, 50,
    51, 262n7, 270n106

National Aid Society 294n21
National (Health) Insurance 2, 146,
    206, 210, 211, 212
    Act (1911) 69, 77
    Committees 186, 278n75
    see also general practitioners
National Health Service 3, 4, 5, 6,
    155, 198, 237, 233, 240, 243,
    247–9, 338n30
National Hospital for Paralysis and
    Epilepsy 46
National Industrial Home for
    Crippled Boys, Kensington 54
National Orthopaedic Hospital 13,
    17, 35, 258n43
National Safety Congress,
    USA 140
National Safety Council 193,
    314n77
National Scheme for Cripples 153–5,
    156, 157, 159, 162, 168, 176
navvies 100
Navvies' Union 102–3
Nelson Street Clinic *see* Liverpool
nerves 28, 52, 106, 127; *see also*
    peripheral nerves
Nethersole, Olga 318n30
Netley 97
neurology 48, 120, 121, 229, 297,
    252n22, 322n73
neurosurgery 47, 183, 243, 265n46,
    304n137, 335n92,
New Jersey 140, 144, 145

New York 31, 34, 43, 70, 130
    Academy of Medicine 70, 140
    ambulance system 95, 96, 98, 99
    Hospital for the Relief of the
        Ruptured and Crippled 34,
        40
    Bellevue Hospital Medical
        College 34
    Chair of Orthopaedic Surgery 26
    Chair of Pediatrics 45
    Employment Center for the
        Handicapped 307n21,
        308n24
    Institute for Crippled and Disabled
        Men 140
    Orthopaedic Dispensary and
        Hospital 264n34
    Post Graduate Medical School
        Clinic 263n20
    Rochester General
        Hospital 283n8
    State Reconstruction Home,
        Haverstraw 279n82
Newbolt, George P. 274n25,
    294n21
Newcastle upon Tyne 46, 132
Newman, George 58, 66, 69, 152,
    157–9, 165–9, 171, 274n25,
    303n134, 317n28
Newsholme, Arthur 159
Nicoll, Ernest Alexander 208, 209,
    210, 211, 217, 231
'no-touch' surgical technique 31,
    110, 351n14
Norfolk and Norwich General
    Hospital 87
Norris, Donald C. (MO) 333n69
North Staffordshire
    Coal Owners'
        Association 307n87
    Cripples' Aid Society 319n42
    Infirmary 307n87
North-West London
    Hospital 264n42
Nottingham 208, 311n47, 345n20
    Harlow Wood Orthopaedic
        Hospital 215
Nuffield Provincial Hospitals'
    Trust 198, 223, 244

Nuffield, Lord 161, 217, 224
nurses 58, 155, 206
  Canadian 347n50
  Queen's Jubilee 72
  training centre, Oswestry 78
nutrition 19, 52, 175

occupational therapy 122, 145, 210,
  227, 228; *see also* rehabilitation;
  vocational training
Ogilvie, William Heneage
  (1887–1971) 211, 294n25,
  328n30, 344n12
Oldham 83, 272n9
Oliver, Thomas 310n45
Ollerenshaw, Robert
  (1882–1948) 300n93, 343n79,
  354n46
Ollier, L. X. E. (1882–1948) 235
open-air therapy 24, 72, 73–5
Openshaw, Thomas H.
  (1856–1929) 128, 261n1,
  270n103, 283n9, 293n9, 294n21,
  295n29, 300n89, 329n33,
  353n36
ophthalmology 2, 5, 44, 240,
  251n15, 354n49
Orr, H. Winnett (1877–1956) 127,
  279n77
orthopaedic medicine 239, 322n71
orthopaedic physicians 170, 228
Orthopaedic Research Society 235
Orthopaedic Service Reconstruction
  hospitals, USA 142
orthopaedics
  ambiguous place in medicine and
    surgery 5, 17, 36, 44, 179
  clinics 31, 77–8, 132, 169, 176,
    178, 244; *see also* outpatient
    departments
  etymology 11–13, 17, 142, 151,
    254n6, 258n40, 200
  and fractures pre-1914 80–1,
    283n11
  in general hospitals pre-1914 107
  and general medicine 244, 245
  and general surgery 5, 131–2,
    142, 242, 245

  in Germany 71–2
  historiography 1, 52
  in London 14–17, 37, 46, 95, 131,
    176, 178, 183, 242, 264n33
  incompetence in 183
  'medical' 121
  and medical education 1, 37,
    130–1, 135, 238–48, 267n64,
    352n32, 353n36
  'mental' 254n7
  modernist construction 5, 14,
    29–30, 34–5, 48, 51, 121,
  'moral' 12
  and National Health Service 240,
    243, 248–9
  'new' (early 19th century) 13, 14
  v. paediatrics 44–5, 349n1
  v. physical medicine 228
  v. physiotherapy 309n37
  and private practice 37, 195, 240,
    241, 243
  and professorships 43, 238
  as psychosocial 245
  reconceptualization, late-19th
    century 34–5, 39
  v. rehabilitation 231
  research and development 234–9,
    266n57
  'social' 179, 236, 245
  and the state 6, 160, 171, 188
  status/image 1, 34–7, 106–7,
    237–8, 241
  and teaching hospitals 29, 43,
    183, 241
  and third-party funding 143, 148,
    149, 213
  v. traumatology 231
  in USA 31, 37, 43, 122, 140, 142,
    170
  and world wars compared 222
  see also hospitals
'orthopaedist', the word 33
orthopaedists
  appointments in general
    hospitals 29, 37
  as 'bone and joint surgeons' 170
  as 'darlings of the Gods' 6
  employment at COHs 155
  isolation 179

orthopaedists (*cont.*)
  manipulative *v.* operative 37–8,
    142
  as medical radicals 196
  opponents *see* general surgeons
  and professionalization 70–1,
    165
  philosophy of organization 5–6
  postgraduate instruction 132,
    241–2
  recruitment problem 36, 39,
    124–7, 130, 240–1, 247
  and regionalism 155, 232
  remuneration 1, 37, 131, 349n2
  and salaried service 43, 188,
    195–7, 231, 326n6, 345n25
  and specialist autonomy 5–6
  see also general practitioners;
    general surgeons; medical
    profession
orthoplasty 270n106
Osgood, Robert B. (1873–1956) 41,
  139, 174, 108, 121, 127, 128, 235,
  245, 262n7, 323n80, 327n22
Osler, William 170, 304n149
Osmond-Clarke, H. 354n46
osteopathy 239, 298n76
osteotomy 18–19, 33, 37, 40–41
Oswestry, Shropshire 31, 72–8, 153,
  161, 178, 321n67, 338n23
otolaryngology 240, 354n49
'outpatient', definition 90
outpatient departments 8, 58, 89
  'abuse' 90–2, 99
  attendances 89–90
  and London orthopaedic
    hospitals 15
  for orthopaedics 31, 178, 244
  and specialty interests 49, 91
  status 37
outpatient dispensaries for
  children 42
Oxford 154, 159, 177, 220, 224, 230,
  300n93, 301n106, 332n63,
  352n32
  orthopaedic training in 242
  Wingfield-Morris Orthopaedic
    Hospital 132, 159, 221,
    301n106, 321n67, 348n55

Packard, George 262n7
Packard, John H. 282n7
paediatrics 29, 44, 45, 53, 64,
  267n70, 335n92, 349n1
Page, C. Max (1882–1963) 328n30,
  348n60
Paget, James 21, 24, 25, 26, 262n2,
  270n106
Papworth 201
parachuting 220
Paris 24, 122
  Faculty of Physick 11
  Hôtel-Dieu 83
  International Exhibition
    (1867) 290n96
  Orthophrenic Institute 12
Parker, Rushton (b. 1847) 26,
  260n64
Passmore-Edwards Settlement 59
paternalism 102, 205
Paterson, A. M. 124
pathology 28, 35, 236, 237
patient relapse 62, 77
Pattison, Dorothy 88
Payr, Erwin (1871–1946) 270n106
Pearson, Maurice G. 295n31
Pediatrics 44
Pennsylvania, University 304n149
People's League of Health 160
Peripheral Nerve Injury, Advisory
  Committee on 229
peripheral nerves 221, 229, 303n137
Perkins, George (1892–1979) 223,
  345n28, 354n51
Petit, Jean-Louis 254n3
Peto, Geoffrey Kelsall, MP
  (1878–1956) 202–3
Philadelphia 130,
  Orthopaedic Hospital 322n73
  Pediatric Society 44
  Polyclinic 282n7
  Children's Hospital 265n51
philanthropists and industrial
  medicine 144
'physiatrists' 228
physical culture 166
'physical deterioration' 57
physical medicine 135, 200, 227–9,
  239, 322n71

Physical Medicine, Ministry of Health's Advisory Committee 227
physiology 5, 47–8, 93, 121, 142, 236, 237
physiotherapy 122, 125, 134, 135, 142, 143, 145, 203, 227, 228, 251n15, 347n47
Pilkington Glass Co. Orthopaedic Hospital 150, 151, 315n82
Pittsburgh, US Steel Corporation 141, 204
plaster of Paris 33, 187, 191
plastic surgery 106, 219, 243, 252n20, 350n2
Platt, Harry (1886–1986) 128, 156, 170, 171, 196, 198, 206, 235, 248, 255n13, 268n79, 269n100, 297n60, 300n93
 early career 42, 181–2
 and fracture campaign 180, 183–4, 185, 191, 193, 194, 211
 and medical education 240, 241–7
 and Second World War 209, 221, 222, 223, 224–5, 229, 232
Poland, John 294n21
police 93, 96, 97
poliomyelitis 44, 54, 67, 170, 174, 175, 178; *see also* infantile paralysis
polyclinics 91
Poor Law 85, 86, 90, 92, 119, 194, 206
 authorities 69, 146, 303n130
 Boards of Guardians 57, 69, 157, 278n75
 infirmaries 69, 87, 88, 93, 184–5, 188
 unions 76, 78
 workhouses 54, 86, 92, 131, 272n9
Poplar Accident Hospital 87, 88, 94, 278n67, 290n100, 291n111, 294n21
Portland, Maine, Children's Hospital, 280n83
Postal Workers' Union 212
Potteries Cripples' Guild 272n14

Potts, Dr William A. 64 , 67
Preston Royal Infirmary 232, 335n1
preventive medicine 77, 96, 140, 164–72
Prince of Wales' Hospital Fund for London *see* King's Fund
Pringle, Hogarth 269n98
private practice 3, 38, 43, 84, 102, 150, 167, 188, 325n6; *see also* orthopaedics
proctology 9, 46
productivity *see* efficiency
professionalization 1–2, 4, 53, 62, 83; *see also* orthopaedists
Providence, Rhode Island 127
Provident Dispensaries 90–1
Provincial Surgeons' Club 50, 116
Psycho-Medical Society 112
psychology and rehabilitation 211
psychology, 'new' 268n88
psychosomatic illness 239
psychotherapy 200
public health 165–8
public health authorities 58, 66, 78, 157
Public Health (Tuberculosis) Act (1921) 77, 157
Pugh, W. T. Gordon (1872–1945) 178
Putti, Vittorio (1880–1940) 306n5

Queen Mary's Hospital, Roehampton 295n29,n30
Quine, Mr 304n145

Rabinbach, Anson 47
Radcliffe, Frank 305n149
radiology 36, 76, 182, 251n15, 335n92
railways 101, 103; *see also* accidents; London Midland Scottish
Ramsbotham, Herwald 203
record keeping *see* medical records
Red Cross 117, 156, 160, 222, 301n102, 317n28, 356n68
Red Cross Clinic for the Physical Treatment of Disabled Officers 305n149

Red Cross and Order of St John, Joint Committee 116–8, 295n37
regionalism, concept and practice 8, 155, 166, 222, 232, 247–8, 316n8
rehabilitation
  and company schemes 209, 211
  and integration of health services 199
  legislation 142–5
  and National Health Service 218, 248
  v. orthopaedics 227–9, 231–2
  of seamen 204
  of soldiers (First World War) 116–17, 119, 135
  of workers 207–9, 211
  vocational/occupational 130, 142, 146, 172, 200–203
  word and concept 135, 137, 199–200, 211, 216, 218, 227, 228, 306n5
  see also occupational therapy; physiotherapy
Rehabilitation of Disabled Persons, Report (1956) 248
Rehabilitation and Resettlement of Disabled Persons, Committee (1942) 219
Rehabilitation of Persons Injured Through Enemy Action, Conference (1940) 226
Rehabilitation Review 142
rest therapy 20, 24, 25, 38
Reverby, Susan 47
rheumatism 68, 176, 216, 321n70
Rhode Island Hospital 127, 329n31
Rhondda, Lord 157
Rhyl, Wales 24, 72, 73, 74
rickets 41, 44, 54, 60, 61, 67, 68, 70, 161, 167, 172, 174, 175,
Riddoch, George 229
Ridlon, John (1852–1936) 26, 35, 262n7, 268n78
Riordan, D. C. 351n23
Road Traffic Act (1934) 190
Roberton, John (1797–1876) 79, 82, 83, 87, 98, 99, 100, 102

Roberts, Lord F. S. 145
Robson, Mayo 97
Rochester, Minnesota, Mayo Clinic 42, 43, 49, 185, 299n84
Rollier, Auguste 169, 320n60
Rosen, George 2
Roth, Mathias 258n43
Rothbaud Scheme 298n77
Rouen, Anglo–Belgian Hospital 118
Royal Air Force 219, 222, 223, 225, 337n22
Royal Army Medical Corps 116, 290n101
Royal Automobile Club 117
Royal College of Surgeons 36, 49, 50, 51, 115, 121, 133, 240, 354n49
Royal Colleges 2, 3, 167, 223, 337n21
Royal National Orthopaedic Hospital 17, 131, 178–9, 131, 181, 260n57, 265n46, 300n93, 346n43, 353n33
Royal Navy 221, 222, 225, 246, 337n22, 345n32
Royal Orthopaedic Hospital 13, 15, 16, 17, 37, 66, 257n34, 263n31
Royal Postgraduate Medical School 241, 332n65, 333n69
Royal Society of Medicine 30, 45–6, 51, 178, 200, 268n79, 304n141, 308n29, 346n41
Royal Victoria Albert Dock Hospital 88, 204, 211, 291n111, 331n62
Ruskin, John 56
Russell, James (1880–1960) 207, 333n78
Russell, Sir Edward 63
Russo-Japanese War 110
Russo-Turkish War 97

Safety First 141
safety of workplace 82, 140
St Bartholomew's Hospital 20, 21, 60, 90, 132, 185, 261n1, 284n26, 327n24, 328n30, 333n69, 334n83, 353n36

orthopaedic department 65, 183, 263n31, 297n62
Policy Committee of the Medical College 242
St James's Hospital, Balham 181, 193, 197
St John Ambulance Association 86, 96, 97
St John Ambulance Brigade 86
St John Clinic for Rheumatism 322n71
St Mark's Hospital 181
St Marthe, Scevale de 12
St Mary's Hospital 290n101, 313n65
St Olave's Poor Law Union Infirmary 274n34
St Paul, Minnesota 70, 77
St Peter's Hospital 181
St Thomas's Hospital 110, 117, 170, 184, 223, 303n134, 321n71, 334n83, 353n36
  Medical School 354n51
  Physical Exercise Department 125, 297n62
Salford
  Greengate Dispensary 61
  motor accidents 330n52
  Ragged and Industrial Schools 275n37
  Royal Hospital 183, 354n46
Salop Royal Infirmary 229
Salvarsan 166
sanatorium movement 201
Sankey Commission 342n73
Sankey, Lord 232, 306n8
Save the Children Fund 160
Sayre, Lewis A. (1820–1900), 26, 34
School Medical Officers 58, 157
School Medical Service 65–70, 169, 175
schools
  day 176, 178
  elementary 173
  for the physically 'defective' ('PD') 66, 67, 68, 172, 278–9n75
  in hospitals 66, 68, 78, 169

residential 61, 65, 68, 69, 176
  see also education
scientific management 48, 92–3, 103, 120–4, 141, 142, 187, 236
scoliosis 68, 256n16
Scotland 43, 207, 221, 229, 300n93, 324n105
  Department of Health 327n17, 347n50
Scott, J. C. 352n32
Scurr, John (MP) 212
Sears-Smith Vocational Rehabilitation Act, USA (1918) 142
Second World War
  Army hospitals 225
  military orthopaedic centres 221, 225
  special hospitals 219, 228–9
  see also Army Medical Service
Secretary of War, USA 130
segregation, concept and practice 91, 114, 121, 134, 154, 183, 187, 192, 225, 226, 229
Selby-Bigge, Lewis Amherst 279n77
settlement movement 57
Sevenoaks Children's Hospital for Hip Diseases 263n26
Sever, James W. 283n8
Shaffer, Newton (1846–1928) 37, 268n78, 279n83
Shaftesbury Ragged School Union Society 156, 160
Sharpey-Schäfer, Edward 47
Sheffield Royal Infirmary 215
Sheffield, Chair of Surgery 269n100
Sheldrake, Timothy 255n14
shellshock 112, 120, 121,
Shepherd's Bush Military Orthopaedic Hospital 106, 117–120, 125, 131–32, 134, 142, 241, 300 n93, 301n106, 305n150, 306n5, 321n71
Sherborne, Dorset 345n32
Sherman, Harry M. 268n78
Sherman, William O'Neill 141, 204
ship building 137
Shrewsbury 76

Shropshire
  County Council 78
  cripples 278n68,n70
  Robert Jones and Agnes Hunt
    Orthopaedic Hospital *see*
    Oswestry
Shufflebotham, Frank 138, 150
Sidcup Hospital for Facial
  Injuries 178
Sieveking, Edward 91, 92, 98
Sinclair, Meurice 181, 197
Sloggett, Sir Arthur 297n56
Smart, Sir Morton 227, 321n71
Smith, Ben 202
Smith, E. Noble, 36, 37
Smith, S. Alwyn 300n93
Smith, S. Maynard 270n103
Smyth, J. L. (1882–1966) 213, 215,
  334n86
social medicine/community care 8,
  172, 176, 246
social/welfare workers 55–56, 77,
  211, 274n34
socialism 56, 57
Socialist Medical Association 212,
  316n8, 321n64
Society for Preventing Street
  Accidents and Dangerous
  Driving 95
Society for the Study of Diseases in
  Children (SSDC) 44–46, 64
Society of Clinical Surgery,
  USA 50
Society of Physiotherapists 347n47
'soldiers's heart' 120–1
Somme, battle of 294n22
Souttar, Henry S.
  (1875–1964) 291n102, 326n16,
  333n77
Spanish Civil War 220, 331n58
Special Hospitals and Physical
  Reconstruction, Division of,
  USA 142
specialists
  in children's medicine 62
  in school clinics 65
  in surgery 251n13, 300n90
  see also general practitioners; and
    under individual specialisms

specialization in medicine
  American *v.* British contexts 2–4,
    38
  cultivation 28, 38, 93, 245
  and First World War 105, 106,
    123
  v. generalism 38, 45, 47, 185–6,
    246, 247, 267n68
  historiography 1–2, 4, 52, 53,
    109, 136, 250–1n9
  impact on patients 9
  intellectual status of 245–6
  in postwar Britain 132
  see also medical education; *and*
    under individual specialisms
specialty boards 4, 240
Speed, Kellogg (1879–1955) 322n75
Spence, James (1892–1954) 335n92
Spencerianism 38
Spens Report 352n26
spinal deformity/diseases 12, 15, 16,
  18, 37, 61–62, 68, 256n20
splints 24, 33, 44, 118, 314n80, 325n4
splints, Thomas 25, 26–27, 106–7,
  111, 220
squints 13, 175
Staffordshire 88, 138
  County Council 319n42
  Biddulph Grange Orthopaedic
    Hospital 162
standardization, principle and
  practice 50, 91, 93, 121, 187,
    232, 314n79
Stanley, Arthur 117, 156, 222,
  255n13, 324n103
Starauss, George (MP) 337n21
State (the) 116, 168–71; *see also*
  under individual departments of
  government; hospitals; local
  authorities; orthopaedics;
  voluntarism
statism 6, 57, 218
Stephenson, Sydney 44
Stevens, Rosemary 4
Stewart, Grainger 297n62
Stiles, Harold (1863–1946) 42–3,
  44, 46, 47, 50, 64, 77, 131, 133,
  269n95, 270n103, 283n12,
  300n93, 350n13

Stockholm 329n39
Stoke-on-Trent War
  Hospital 307n8
Stopford, John 244, 297n60
Stromeyer, G. F. Louis
  (1804–76) 14, 255n13, 258n40
Sunderland 86
sunlight 170, 320n60
surgery
  bloodless 22, 264n35
  and meritocratic values in 48
  bone and joint 25, 40–1, 42; *see
    also* general surgeons
  cardiovascular 243, 252n20,
    350n2
  children's 43, 47
  French 12
  fragmentation of 243, 245
  and generalism 33, 39, 43, 48
  hero cult 22, 222, 345n22
  and image-making 70
  mechanical 124
  industrial 143, 203
  operative *v.* conservative 20–3,
    36–7, 171, 259n56
  reconstructive 139
  rhetoric of unity 17, 38, 47
  textbooks 80
  of war 344n12
  see also abdominal; conservative;
    consultants; general surgeons;
    medical profession;
    neurosurgery; orthopaedics;
    plastic; specialization
*Surgery, Gynecology and
  Obstetrics* 50
'surgical tuberculosis' 69, 75, 78,
  173, 177
Surgical Union 270n104
surgical wards, organization of 82,
  84
Surrey Voluntary Association for the
  Care of Cripples 172
Sutherland, Duchess of 272n14
Swain, William Paul 81
Sydney, Royal Alexandra
  Hospital 265n46
Syme, James (1799–1870) 20, 24
syphilis 54, 166

Tait, Lawson (1845–99) 49, 88
talipes *see* club foot
Taylor, Charles F. (1827–99) 26,
  267n65
Taylor, Frederick 93, 122
Taylor, James 284n26
Taylorism 122, 123
team work 121–2, 124, 154, 169,
  177, 183, 187–8, 191, 206, 220,
  236, 239
technological determinism 52
technological fixes 18, 166
Telford, E. D. (1876–1961) 42, 43,
  44, 335n92
tendons 13, 14, 44, 52, 127, 350n6
Tennant, H. J. (MP) 119
Tennyson, Lord Alfred 21
tenotomy 13, 15, 17, 18, 51, 258n40
thermaltherapy 227
Thomas, Hugh Owen
  (1834–91) 23–8, 29, 30, 31, 33,
  38, 72, 73, 74, 235, 299n81
Thompson, Alexis 269n98
Thomson, John 64
Thorburn, Sir William (1861–
  1923) 270n103, 337n22,n23,
  338n26
Tillett, Ben 212
Times (The) 59, 62, 134, 202, 209
Tissot, S. A. S., *Advice to the People*
  (1767) 81
Titmuss, Richard 145, 218
Tomlinson Committee 228
Tomlinson, George, MP
  (1890–1952) 209, 219
Townsend, Ursula (Secretary
  CCCC) 156, 274n25, 324n103,
  316n13
trade unions 6, 102, 149, 180, 187,
  202, 211
Trades Union Congress (TUC) 8,
  146, 180, 188, 196, 212, 215,
  338n30
  Annual Congress at Blackpool
    (1938) 214
  BMA alliance 213–16
  Scottish 327n17
  Social Insurance Committee 213
  see also Labour Party

Transport and General Workers'
    Union 212
trauma/accident centres 193, 206,
    230; *see also* accident hospitals;
    fracture clinics; industrial
    injuries
traumatology, as specialism 231
Treasury 161
Treloar, Sir William Purdie 56, 65,
    275n37
Trethowan, William Henry
    (1882–1934) 128, 132, 156,
    261n1, 263n31
Trevelyan, George Macaulay 59
Treves, Frederick 26, 37
Trinity College, Dublin 351n21
Trueta, Josep (1897–1977) 220,
    306n5, 348n60, 332n63
Tubby, Alfred H. (1863–1930) 35,
    36, 37, 44, 45, 46, 64, 169,
    267n64, 270n103, 270n107,
    351n16
tuberculosis 67, 68, 69, 112, 166,
    167, 175
  of bones and joints 21, 22, 24, 28,
      41, 44, 54, 58, 63, 67, 161,
      172–3, 174, 259n56, 320n60;
      see also surgical tuberculosis
  decline 174
  in Royal Navy 246
  mortality 173
  research 266n57, 350n13
  schemes 155, 198
  see also clinics; hospitals
tuberculous, lesions 170
Turko-Servian War 294n21
Turner, George Grey
    (1877–1951) 269n98

Uddingston, Glasgow 207, 210
Ulster 124, 127
unemployment 175, 203
uniformity, concept and
    practice 82, 91, 92, 93, 121,
    160, 182, 187, 189, 218
United States Public Health
    Service 140

unity of control, in patient care 100,
    134, 188, 206, 226, 228–30, 239,
    247, 248
University College 351n16
University Grants Committee 238
unorthodox practitioners 135; *see
    also* bonesetting; homoeopathy;
    osteopathy
urology 46, 243

Varrier-Jones, P. C. 201
venereal disease 68, 69, 246
Vienna Accident Hospital 190–3,
    204, 205
vis medicatrix naturae 20
vitalism 211
vitamin D 167, 175
Voisin, Felix 12
Volkmann, Richard von
    (1830–89) 257n36
voluntarism and the state 57,
    157–9, 161, 194–6
Vulpius, Otto 263n26

Wade, Sir Robert B.
    (1874–1954) 265n46
Waldenström, Henning 329n39
Wales, schemes for cripples
    in 317n19
Walsall Cottage Hospital 88, 89
Walsham, W. J. 36
Walter Reed General Hospital,
    Washington, DC 130
Walter, Albert G. 259n52
war and medicine, alleged
    relations 7, 105, 108
War Measures
    Committee 305n149
War Office 106, 112, 114, 115,
    116–17, 118, 125, 127, 154, 168,
    221, 222
War Pensions Committees 137
Ward, Mrs Humphry
    (1851–1920) 59, 64, 66, 69, 156,
    277n57
Warner, Francis 65, 66, 275n44,
    277n61

Warren, Guy de Grouchy 208
Watson, Frederick 260n64
Watson-Jones, Reginald
   (1902–72) 206, 209, 210, 211,
   242, 243, 244, 248, 327n24,
   335n3, 337n22, 338n30, 339n34,
   340n53, 353n33, 355n55
   and rehabilitation 213, 215, 217,
   219, 223,
   and accident services 230, 241
Webb, Sidney 316n8
*Weekly Welsh Orthopaedic
   News* 127
Weir Mitchell Treatment 258n43
welfare agencies 57–8, 150, 160–1
Wells, H. G. 119
Wells, Spencer 287n65
West Derby Board of
   Guardians 144
West Kensington, Poor Law
   Guardians 73
West London Hospital,
   Hammersmith 72, 73
West London Society for Sick
   Children 61
Western Pennsylvania
   Hospital 259n52
Westminster Hospital 35, 255n14,
   267n64, 334n83, 353n36
Westminster Review 23, 38
Wheeler, Sir William Ireland de
   Courcy (1879–1943) 50, 221,
   270n103, 318n29
Wheelhouse, C. G. 25, 257n33
Whitley, J. 202
Whitman, Royal (1857–1946) 35
whole-person medicine 44, 170; *see
   also* holistic philosophy
Wigan, Royal Albert Edward
   Hospital 314n80
Willard, De Forest (1846–1910) 26,
   279n83
Williams, Dawson 44, 77, 186
Williams, Gwynne 223
Wilson, Arnold 216, 335n6
Wilson, J. Greenwood
   (MOH) 329n39

women in welfare 55, 58–9, 156,
   160
Women's National Anti-Suffrage
   League 59
Women's University Mission 76
Woods, Sir Robert Stanton
   (1877–1954) 227, 347n44
workers
   health care, history of 151
   medical examination 140
   and welfare schemes 209–11
   see also hospitals
workhouse *see* Poor Law
workmen's compensation 8, 138,
   141, 143, 146, 148, 149, 201
   abuse 145
   Acts 86, 145, 214, 342n71
   bogey 215
   costs 202, 209
   and medical
      rehabilitation 342n71
   and specialist services 146
   insurance 192
   cuts in 211
   legislation 80
   medicalization 149
   payments 203
   Royal Commission
      (1939–40) 204
   USA 145
workmen's councils 338n26
wounds, closed plaster
   treatment 344n13
Wright, G. A. (1851–1920) 42, 43,
   260n64, 275n37
Wright, Sir Almroth 297n56

X-rays 31, 51, 76, 187, 284n20,
   325n4

*Year Book of Industrial and
   Orthopedic Surgery* 348n65
Yvetot, France, Hôpital de
   l'Alliance 150, 151

Zeitschrift für Krüppelfürsorge 71
Zelizer, Viviana 55

Printed in the United States
By Bookmasters